Sustainable Water Resources Management XI

WIT*PRESS*

WIT Press publishes leading books in Science and Technology.
Visit our website for new and current list of titles.
www.witpress.com

WIT*eLibrary*

Home of the Transactions of the Wessex Institute.
Papers published in this volume are archived in the WIT eLibrary in volume 250 of WIT Transactions on Ecology and the Environment (ISSN 1743-3541).
The WIT eLibrary provides the international scientific community with immediate and permanent access to individual papers presented at WIT conferences.
Visit the WIT eLibrary at www.witpress.com/elibrary.

ELEVENTH INTERNATIONAL CONFERENCE ON SUSTAINABLE WATER RESOURCES MANAGEMENT: EFFECTIVE APPROACHES FOR RIVER BASINS AND URBAN CATCHMENTS

SUSTAINABLE WATER RESOURCES MANAGEMENT XI

Conference Chairman

S. Mambretti
Polytechnic of Milan, Italy
Member of WIT Board of Directors

International Scientific Advisory Committee

Organised by

Wessex Institute, UK
Polytechnic of Milan, Italy

Sponsored by

WIT Transactions on Ecology and the Environment
International Journal of Environmental Impacts

Sustainable Water Resources Management XI

WITPRESS
WIT Press publishes leading books in Science and Technology.
Visit our website for new and current list of titles.
www.witpress.com

WITeLibrary
Home of the Transactions of the Wessex Institute.
Papers published in this volume are archived in the WIT eLibrary in volume 250 of WIT Transactions on Ecology and the Environment (ISSN 1743-3541).
The WIT eLibrary provides the international scientific community with immediate and permanent access to individual papers presented at WIT conferences.
Visit the WIT eLibrary at www.witpress.com/elibrary.

ELEVENTH INTERNATIONAL CONFERENCE ON SUSTAINABLE WATER RESOURCES MANAGEMENT: EFFECTIVE APPROACHES FOR RIVER BASINS AND URBAN CATCHMENTS

SUSTAINABLE WATER RESOURCES MANAGEMENT XI

Conference Chairman

S. Mambretti
Polytechnic of Milan, Italy
Member of WIT Board of Directors

International Scientific Advisory Committee

Organised by

Wessex Institute, UK
Polytechnic of Milan, Italy

Sponsored by

WIT Transactions on Ecology and the Environment
International Journal of Environmental Impacts

WIT Transactions

Wessex Institute
Ashurst Lodge, Ashurst
Southampton SO40 7AA, UK

Editorial Board

Sustainable Water Resources Management XI

Effective Approaches for River Basins and Urban Catchments

EDITOR

S. Mambretti
Polytechnic of Milan, Italy
Member of WIT Board of Directors

WIT*PRESS* Southampton, Boston

Editor:
S. Mambretti
Polytechnic of Milan, Italy
Member of WIT Board of Directors

Published by

WIT Press
Ashurst Lodge, Ashurst, Southampton, SO40 7AA, UK
Tel: 44 (0) 238 029 3223; Fax: 44 (0) 238 029 2853
E-Mail: witpress@witpress.com
http://www.witpress.com

For USA, Canada and Mexico

Computational Mechanics International Inc
25 Bridge Street, Billerica, MA 01821, USA
Tel: 978 667 5841; Fax: 978 667 7582
E-Mail: infousa@witpress.com
http://www.witpress.com

British Library Cataloguing-in-Publication Data

A Catalogue record for this book is available
from the British Library

ISBN: 978-1-78466-421-3
eISBN: 978-1-78466-422-0
ISSN: 1746-448X (print)
ISSN: 1743-3541 (on-line)

The texts of the papers in this volume were set individually by the authors or under their supervision. Only minor corrections to the text may have been carried out by the publisher.

Sustainable Water Resources Management XI

Effective Approaches for River Basins and Urban Catchments

EDITOR

S. Mambretti
Polytechnic of Milan, Italy
Member of WIT Board of Directors

WIT PRESS Southampton, Boston

Editor:
S. Mambretti
Polytechnic of Milan, Italy
Member of WIT Board of Directors

Published by

WIT Press
Ashurst Lodge, Ashurst, Southampton, SO40 7AA, UK
Tel: 44 (0) 238 029 3223; Fax: 44 (0) 238 029 2853
E-Mail: witpress@witpress.com
http://www.witpress.com

For USA, Canada and Mexico

Computational Mechanics International Inc
25 Bridge Street, Billerica, MA 01821, USA
Tel: 978 667 5841; Fax: 978 667 7582
E-Mail: infousa@witpress.com
http://www.witpress.com

British Library Cataloguing-in-Publication Data

A Catalogue record for this book is available
from the British Library

ISBN: 978-1-78466-421-3
eISBN: 978-1-78466-422-0
ISSN: 1746-448X (print)
ISSN: 1743-3541 (on-line)

The texts of the papers in this volume were set individually by the authors or under their supervision. Only minor corrections to the text may have been carried out by the publisher.

Preface

The present book includes the proceedings of the 11th International Conference on Sustainable Water Resources Management: Effective Approaches for River Basins and Urban Catchments, held online due to the COVID pandemic, organised by the Wessex Institute, UK and the Politecnico di Milano, Italy. The Conference was sponsored by the International Journal of Environmental Impacts and WIT Transactions on Ecology and the Environment. The Conference combines the topics covered by the former:

- **International Conference on Water Resources Management**, which follows those held in Halkidiki, Greece in 2001, followed by Las Palmas, Gran Canaria (2003); the Algarve, Portugal (2005); Kos, Greece (2007); Malta (2009), Riverside; California (2011) the New Forest, UK, home of the Wessex Institute (2013); A Coruña, Spain (2015); Prague, Czech Republic (2017); and Alicante, Spain (2019).
- **International Conference on River Basin Management**, which follows those held in Cardiff, UK in 2001, followed by meetings in Las Palmas, Gran Canaria (2003); Bologna, Italy (2005); Kos, Greece (2007); Malta (2009); Riverside, California (2011); The New Forest, UK (2013), home of the Wessex Institute; A Coruña, Spain (2015); Prague, Czech Republic (2017); and Alicante, Spain (2019).

Water is essential for sustaining life on our planet and its uneven distribution is a source of permanent conflict. The growth of human population combined with the irregularity in precipitation and water availability may restrict even further the access to water in certain regions of the world. This problem is made more severe by anthropogenic activities that affect its quality.

Moreover, changes in the landscape, use of the land and climate conditions lead to a continuous revaluation of river basin management objectives. This requires the development of better measuring tools as well as the use of increasingly accurate computer software.

Catastrophic events such as floods and associated landslides, erosion and sedimentation can have serious effects not only on life and property but also on the basin ecology. Frequently these problems are aggravated by the unforeseen consequences of man made changes in the river basin. This has led in recent years to work on river restoration and rehabilitation with various degrees of success.

As we emerge from the global pandemic, we have the opportunity to adapt and improve our approach in order to reduce risks, to produce fewer carbon emissions and, generally speaking, to have a more sustainable way to manage water resources.

The objective of these series of conferences is to bring together practitioners and researchers in academia and industry in the hope that their interaction will foster mutual understanding and lead to better solutions for water resources and river basin management. By necessity, the theme of the conference is designed to be interdisciplinary and draws expertise from a wide range of fields including engineering, computing and the social sciences. All aspects of Hydrology, Ecology, Environmental Management, Flood Plains and Wetlands are included in the Conference.

To this end, papers are written by professionals and academics involved in the field to exchange knowledge and gain an insight into the state of the art in the current technology, techniques and solutions as they have been developed and applied in different Countries. Contributions include a wide variety of stakeholders from research and academia, to industrial sectors as well as government organisations.

Papers presented at the Conferences are permanently available to the international community in digital format at http://www.witpress.com/elibrary.

The Editor 2021

Contents

Section 1: Water management

Updated rainfall series and their trends for mainland Portugal (1913–2019)
Maria Manuela Portela, Luis Angel Espinosa & Martina Zelenakova 3

Annual vapor, precipitation, groundwater, and transpiration fluctuations in simple water cycle processes
Koji Tamai 13

Water governance performance assessment of riparian communities: A case study of the Volta River Basin
Abdul-Razak Zakaria & Kenichi Matsui 23

River restoration for the replacement of lost spawning grounds due to dam construction
Regina Santos, Luís Sanches Fernandes, Fernando Pacheco, Mariza Monteiro & Joaquim Jesus 35

Section 2: Ecological and environmental impact

Reconditioning the sewerage and drinking water system within the Faculty of Mathematics, University of Guayaquil, Ecuador
Armando Saltos, Patricia Villa, Jacinto Rojas & Manuel Gomez de la Torre 47

Heavy metals in rivers affected by mining activities in Ecuador: Pollution and human health implications
Kenny Escobar-Segovia, Samantha Jiménez-Oyola, Daniel Garcés-León, Daniela Paz-Barzola, Eduardo Chavez Navarrete, Paola Romero-Crespo & Bryan Salgado 61

Filtration of dissolved organic nutrients from fish farm wastewater using a macroalgae biofilter
Jack R. Hall & Georg Martin 73

Use of agro-inputs in agricultural production and water resources pollution: A case study of Chu-mango in the Mekong River Delta, Vietnam
Truong Hong Vo Tuan Kiet, Pham Thi Nguyen & Nguyen Thi Kim Thoa 81

Section 3: Water and the community

Valuation of a community company and its impact on development strategies
Bethy Merchan, Romina Torres, Jannely Benalcazar, Paul Carrión-Mero, Gricelda Herrera-Franco, Edgar Berrezueta & María Jaya-Montalvo 95

Communication methods on water care during the COVID-19 pandemic and its impact on the resilience of the rural community of "Libertador Bolívar", Ecuador
Gricelda Herrera-Franco, Joselyn Alvarado, Pablo Gordillo, Lenin Veintimilla, Bethy Merchan, Paúl Carrión-Mero & Edgar Berrezueta 109

Section 4: Hydraulic modelling

Graph theory and community detection for elementary DMA design
Stefano Mambretti, Anita Raimondi & Fernando F. Stroppa 121

Comparison of urban retrofitting scenarios for sustainable stormwater control: A case study in Milan, Italy
Franco Raimondi, Mariana Lobo Marchioni, Anita Raimondi & Gianfranco Becciu 131

Modeling the behavior of rain gardens
Harsho Sanyal, Yara Yasser Elborolosy & Joseph Cataldo 143

Section 5: Erosion and sediment transport

Estimating storage capacity loss due to sedimentation for Murum reservoir in Sarawak, Malaysia
Joyce Janggu, Susie Nadya & Mubasher Hussain 157

Semiautomatic algorithm for the interpretation of the bedforms and statistical analysis
Leonardo Schippa & Irene Cavalieri 169

Coastal erosion hazard and vulnerability: Case study of Porticello, South Calabria, Italy
Giuseppina Chiara Barillà, Giuseppe Barbaro, Giandomenico Foti, Pierluigi Mancuso, Vincenzo Fiamma, Agnieszka Malesinska, Pierfabrizio Puntorieri & Maria Mandalari 181

Section 6: Flood risk management

Towards an absorption strategy for the management of flood risk: Study of the strategy of resistance and the artificialization of soils in the Franco–Belgian cross-border territory
Emeline Coszach, Vincent Becue, Jérémy Cenci, Isabelle de Smet, Smail Khainaar, Patrizia Laudati & Cédric Rivière 197

Nature-based solution as flood protection strategies: The case of Carahá River Park, Brazil
Renata V. de Oliveira Gobetti, Ornella Salimbene & Deborah Panepinto 209

Development of flood tracking technique for flood disaster management in urban areas
Suk Ho Lee, Dong Ho Kang & Byung Sik Kim 221

Author index 235

SECTION 1
WATER MANAGEMENT

UPDATED RAINFALL SERIES AND THEIR TRENDS FOR MAINLAND PORTUGAL (1913–2019)

MARIA MANUELA PORTELA[1], LUIS ANGEL ESPINOSA[1] & MARTINA ZELENAKOVA[2]
[1]Instituto Superior Técnico, Lisbon University (IST/UL, CERIS), Portugal
[2]Technical University of Kosice, Slovakia

ABSTRACT

This paper summarises the main findings from a comprehensive study on rainfall trends based on updated long-running rainfall series from more than 500 stations uniformly dispersed over mainland Portugal (89015 km^2, i.e., almost 60 stations each 10,000 km^2) considering a total period of 106 hydrological years, from October 1913 to September 2019. To understand the rainfall behaviour at different time scales, cumulative intra annual (monthly, quarterly and half-yearly), and annual rainfall series were analysed. To identify possible shifts in the long-term rainfall pattern, two additional contiguous periods were analysed along with the 106-year total period: the initial period, with 55 years and, followed by the final one, with the last 51 years. The detection of the rainfall trends at the different time scales and periods were addressed via the Mann–Kendall (MK) test together with the Sen's slope estimator for measuring the magnitude of the trends. The majority of results achieved revealed that the generalised positive trends in the initial 55 years were followed by negative trends in the last 51 years of the total period clearly outweighing the trends observed in the initial years. This counterbalance resulted in widespread downward trends for the total period. The spatial representation of the magnitude of the trends clearly showed an increasing asymmetry between northern and southern conditions, the latter becoming progressively and remarkable more arid. Such behaviour is also consistent with the previously identified pattern of the "dry gets drier" paradigm for the south western region of the Iberian Peninsula. It was also found that, closely to the year 1968, the intra annual rainfall pattern changed. Since about that year, the rainfall in the last wet months of the hydrological year (January, February, and March) has showed a sustained decrease of its relative weight, thus diminishing its relative contribution to the groundwater refill and to the water needs during the forthcoming dry months.

Keywords: climate change, mainland Portugal, rainfall trends, Mann–Kendall test, Sen's slope test

1 INTRODUCTION

Through the regular scientific assessments on climate change of the Intergovernmental Panel on Climate Change [1], it is pointed out that changes are occurring in the amount, intensity, frequency and type of precipitation. According to these assessments, specifically for rainfall, trends would be positive in some parts of the globe [2]. However, at smaller scales an opposite pattern have been identified with sustained decreasing rainfalls [3]. Despite these patterns are still uncertain, there is enough evidence to still considering them as very likely scenarios as, for instance, for the Iberian Peninsula. According to the reports by [4]–[7] all over the 21st century the Iberian rainfall is likely to decrease at longer time scales, and be grouped in more intense extreme events.

In the case of the Mediterranean region, several studies have addressed climate variables' trends (such as precipitation, specifically its liquid state, i.e., rainfall) based on ground-based measurements during the 20th century, but yet a small number of studies ascertaining the current and future conditions [8]–[10]. This fact highlights that such trend analyses need to be continuously updated since they describe current trends to predict future ones and that almost all the climate variables come from a complex system with changes that may or may not be identified subject to the length and time window of the utilised series. Under such understanding and having in mind the major role of the rainfall in the hydrological cycle, long-term and recent rainfall trends, including their variability and uncertainty over time,

WIT Transactions on Ecology and the Environment, Vol 250,
www.witpress.com, ISSN 1743-3541 (on-line)
doi:10.2495/WRM210011

were analysed for mainland Portugal at different time scales, from month to year, by means of the Mann–Kendall (MK) test together with the Sen's slope estimator. The results achieved clearly indicate largely predominant negative MK values, i.e., decreasing rainfalls.

2 CASE STUDY: RAINFALL DATA

The study area is mainland Portugal, with approximately 89,015 km^2, located on the Iberian Peninsula, in the south westernmost part of Europe (centred at latitude 39°23′59.5″N and longitude 8°13′28″W). Portugal faces the North Atlantic Ocean (Fig. 1), and is characterised by a Mediterranean climate with warm and dry summers and cool and wet winters. The major natural factors that substantially determine the Portuguese climate are the latitude, the sharp hilly terrain, ranging from 0 m.a.m.s.l. at the Atlantic Ocean to 1993 m.a.m.s.l. in the Serra da Estrela, and also the weak ocean current that flows south along the coast of the country. The average annual rainfall is in close connection to the relief following a predominantly complex north to south pattern, and thus, varying considerably throughout the country ranging from values above 2,800 mm to less than 400 mm, as shown in Fig. 1.

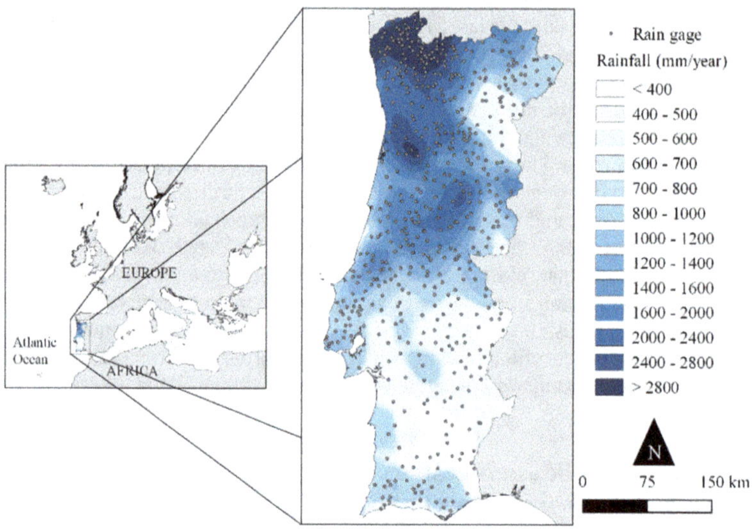

Figure 1: Average annual rainfall of mainland Portugal during the total 106-year period (from 1913/1914 to 2018/2019), and location of the 532 rain gauges analysed in the study.

The monthly rainfall records used in the study were acquired from the public database of the Portuguese Environment Agency [11]. Because most of the rain gauges had missing data, the fill-in procedure described in [12] was applied to the original monthly records, resulting in 532 rain gauges (Fig. 1) with updated long-running monthly rainfall data accounting for 106 hydrological years, each starting October 1 (from October 1913 to September 2019).

3 MODELS

The Mann–Kendall test was used to detect consistent upward and downward trends (i.e., monotonic trends) in the monthly and aggregated rainfall time series [13], [14] for the 95%

confidence interval. Additionally, the Sen's slope estimator [15] was utilised to determine the magnitude of the trends.

4 RESULTS

The trends at the selected 532 rain gauges were characterised at the annual level but also at the following shorter time scales aiming at ascertaining the within-the-year changes in the rainfall pattern [10], [12]: months, quarters (Q1 to Q4 starting in October, January, April and July, respectively) and semesters (S1, from October to March, and S2, from April to September).

Also, to examine the spatiotemporal dynamics of rainfall in the context of changing climate, the total period of 106 years (from October 1913 to September 2019) was divided into two contiguous periods, i.e., the initial 55 hydrological years (1913/1914 to 1967/1968) and the last 51 hydrological years (1968/1969 to 2018/2019). This was done based on the identified accelerated downward rainfall trends since late 1960s for southern Portugal [10].

The rainfall trend magnitudes obtained from the models were mapped including both statistically significant and non-significant trends (Figs 2–4). The figures considered the three periods, thus, enabling to assess the rainfall changes over time.

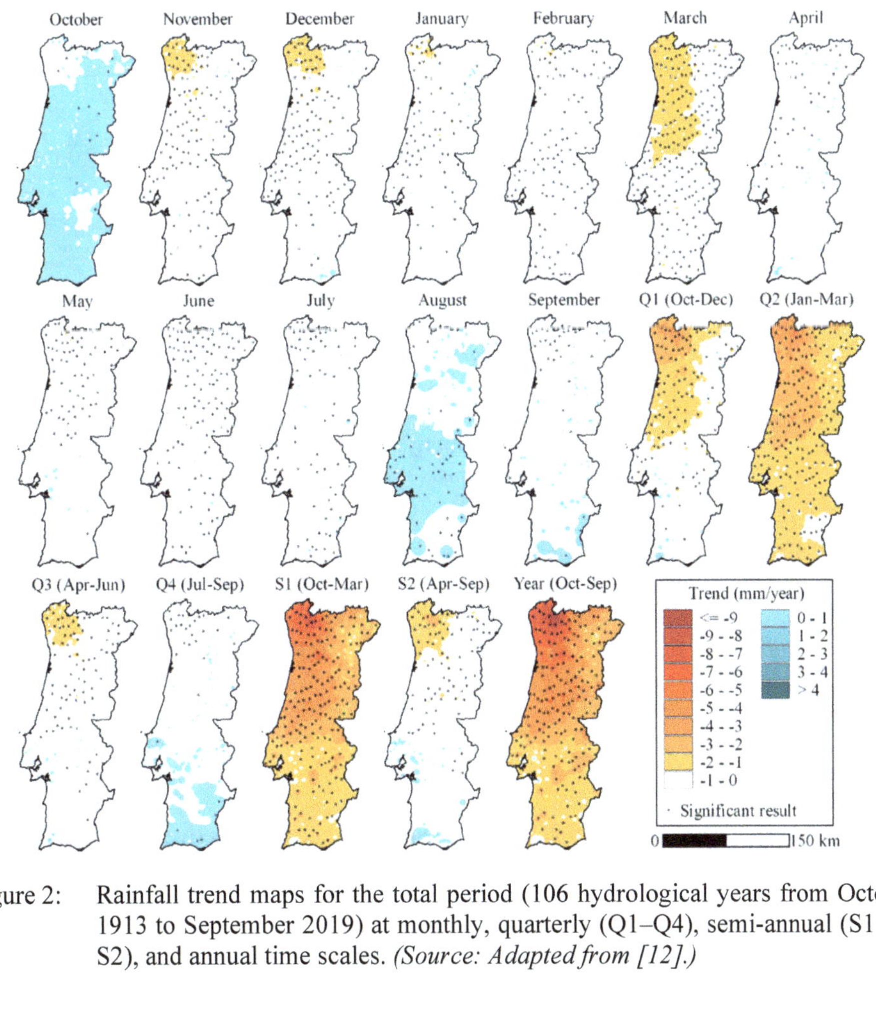

Figure 2: Rainfall trend maps for the total period (106 hydrological years from October 1913 to September 2019) at monthly, quarterly (Q1–Q4), semi-annual (S1 and S2), and annual time scales. *(Source: Adapted from [12].)*

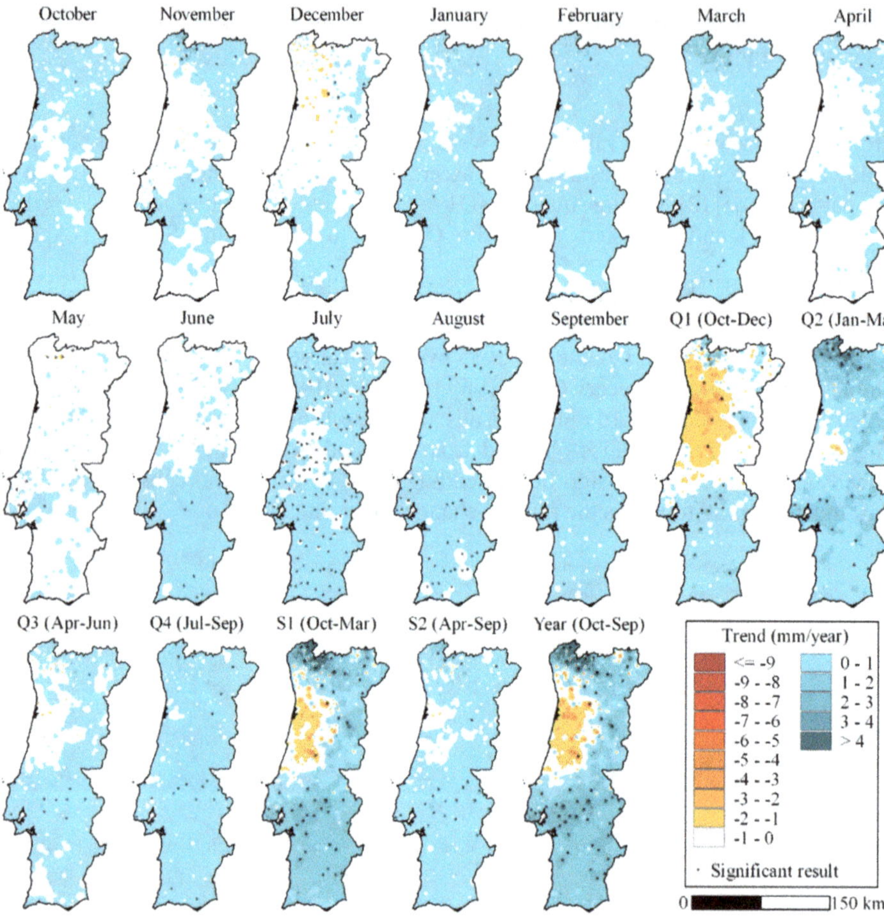

Figure 3: Rainfall trend maps for the initial period (55 hydrological years from October 1913 to September 1968) at monthly, quarterly (Q1–Q4), semi-annual (S1 and S2), and annual time scales. *(Source: Adapted from [12].)*

For the total period, the trend analysis showed generalised decreases in rainfall for longer time scales (e.g., Q2, S1, and the year in Fig. 2). However, these magnitudes were determined from a linear model. Therefore, an improved interpretation was done for the division of the 106 years. During the first 55 years (Fig. 3) large parts of Portugal denoted upward trends. In the following 51 years (Fig. 4), these patterns swung into markedly downward trends except July and August which, either way, provide negligible amounts of rainfall. Hence, the latter period counterbalanced the trends observed in the initial years and resulted in widespread, but less marked, downward trends for the total period.

The comparison on the trend maps of Figs 3 and 4 denoted a relatively rapid "step" change from the initial upward trends to the final downward trends, respectively. This contrast is highly marked for Q1, Q2, S1, and the year with more steep declines in the earlier years. The comparison now including Fig. 2 reveals some relevant rainfall distribution changes at monthly scale. For instance, March is the month with the highest number of significant and most pronounced trends for the total period. Nevertheless, by inspecting Fig. 4, apparently the most significant changes take place in the two previous months. This suggests that the notorious monthly rainfall decline is shifting "backwards" from March to February and January – note that these months have a determining effect on the groundwater and artificial reservoirs refill and in the water budgets in mainland Portugal.

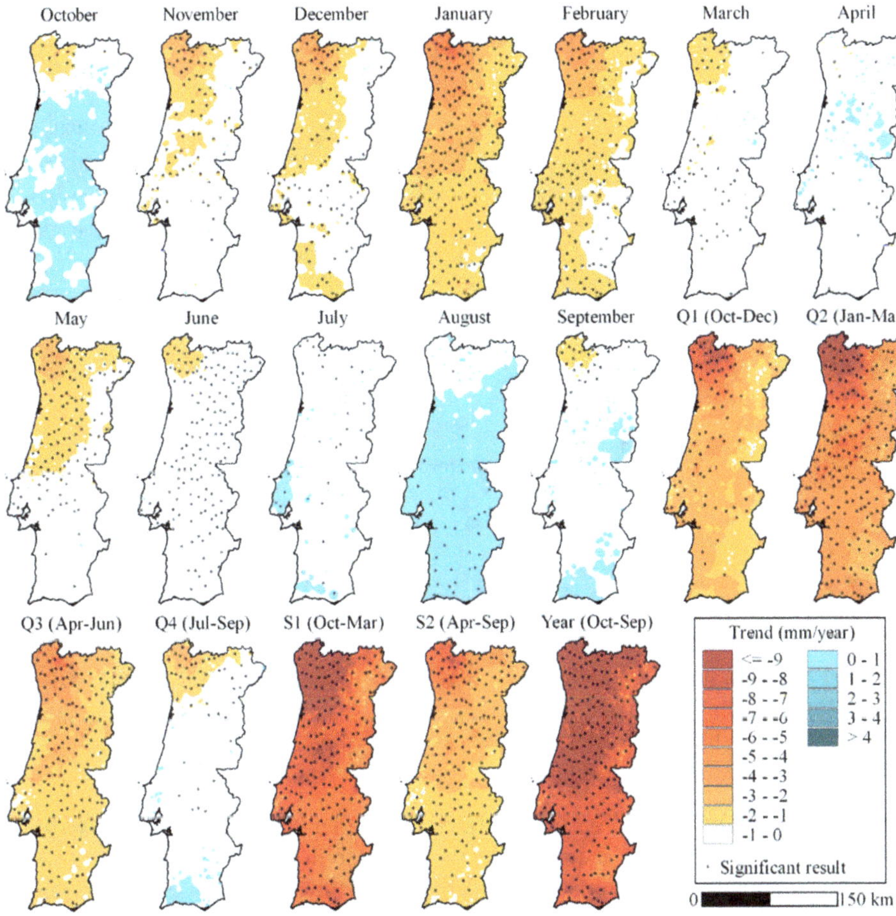

Figure 4: Rainfall trend maps for the last period (51 hydrological years from October 1968 to September 2019) at monthly, quarterly (Q1–Q4), semi-annually (S1 and S2), and annually time scales. *(Source: Adapted from [12].)*

Furthermore, a homogeneous spatial distribution of downward rainfall trends was recognised at seasonal scale over the country, especially in the wet season of the hydrological year (S1, from October to March). This is of great importance given S1 may represent three-quarters or even more of the annual rainfall. However, October denoted only few areas with negative trends compared to the rest of the rainy months (as shown in Fig. 4). This highlights a greater vulnerability to such negative pattern since, based on these results, wetter conditions which could turn round the already dry conditions are not expected to occur.

Analogously, long-term averages at annual level were calculated over the three adopted periods using aggregates of the updated and filled-in rainfall series [12]. The average annual rainfalls are mapped in Fig. 5 including a weighted mean annual value in each case. From the figure, wetter conditions can be observed in the initial 55 years which contrast with the drier ones in the last 51 years given the general rainfall decreases from one period to the other. Such decrease is about 170 mm, also reflecting a more pronounced asymmetric with a relatively still wet north and a conclusively drier south.

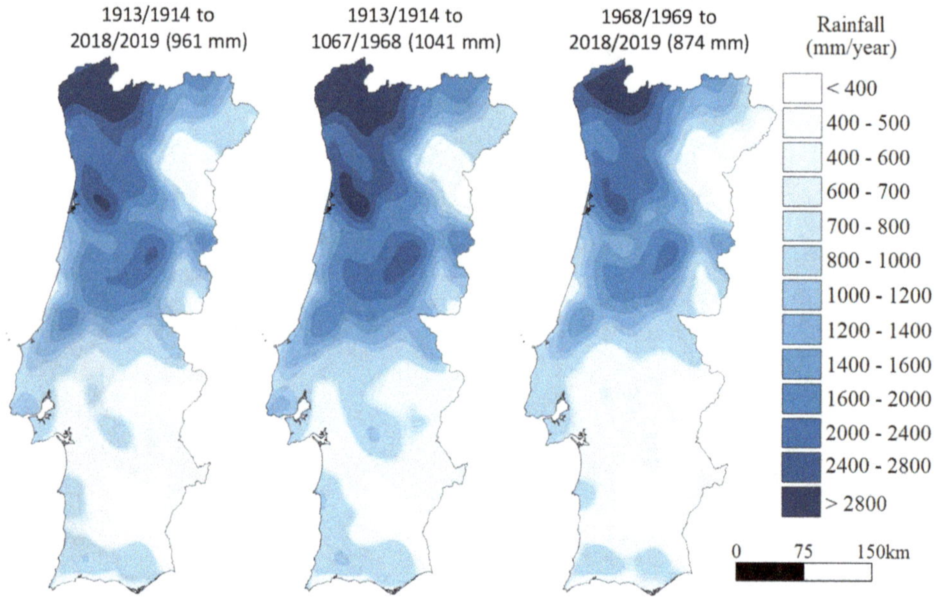

Figure 5: Average annual rainfall maps over the period from 1913/1914 to 2018/2019, and over the initial 55 and last 51 years. The respective weighted annual average is between brackets scales. *(Source: Adapted from [12].)*

In conjunction with the rainfall trend estimates, long-term rainfall anomalies were also calculated to provide a clearer indication of the vulnerability of Portugal to the declining rainfall. The variability of rainfall, i.e., anomalies, were defined, for each of the 532 stations, as the difference between averages of annual rainfall in the last years and in the initial years. The interpolated anomalies both in absolute and relative terms are depicted in Fig. 6 left and right side, respectively. The dimensionless anomalies were obtained by dividing the absolute values by their respective annual average in the initial 55-year.

As seen in Fig. 6 left side, some regions of the country denoted a substantial decrease of more than 500 mm in the average annual rainfall from the initial 55-year period to the last period. Fig. 6 right side on the other hand, evidences a more homogenous rainfall reduction in relative terms affecting the wet (northern) and dry (southern) regions. This suggests that the already dry south is getting even drier. When coupling these patterns with those already described for the rainfall trend and with the average annual rainfall maps, the difference between a drier northern Portugal (but still wet) and a certainly drier arid south is reinforced. This coincides with the claims of [16] who, based on a trend analysis of various hydrological data sets for the reference period from 1948 to 2005, pinpointed the southwest of the Iberian Peninsula as one hot spot of the "dry regions dry out further" pattern, i.e., of the DD paradigm.

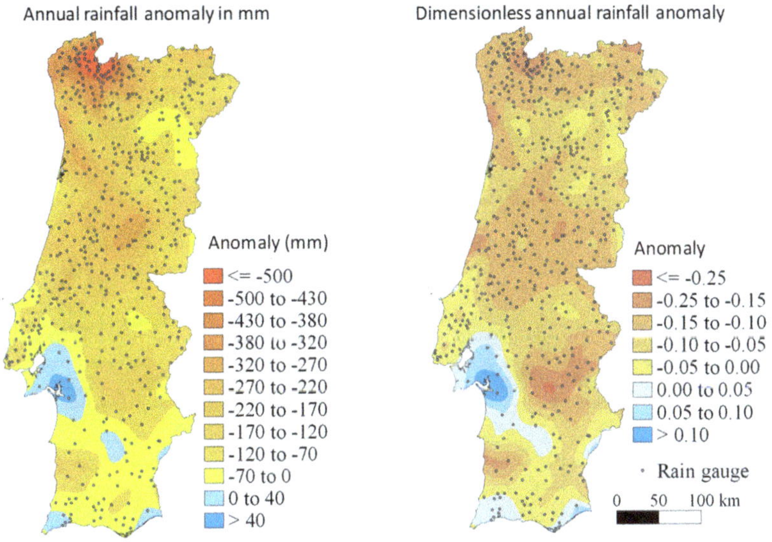

Figure 6: Differences between average annual rainfalls in the last 51-year period (from 1968/1696 to 2018/2019) minus the initial 55-year period (from 1913/1914 to 1967/1968) expressed in absolute terms (left side). The previous differences made relative to their respective average annual rainfalls in the 55-year initial period (right side).

Aiming at understanding the long-term within-the-year modification and uncertainty of the detected rainfall trends for the second trimester of quarter of the hydrological year, a sequential trend analysis was performed for the total period based on rainfall moving averages of the related months of Q2, from January to March. A period of 30 years was adopted as moving time window ($n = 30$ years). For comparison purposes, the rainfall values in each month at the 532 stations were made dimensionless regarding the weighted average annual rainfall of 961 mm (from 1913/1914 to 2018/2019 in Fig. 5).

Each monthly value was assigned to the year in the middle of the 30-year time window. For instance, the first moving average considered the years from 1913/1914 to 1942/1943 and was assign to 1928/1929. In each rain gauge and month, the application of this technique resulted in 77 moving averages for the 106-year period (with 532 values each), being the last moving average, from 1989/1990 to 2018/2019, assigned to 2004/2005. This allowed examining fluctuations of rainfall trend [17] by adopting a commonly used technique for smoothing out climate data series. Time windows of 30 years are also adopted in the computation of the climatological standard normals aiming at to ascertain the climatic conditions likely to be experienced in a given location [18], [19].

For each month of Q2, Fig. 7 shows some of the results from the moving average technique, related to the medians and empirical quantiles (Q) for non-exceedance probabilities between 5% and 95% (Q5% and Q95%, respectively) for the set of 532 rainfall series. For March the three curves thus obtained denoted an upward trend from the start of the total period with turning point in trend around late 1950s, reaching very low values regarding the long-term average. This dynamical pattern apparently "stabilised" from 1970 on with no upward or downward trend. This stabilised pattern was not detected for February nor January which could mean that the rainfall is still going through a somewhat active process of rainfall decrease. The figure also illustrates that the range between the quantiles Q5% and Q95% is getting narrower meaning less rainfall variability among the set of series from the 532 rain gauges, therefore, less uncertainty regarding the detected trends, particularly of the downward trends over mainland Portugal in earlier years.

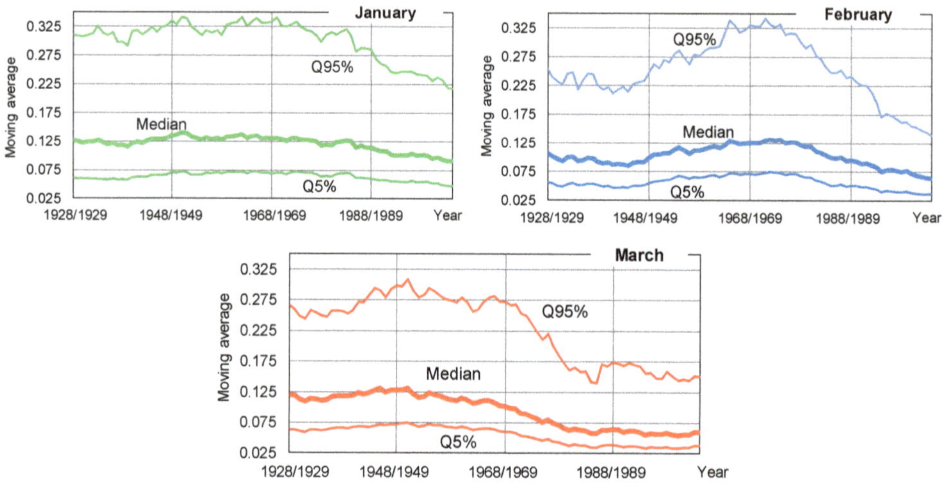

Figure 7: Moving average results, based on monthly rainfalls from the 532 rain gauges, characterised by the median and empirical quantiles for the non-exceedance probabilities of 5% (Q5%) and 95% (Q95%) in relative terms. *(Source: Adapted from [12].)*

5 FINAL REMARKS

Mainland Portugal, as for other regions in the globe, is susceptible to rainfall variability and change. Fluctuations in rainfall is a recurring phenomenon in the study area which has had a more negatively sustained and statistically significant behaviour in recent years as addressed in this work based on densely, updated and long-running rainfall series from 532 rain gauges (from 1968/1696 to 2018/2019 and two subperiods). Therefore, understanding past rainfall variations and updating rainfall trends at relevant time scales play a critical role in water resources management addressing, among different water-related challenges, the water security in the country in a well-recognised changing climate.

Although obtaining a single rainfall pattern for Portugal was not possible – due to the country's complex physical features and different climatic drivers involved in the rainfall process – the results showed strong evidence about the generalised, less uncertain, and widespread rainfall decrease in the country from early 1970 on. This negative trend in rainfall has been affecting the months that contribute the most to the annual rainfall (e.g., January and February), thus strongly determining the within-the-year rainfall pattern. These findings are backed, for instance, by the global evidence of a near future with lower freshwater availability in subtropical latitudes. Overall, this stresses the urge of establishing updated rainfall records to evaluate the change of rainfall and to provide useful information for future rainfall assessments. Furthermore, public participation is a key factor, since this practice calls for updated information and political commitment towards the implementation of mitigation and adaptation measures and of new water resources planning and management policies in the face of current and future severe climate change patterns. Both information and political issues have not yet been fully addressed for mainland Portugal [20].

ACKNOWLEDGEMENTS

For the first and third authors, this work is supported by the FCT – Scientific Cooperation Agreement between Portugal and Slovakia 2019/2020, SK-PT-18-0008, and for the second author by the FCT – grant number PD/BD/128509/2017.

REFERENCES

[1] Barros, V. et al., *Climate Change 2014: Impacts, Adaptation, and Vulnerability – Part B: Regional Aspects Contribution of Working Group II to the Fifth Assessment Report of the Intergovernmental Panel on Climate Change.* Cambridge University Press, 2014.

[2] Khan, M.Z.K., Sharma, A. & Mehrotra, R., Global seasonal precipitation forecasts using improved sea surface temperature predictions. *Journal of Geophysical Research: Atmospheres*, **122**, pp. 4773–4785, 2017.

[3] López-Moreno, J.I., Vicente-Serrano, S.M., Gimeno, L. & Nieto, R., Stability of the seasonal distribution of precipitation in the Mediterranean region: Observations since 1950 and projections for the 21st century. *Geophysical Research Letters*, **36**, 2009.

[4] Muñoz, J.M.H., Assessing the impact of climate variability on seasonal streamflow forecasting in the Iberian Peninsula. PhD thesis, Universidad de Granada, 2015.

[5] Ojeda, M.M.d.V.G.V., Climate-change projections in the Iberian Peninsula: A study of the hydrological impacts. PhD thesis, Universidad de Granada, 2018.

[6] Soares, P.M., Cardoso, R.M., Ferreira, J.J. & Miranda, P.M., Climate change and the Portuguese precipitation: ENSEMBLES regional climate models results. *Climate Dynamics*, **45**, pp. 1771–1787, 2015.

[7] Santos, F.D., Forbes, K. & Moita, R., *Climate Change in Portugal: Scenarios, Impacts and Adaptation Measures: SIAM Project*, Gradiva, 2002.

[8] Andrade, C. & Belo-Pereira, M., Assessment of droughts in the Iberian Peninsula using the WASP-Index. *Atmospheric Science Letters*, **16**, pp. 208–218, 2015.

[9] Páscoa, P., Gouveia, C., Russo, A. & Trigo, R., Drought trends in the Iberian Peninsula over the last 112 years. *Advances in Meteorology*, **2017**, 2017.

[10] Portela, M.M., Espinosa, L.A., Studart, T. & Zelenakova, M., Rainfall trends in Southern Portugal at different time scales. *International Congress on Engineering and Sustainability in the XXI Century*, Springer, pp. 3–19, 2019.

[11] SNIRH, *Sistema Nacional de Informação de Recursos Hídricos*, APA – Agência Portuguesa do Ambiente, 2020.

[12] Portela, M.M., Espinosa, L.A. & Zelenakova, M., Long-term rainfall trends and their variability in mainland Portugal in the last 106 years. *Climate*, **8**, p. 146, 2020. DOI: 10.3390/cli8120146.

[13] Mann, H.B., Nonparametric tests against trend. *Econometrica: Journal of the Econometric Society*, pp. 245–259, 1945.

[14] Kendall, M.G., Rank correlation methods. 1948.

[15] Sen, P.K., Estimates of the regression coefficient based on Kendall's tau. *Journal of the American Statistical Association*, **63**, pp. 1379–1389, 1968.

[16] Greve, P., Orlowsky, B., Mueller, B., Sheffield, J., Reichstein, M. & Seneviratne, S.I., Global assessment of trends in wetting and drying over land. *Nature Geoscience*, **7**, pp. 716–721, 2014. DOI: 10.1038/ngeo2247.

[17] Kenney, J., Moving averages. *Mathematics of Statistics*, **1**, pp. 221–223, 1962.
[18] WMO, International meteorological vocabulary. Technical Report, WMO/OMM/ IMGW, No. 182, Geneva, 1992.
[19] WMO, WMO guidelines on the calculation of climate normals. Technical Report, World Meteorological Organization (WMO)/OMM/IMGW, 182, Geneva, 2017.
[20] Carvalho, A., Schmidt, L., Santos, F.D. & Delicado, A., Climate change research and policy in Portugal. Wiley Interdisciplinary Reviews. *Climate Change*, **5**, pp. 199–217, 2014.

ANNUAL VAPOR, PRECIPITATION, GROUNDWATER, AND TRANSPIRATION FLUCTUATIONS IN SIMPLE WATER CYCLE PROCESSES

KOJI TAMAI
Forestry and Forest Products Research Institute, Japan

ABSTRACT

Cambodia has been actively developing its land use from forests to farmland or plantations. As a result, population, plantations, and agricultural land are increasing and the water resources must depend on groundwater. Therefore, the importance of groundwater as a water resource is increasing. Annual fluctuations of simple water cycle processes, with transitions from vapor in the atmosphere to groundwater, and transpiration processes that release soil water to the atmosphere are interesting to consider when assessing stable use of groundwater. The results of the monitoring data of simple water cycle processes I Kampong Thom, central Cambodia, show that the annual precipitation depends on the number of high-precipitation events rather than on the number of annual precipitation observation days and the dry evergreen forest examined for this study was judged as consuming water to the greatest extent potential with no restrictions. It is concluded that groundwater is useful for stability even if dry evergreen forests are developed into plantations and agricultural lands.

Keywords: monitoring data, dry and rainy monsoon, dry season, dry evergreen forest.

1 INTRODUCTION

In recent years, Cambodia has been actively developing its land use from forests to farmland or plantations. In Kampong Thom Province, the momentum is especially remarkable. In this area, the sediment carried by the Mekong River is deposited thickly, in many places reaching around 10 m (Ohnuki et al. [1]). Because large amounts of water can be stored in the soil, it is an area with few natural rivers where water flows on the surface. As a result, population, plantations, and agricultural land are increasing, even in areas (Fig. 1) that are several tens of kilometres distant from natural rivers. In such cases, the water resources must depend on groundwater. Therefore, the importance of groundwater as a water resource is increasing.

The tendency is even stronger in the dry season, when there is little precipitation from December through April. Precipitation, which is the source of groundwater, is brought about by southwest monsoons during the rainy season. Water in Cambodia becomes soil water and groundwater through vapor and precipitation that are borne with the southwest monsoon. Furthermore, part of the soil water is consumed by the transpiration activity of the remaining forest trees. Annual fluctuations of simple water cycle processes, with transitions from vapor in the atmosphere to groundwater, and transpiration processes that release soil water to the atmosphere are interesting to consider when assessing stable use of groundwater. Therefore, comparisons of annual fluctuations on meteorologically significant dates (e.g. the beginning of the monsoon, the rainy season, and the dry season), annual precipitation, evapotranspiration from forests, groundwater level, and other matters are reviewed and discussed herein to assess groundwater stability as a water resource.

2 OBSERVATION

For this study, meteorological data and the groundwater level were observed at a 60-m weather observation tower site in the O Thom I watershed (12°44′N, 105°28′E, 88 m asl) in

WIT Transactions on Ecology and the Environment, Vol 250, © 2021 WIT Press
www.witpress.com, ISSN 1743-3541 (on-line)
doi:10.2495/WRM210021

Kampong Thom Province, central Cambodia. Meteorological data have been stored there every 10 min. The groundwater level was measured manually once daily at the well with 10 m depth. The forest around the tower, categorized as "dry evergreen forest", consists of evergreen trees with tree density of 1600 stems ha^{-1}. The average canopy surface height (h) is 27.2 m. However, there are several emergent trees taller than 35 m occur per hectare (Shimizu et al. [2])

Figure 1: Conversion from dry evergreen forest to rubber plantations (left) and cassava farmland (right) in Kampong Thom Province, Cambodia.

3 METHOD

3.1 Determination of the beginning dates of dry and rainy monsoons and dry and rainy seasons

Wind direction data were used to ascertain the beginning dates of dry and rainy monsoon seasons as followings (Nang et al. [3]). Fig. 2 presents aggregated results for 2007 as an example.

1) Wind direction data were categorized as dry monsoon (northeast–northwest), wet monsoon (west–south), and others (southeast–east).
2) The proportion of each category was calculated for each day.
3) Proportions over 50% for the dry monsoon (□) and wet monsoon (●) are shown in Fig. 2.
4) Periods with high density of □ and ● respectively show that the dry monsoon and wet monsoon are predominant.
5) In Fig. 2, the beginning date of the rainy and the dry monsoon were determined from the distribution of □ and ●; they are indicated respectively by a solid arrow and a broken arrow.

Precipitation data were used as follows to ascertain the initial dates of dry and rainy seasons.

1) Accumulated precipitation lines are depicted in Fig. 2.
2) Lines with a steep decline (thick line) and a gentle decline (thin line) are inferred as respectively representing the dry and rainy seasons.
3) Right and left ends of the thick line were regarded respectively as the beginnings of the rainy and the dry seasons.

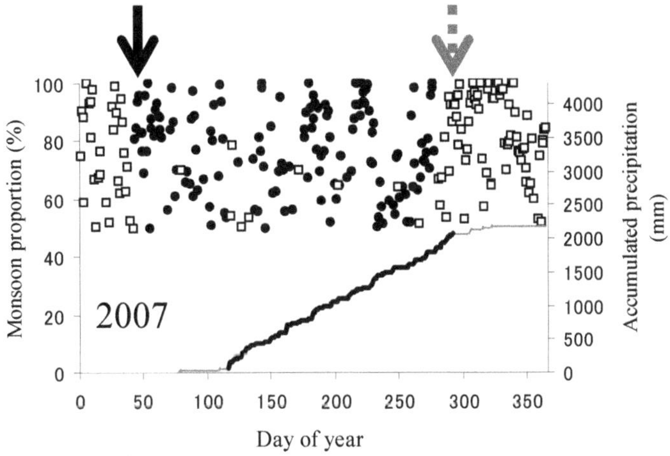

Figure 2: Wind direction and precipitation data analysis to judge the periods of dry and rainy seasons and monsoons in 2007 [3].

3.2 Estimation of evapotranspiration and potential evaporation rates from the evergreen forest

Evapotranspiration rate (E) is obtained using the Bowen ratio method expressed as

$$\lambda E = (R_n - G)/(1 + \text{Bowen}) \tag{1}$$

$$\text{Bowen} = \gamma(TD_1 - TD_2)/C \tag{2}$$

$$C - (\Delta + \gamma)(TW_1 - TW_2) - \gamma(TD_1 - TD_2) \tag{3}$$

where λ represents the heat of vaporization of water, Bowen is the Bowen ratio, γ stands for the psychrometer formula to be 0.66 hPa°C^{-1} and Δ denotes the slope of the saturated water vapor pressure curve when the vertical axis is the saturated water vapor pressure and the horizontal axis is the temperature. Also, TD_i and TW_i respectively represent the ventilated dry and wet bulb temperatures (i = 1 and 2 respectively denote the higher and lower observation height, R_n expresses the net radiation, and G stands for the soil heat flux observed as the meteorological data.

The potential evaporation rate (E_p), indicating the climatic conditions that enforce evaporation from the water surface, is calculated with the Penman [4] equation expressed as

$$lE_p = \frac{\Delta(R_n - G)}{(\Delta + \gamma)} + \frac{\Delta}{\Delta + \gamma} f(u)VPD \tag{4}$$

$$f(u) = 0.26(1.0 + 0.537u) \tag{5}$$

where u denotes the wind speed and VPD represents the vapor pressure deficit calculated with TW_1 and TD_1. Also, E_p is an index of the temporal variation of the climatic conditions that force evaporation.

4 RESULTS AND DISCUSSION

4.1 Seasonal course of absolute humidity

Fig. 3 shows the seasonal course of absolute humidity in 2004–2011. In each figure, ◆, ◇, ▲, and △ respectively denote the beginning date of rainy monsoon, rainy season, dry monsoon, and dry season, with the method explained in III-1. The daily average absolute humidity in the rainy season was not only larger than that in the dry season, but also day-to-day fluctuated less. Even more interesting is that the diamond or triangle dots are not located at the trend change points of seasonal course, but in the middle of an increasing or decreasing trend of absolute humidity: results show that the amount of water vapor in the air does not change drastically depending on the change of the monsoon and the presence or absence of precipitation.

4.2 Year-to-year fluctuations at the beginning date of rainy and dry monsoon and rainy and dry seasons

Fig. 4 shows the beginning dates of rainy and dry monsoon and rainy and dry seasons, as reported by Nang et al. [3]. Symbols of dots have the same meaning as those of Fig. 2. At the change from the dry to rainy season, ◆ and ◇ occur respectively over widely various DOY 25–95 and DOY 85–120. The time lags between ◆ and ◇ are also long, ranging from 10–80 days.

In contrast, at the changes from the rainy season to the dry season, ▲ and △ occur respectively over a much narrower range of DOY 275–295 and DOY 285–310. Time lags between the beginning date of dry monsoon and dry season are also short or are absent, as shown for 2010 and 2011. The results presented above indicate that the change from the dry to the rainy season occurs in steps. The date of the change in season varies widely. However, the change from the rainy to the dry season is clearer. It occurs over a shorter time period. The date of the change in season determined from precipitation data was fixed at around DOY 290.

4.3 Year-to-year fluctuations of the precipitation amount

Year-to-year fluctuations in annual precipitation were compared for four years. The annual precipitation was high at 2100.0 mm and 1885.4 mm in 2007 and 2011, respectively, and low at 1138.0 mm in 2010, respectively.

Fig. 5 shows the number of precipitation observation days (x-axis) and the change in cumulative precipitation (Y-axis) as the number of days increases. The right end of each line shows the number of annual precipitation observation days (X-axis) and the annual precipitation (y-axis). Both the number of annual precipitation observation days and the annual precipitation in 2007 are the lowest in the four years. However, the annual precipitation in 2008, when the number of annual precipitation observation days was the highest, was lower than in 2007 and 2011.

Figure 6 shows values of daily precipitation arranged in ascending order. Therein, ◇ stands for the value of daily precipitation that was 50 mm day-1 or more. The numbers of days when the daily precipitation was 50 mm or more were 1 day in 2010, 3 days in 2008, 9 days in 2007, and 7 days in 2011, which are of almost identical order to the amount of annual precipitation. Especially in 2011, when the annual rainfall was the second highest,

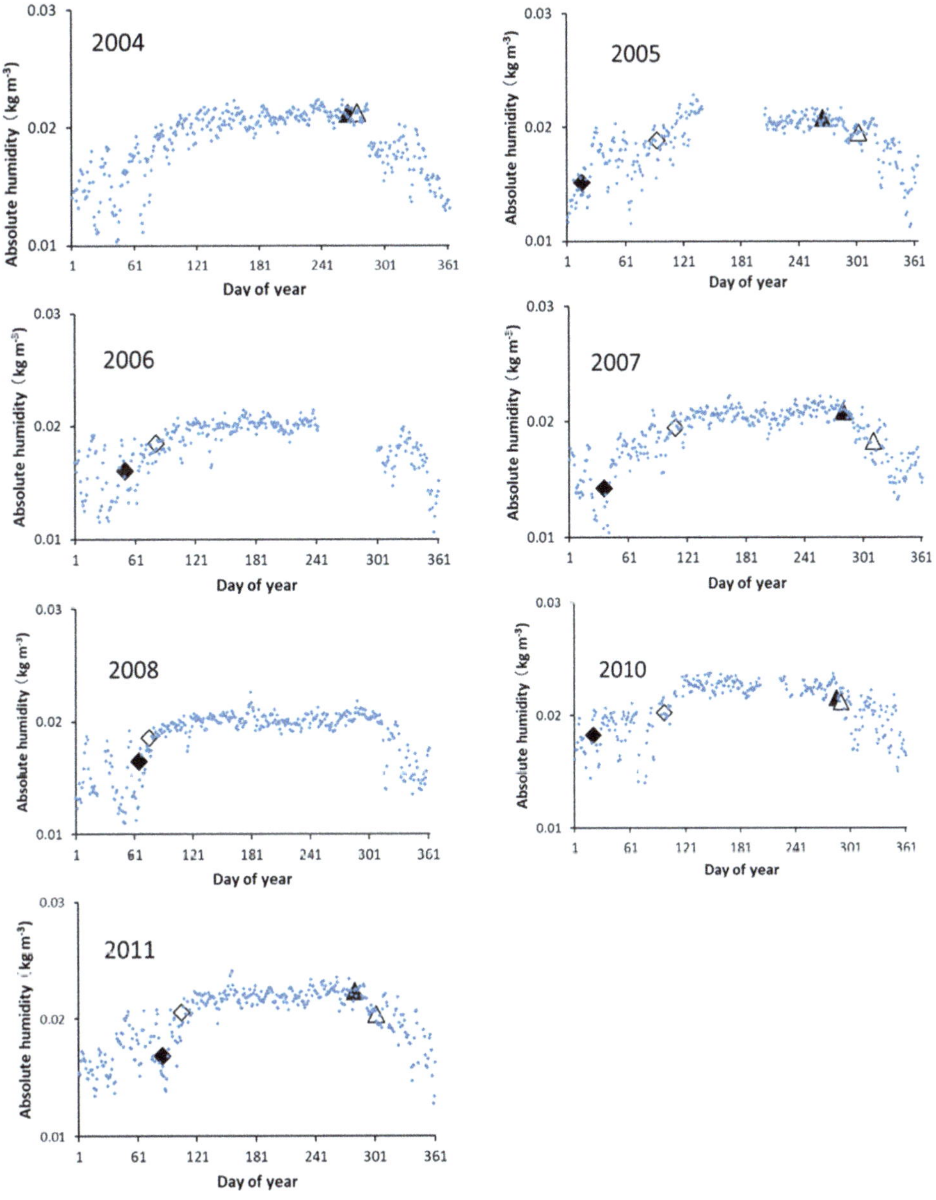

Figure 3: Seasonal course of absolute humidity in 2004–2011. In each figure, ◆, ◇, ▲, and △ respectively denote the beginning date of rainy monsoon, rainy season, dry monsoon, and dry season, with the method explained in Section 3.1.

there was a day with a particularly high daily rainfall of 138.4 mm. Based on the data presented above, results show that the annual precipitation depends on the number of high-precipitation events rather than on the number of annual precipitation observation days.

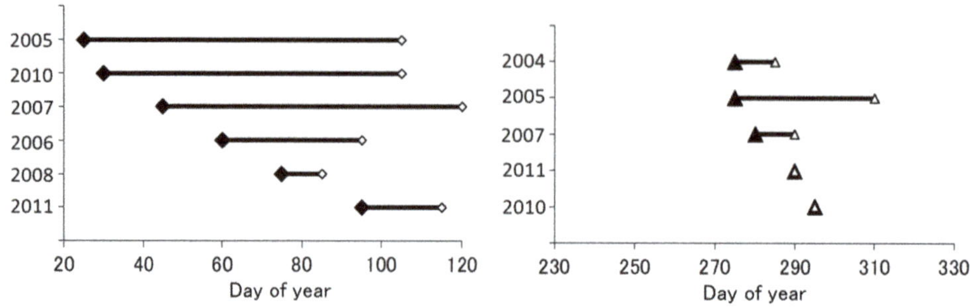

Figure 4: Beginning dates of rainy monsoon and rainy season (left) and dry monsoon and dry season (right). ◆, ◇, ▲, and △ respectively denote the beginning date of rainy monsoon, rainy season, dry monsoon, and dry season [3].

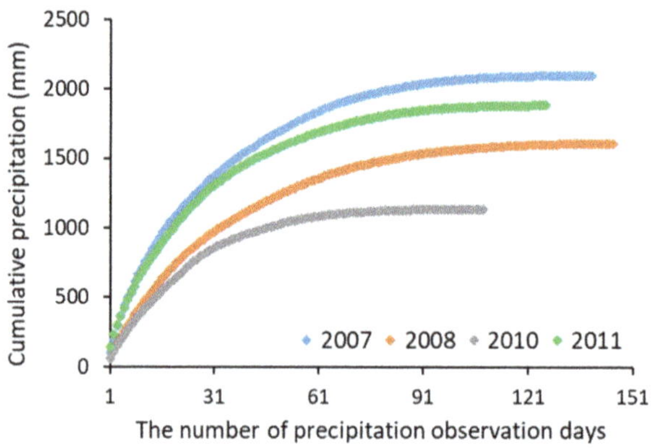

Figure 5: Change in cumulative precipitation as the precipitation observation days increases.

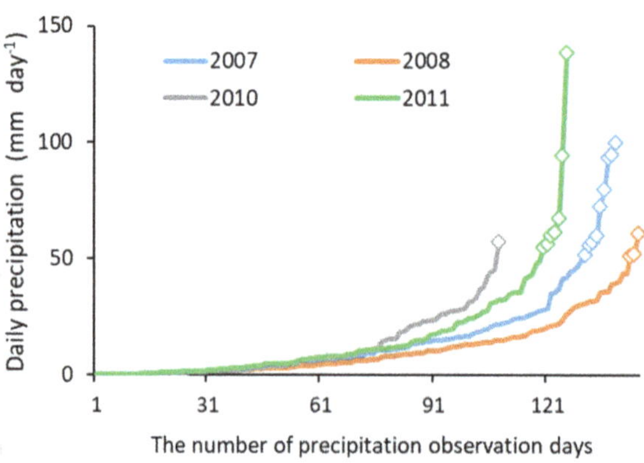

Figure 6: Values of daily precipitation arranged in ascending order. White diamonds stand the value of daily precipitation that was 50 mm day-1 or more.

4.4 Seasonal course of precipitation and groundwater level

The monthly precipitation, monthly averaged groundwater level, monthly averaged Ep and monthly averaged E are shown respectively in Fig. 7 for 2004–2007. Moreover, the averaged rates (ARs) and standard deviations (SDs) were calculated with these monthly rates in each month. The year-to-year variation in each monthly value was compared with the AR and SD. In each panel of Fig. 7, three lines are drawn on the plots. The squares represent the monthly values each year, whereas the thick line shows the average rate for the 4 years. The two thin lines show SD. Values shown above and under the thick lines respectively represent greater and smaller monthly rates. Values shown above the upper thin line represent much larger monthly rates than usual, whereas values shown below the lower thin line denote much smaller monthly rates than usual.

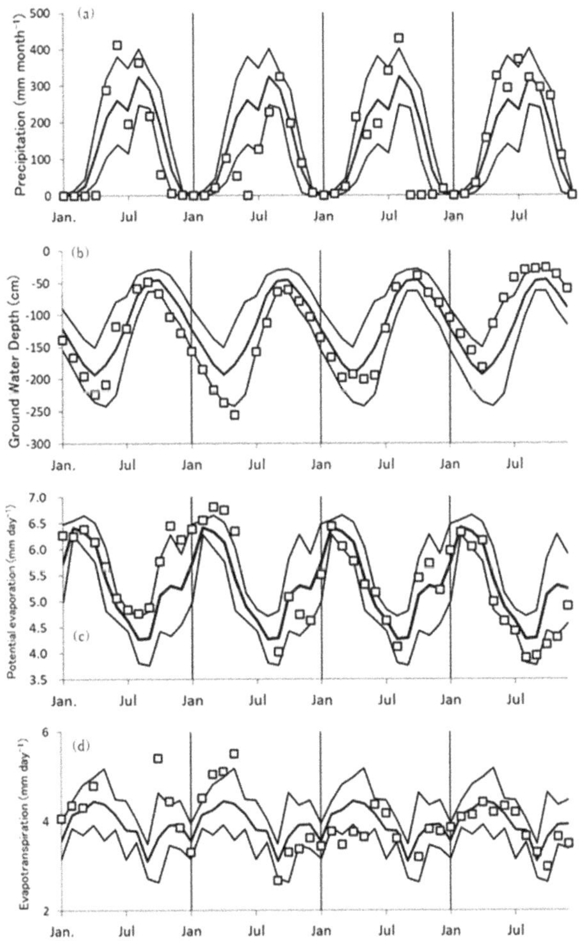

Figure 7: (a) The monthly precipitation; (b) Monthly averaged groundwater level; (c) Monthly averaged Ep; and (d) Monthly averaged E. The averaged rates (Ars; thick line) and standard deviations (SDs; thin lines) were calculated with these monthly rates in each month [5].

4.5 Comparing seasonal fluctuations in precipitation and groundwater level

Precipitation is mainly observed during the rainy season: April–October. The lowest groundwater levels of the year are observed in April and May, which mark the beginning of the rainy season. Therefore, water resources are regarded as tightest in April and May. The amounts of water resources at that time are expected to depend on the amount of precipitation in the rainy season of the prior year. For example, the lowest monthly average groundwater level was observed in May 2005 during the four years depicted in Fig. 7(b) because precipitation tended to be low in September–December 2004.

4.6 Comparing groundwater level and potential evaporation and evapotranspiration

As described in the preceding section, the annual lowest monthly averaged groundwater levels were observed in April of 2004 and 2007 and in May of 2005 and 2006. These months are regarded as the most severe water environment not only for human activities but also for forest trees. Therefore, we compared the monthly averaged daily potential evaporation and the monthly averaged daily evapotranspiration during the month when the groundwater level dropped the most in each year (Fig. 8). Potential evaporation is an index of the temporal variation of the climatic conditions that force evaporation. Trees, by contrast, close their stomata to prevent water loss when they run out of water. Therefore, when the soil is dry and trees cannot absorb a sufficient amount of water, the evapotranspiration rate decreases: when the tree cannot secure sufficient water from the soil, the amount of evapotranspiration decreases to a relative degree. However, in Fig. 8, four points are arranged in a straight line. Therefore, even during a month when the annual minimum monthly average groundwater level is observed, trees can secure sufficient water from the soil, so that photosynthesis and evapotranspiration can be performed without restrictions without closing the stomata. Fig. 9 presents a comparison of the annual minimum monthly average groundwater level with average daily evapotranspiration of the

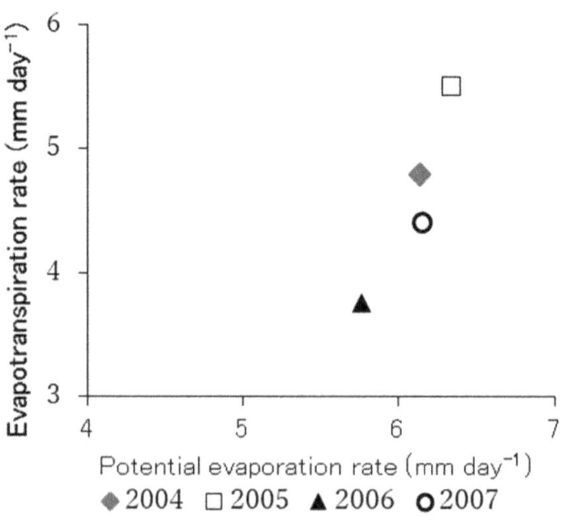

Figure 8: Comparison between the monthly averaged potential evaporation and the monthly averaged daily evapotranspiration during the month when the groundwater level dropped the most in each year.

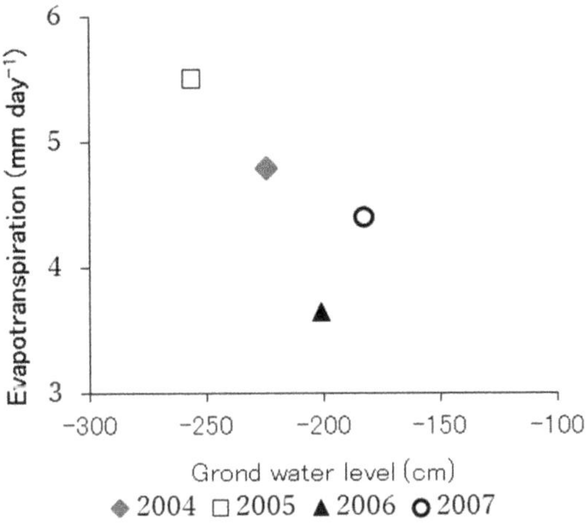

Figure 9: Comparison of the annual minimum monthly average groundwater level with averaged daily evapotranspiration of the same month.

same month. The monthly average daily evapotranspiration in May 2005, for which groundwater level was observed and which is apparently the most severe water environment, is higher than the monthly average daily evapotranspiration in the other three years. As discussed above, the dry evergreen forest examined for this study was judged as consuming water to the greatest extent potential with no restrictions.

5 CONCLUSION

This study examined the sustainability of groundwater as a water resource. Many agricultural areas around the world depend on groundwater for their water resources. The Great Plains of the United States are threatened because the groundwater level continues to drop as a result of excessive pumping of groundwater from the Ogallala Aquifer, which supplies water resources. In Kampong Thom, Cambodia, the subject of this report, dry evergreen forests consume water without limitation in the natural environment because, even though there is a long dry season, rainfall of several thousand millimetres invariably comes during the rainy season. A thick soil layer exceeding 10 m can store vast amounts of groundwater. Water consumption by plantations and crops is unlikely to exceed that of dry evergreen forest considerably. Therefore, we infer groundwater as a sustainable resource that is useful for stability even if dry evergreen forests are developed into plantations and agricultural lands. However, when groundwater is used for production other than plants, such as for industry and livestock, much more water may be consumed than by dry evergreens. In such cases, there is a possibility of groundwater depletion.

REFERENCES

[1] Ohnuki, Y., Shimizu, A., Chann, S., Toriyama, J., Kimhean, C. & Araki, M., Seasonal change in thick regolith hardness and water content in a dry evergreen forest in Kampong Thom Province, Cambodia. *Geoderma*, **146**, pp. 94–101, 2008.

[2] Shimizu, A. et al., Runoff characteristics and observations on evapotranspiration in forest watersheds, central Cambodia. *Forest Environments in the Mekong River Basin*, eds H. Sawada et al., Springer-Verlag: Tokyo, pp. 135–146, 2007.

[3] Nang, K. et al., Determination of seasonal change from dry to rainy using wind and precipitation data – Case study of Kampong Thom, central Cambodia. *Proceedings of 8th International Workshop on Forest Watershed Environment Research in Cambodia*, pp. 11–14, 2012.
[4] Penman, H.L., Natural evaporation from open water, bare soil and grass. *Proceedings of the Royal Society of London A*, **193**, pp. 120–145, 1948.
[5] Chann, S. et al., Interannual variation in the evapotranspiration of an evergreen forest in Kampong Thom, central Cambodia. *Proceedings of International Workshop on Forest Research in Cambodia*, pp. 21–24, Nov. 2011.

WATER GOVERNANCE PERFORMANCE ASSESSMENT OF RIPARIAN COMMUNITIES: A CASE STUDY OF THE VOLTA RIVER BASIN

ABDUL-RAZAK ZAKARIA & KENICHI MATSUI
University of Tsukuba, Japan

ABSTRACT

Governance inefficiencies can exacerbate flooding in developing countries and regions where flood protection infrastructures are particularly weak. This is the case in the Volta Basin of western Africa where irregular flood–drought cycles have devastated the livelihoods of riparian communities. It is critical to know how water governance can better respond to floods. This paper assesses water governance performance in the Volta Basin, especially Burkina Faso–Ghana border areas. The discussion is largely based on our original questionnaire survey that sought to understand riparian perceptions and needs regarding four key water governance principles: public participation, gender equity, transparency and regulatory quality. With a five-point Likert-scale, we sought respondents' level of agreement and or satisfaction regarding statements about these governance principles. Regarding public participation, the results show that 49% of the respondents in Ghana and 42% of those in Burkina Faso thought that their communities were not sufficiently engaged in program implementation. Concerning transparency, 54% and 55% in Ghana thought that it was easy to access information and that decision-making processes were easy to understand. In Burkina Faso, 59% felt it difficult to access information, and 63% was not sure whether decision making processes were easily understood. With regards to regulatory quality, 58% of the respondents in Ghana and 54% of those in Burkina Faso were satisfied with regulatory quality. On gender equity, 72% in Ghana and 78% in Ghana Burkina Faso found it challenging. These findings are discussed in a socio-economic context to illustrate the current status of water governance in the study areas. Based on these findings, we discuss how transboundary water governance in Africa may become more efficient in the future.

Keywords: water governance, public participation, gender equity, transparency, regulatory quality, Volta River Basin, riparian communities' perceptions, assessment, Burkina Faso, Ghana.

1 INTRODUCTION

Amid a global trend of increasing water demand FAO [1] and water related disasters, governance solution has gained much attention [2]–[4]. UN-Water stated that "water crisis is essentially a crisis of governance" [5]. The FAO also included water governance as one of four components of food security [6].

Global Water Partnership [7] defined water governance as the "spectrum of political, social, economic and management structures in the development and management of water resources and water services across different levels of society" [8]. Other definitions of water governance highlight the importance of how decisions are made, by whom, and under what conditions [6], [9]–[12].

Kaufmann et al. [13] reported strong causal relationships between better governance and better development outcomes. Hanasz [14] maintained that good water governance encourages stakeholder participation in decision-making and engenders accountability. Rogers and Hall [8] argued that effective or good water governance can be achieved when all relevant sectors actively participate in decision-making.

Among the most contentious water governance issues is related to the development and operation of large dams that often-deprived downstream regions or countries of reasonable water flow. Lower stream communities also suffered from irregular flood damage due to

WIT Transactions on Ecology and the Environment, Vol 250, © 2021 WIT Press
www.witpress.com, ISSN 1743-3541 (on-line)
doi:10.2495/WRM210031

spillage from dams [15]–[18]. Because of these problems, in the last ten years rising political tensions have been observed between riparian communities in Ghana and Burkina Faso. Although this governance issue can affect more than 20 million people in western Africa by 2050, we do not know much about how governance can be a solution to these rising tensions. This paper, therefore, attempts to assess water governance performance in the Volta River Basin, especially between these two countries. In particular, as rising tensions have been incubated among local communities, it focuses on perspectives of riparian communities with particular focus on such salient governance aspects as public participation, transparency, gender equity and the quality of regulations.

2 CRITERIA FOR GOOD WATER GOVERNANCE

Our examination of these four principles for the purpose of assessing water governance performance was based largely on rich sources of past studies. Rogers and Hall [8] examined operational, organizational, and constitutional levels of governance. The organizational level coordinates and reduces conflicts between competing water interests. It also enforces rules and polices for water usage and users. Here they identified public participation, transparency, gender equity and the quality of regulations as essential components.

International water governance organizations and scholars have recognized the importance of appropriate public participation and transparency in water governance practices. The International Commission for the Protection of the Rhine (ICPR), for instance, noted that good transboundary cooperation in the Rhine was due partly to the long tradition of cooperation among stakeholders [19]. Paul et al. [20] reiterated that public participation offers a voice for all actors in decision making. Good governance requires transparent decision-making processes that ensure traceability of the process by internal and external actors [8]. Transparency enhances public knowledge and access to information. It also promotes accountability [9].

Equal representation and benefit sharing among various stakeholders in the policy development and implementation process ensures fairness [8]. Good water governance must address gendered issues [9].

Regulatory quality determines the accountability of governance practices. Good accountability reassures the public about institutional and administrative framework to address fair, equitable, and transparent processes. Under good regulations, water administrations are required to act to address complaints and requests from stakeholders [9], [20].

Scholars and international organizations have attempted to assess the level of good governance by employing a set of criteria. Pre-requisite conditions of good governance for some scholars included inclusiveness, consensus-making arrangement, accountability, participation, transparency, responsiveness, equity and inclusiveness, effectiveness and efficiency, and lawfulness [8], [14]. The UNDP (1997) proposed that good governance should be assessed by 10 criteria: public participation, consensus orientation, strategic vision, responsiveness, effectiveness, efficiency, accountability, transparency, equity, and rule of law [20]. The Global Water Partnership similarly emphasized openness, transparency, participation, accountability, effectiveness, coherence, efficiency, communication, equity, integration, sustainability, and ethics [12]. Mahon et al. [21] proposed to assess transboundary water governance by governance architecture, governance processes, social justice, ecosystem pressure, stakeholder engagement, ecosystems state, and human well-being.

Some of these criteria were further examined in past studies. Public participation, for example, was assessed by local needs, facilitating conditions, legal frameworks, equity, and decision-making. Equity in water governance was assessed by gender-roles, the performance of water institutions, and justice. The assessment of transparency was done on the basis of public knowledge dissemination, information access for decision-making, openness about processes, and information sharing. Accountability was assessed by regulatory reform (e.g., anti-corruption measures), institutional actions and performance, and public opinion about responsible authorities [9], [20], [22].

The performance level of water governance was assessed by indicators. For instance, Mahon et al. [21] proposed a new Transboundary Waters Governance Assessment Framework. It contains seven somewhat overlapping indicators: governance architecture, governance process, stakeholder engagement, social justice, ecosystem pressure, ecosystem state and human well-being. It then determined the level of governance architecture by the existence of governance procedures. Governance processes were evaluated by plans and function. Stakeholders' engagement was strongly linked to the capacity development of the public for effective engagement. Social justice meant equity. Ecosystem pressure was measured by the extent to which some policies resulted in behavioral changes in favor of environmental conservation. Ecosystem's state was described in terms of whether changes in behaviors resulted in environmental conditions. Human well-being was linked to both social justice and the sustainability of ecosystem goods and services that, in turn, benefit people.

3 STUDY AREA

In order to assess water governance performance, this paper provides a case study on a transboundary region of the Volta River Basin, one of the largest and most important river basins in west Africa (Fig. 1). We chose three riparian communities each along the Volta River that runs through Burkina Faso and Ghana. In Ghana, we looked at the Bawku zone in the Upper East Region. In Burkina Faso, we examine Bagre district in the Eastern Central Region where Bagre Dam is located. Our preliminary field visit to these areas made us choose these communities to focus on flood damage due to the dam spillage.

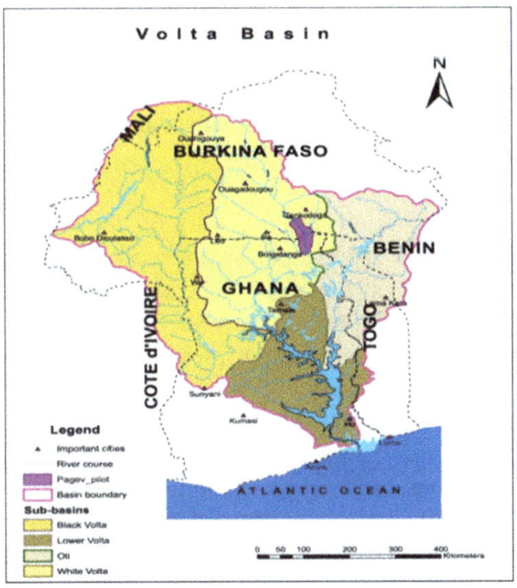

Figure 1: Volta River Basin [30].

The Bawku zone consists of six administrative districts, of which four districts are directly affected by Volta River flows. These districts are Binduri, Garu, Bawku Municipality and Bawku West districts. We selected three communities from these districts as these were considerably affected by dam spillage in the past. These three districts also have similarities in terms of climate, soil type, farming system, culture, language and the crops grown [23]–[25]. Here the climate is characterized by the wet season from May to October and the dry season from November to April. The total population in the three selected districts was 290,117 according to the 2010 population and housing census. Average annual rainfall ranges from 950 mm to 1,100 mm. Vegetation here is the Sahel Savannah type with scattered shrubs, short grasses, and trees. Agriculture is the mainstay with such major food crops as maize, rice, sorghum, pepper, watermelon, and onion. Farmers draw water from the Volta River [23]–[26].

Bagre is one of administrative districts of Burkina Faso's Eastern Central Region. It is located below the Bagre dam area [27]. The climate is characterized by dry (harmattan) and wet (rainy) seasons. Annual rainfall ranges between 850 to 1,100 mm [28]. It is home to 19,164 people with various ethnicities. Due to agribusiness opportunities here, people from various parts of the country come to this area. Agriculture employs about 80% of the population. Crop farming, livestock breeding, and fish culture mainly constitute their livelihood. Rice, maize, vegetable farming and fruits (e.g., mangoes) are produced all year round by irrigation from Bagre dam reservoir [29].

4 MATERIALS AND METHODS

Much of our discussion below is based on the data we collected from our questionnaire survey in Ghana and Burkina Faso. We conducted the field survey between 1 December 2019 and 30 January 2020. We applied simple random sampling and administered the questionnaire among 150 individuals (50 from each community) in Ghana. In Burkina Faso, 90 individuals (30 from each community) were targeted. Altogether, we collected effective answers from 238 respondents. The response rate was 100% in Burkina Faso and 99% in Ghana. Due to the limited reading and writing skills of our respondents, we obtained cooperation from the Ministry of Food and Agriculture personnel in local stations as enumerators. Also, to overcome language barrier, we translated our questions from English into Mossi and Kussal for our respondents in Burkina Faso and Ghana, respectively.

Israel [31] provides a simplified equation for calculating sample sizes. For a 95% confidence level and an estimated proportion of an attribute present in the population,

$$p = 0.5;\ n = N/1 + N(e)^2 \tag{1}$$

where n = sample size, N = population size, and e = 0.05, the level of precision.

For a total population of 309,281 people in our study area, the sample size, n = 399 respondents. The sample size was, however, dependent on the number needed for data analysis. Israel [31] notes that a sample size between 200 and 500 is necessary for multiple regression, an analysis of covariance and log-linear analysis but nearly any sample size suffices for descriptive statistics like frequencies, mode and Chi-squared.

The questionnaire had two parts. The first part was designed to clarify socio-demographic characteristics of the respondents. The second part focused on water governance perceptions by using a set of Likert-type statements and questions. Boone and Boone [32] distinguished Likert-type statements from Likert-scale statements. Likert-type statements are single questions without an intension to combine responses into a composite scale. A Likert-scale, however, comprises four or more Likert-type statements that are to be combined into a single composite score. In total 10 items comprising five statements and five questions constituted the second part of the questionnaire (Table 1). These statements were revolved around four key governance principles: public participation, transparency, gender equity and regulatory quality. The five-point Likert-scale we used to assess the degree of public participation and gender equity in decision-making asked the respondents to select the degree of their agreement or opinion to given statements. To assess the level of transparency, five-point Likert-scale questions had choices that ranged from "not easy at all" to "very easy". To rate efforts made by water managers, five-point Likert-scale questions asked the respondents to choose from "very low" to "very high". In assessing regulatory quality, we first assessed respondents' knowledge about local governance authorities responsible for water governance by using list of institutions to choose. We then asked respondents to rate their level of satisfaction.

Table 1: Governance principle and Likert-type statements.

Governance principles	Questions and statements
Public participation	(1) My community is sufficiently involved in the decision-making
	(2) Customs, religion, and norms of my community are recognized
	(3) My community has an avenue for receiving complaints or raising suggestions about basin
Gender equity	(1) Women and men have equal participation in decision making
	(2) Women and men have equal access to benefits from this basin
Transparency	(1) How easy to access information about spillage of excess Bagre dam water?
	(2) How easy institution makes processes of decision making about water governance in the basin?
	(3) How will you rate water manager's efforts to make stakeholders understand decision making process?
Regulatory quality	(1) Who manages the Volta River Basin?
	(2) How satisfied are you with the regulatory performance of enforcing agencies?

Largely agreeing with Boone and Boone [32], we use descriptive statistics as an appropriate method to discuss the results from Likert-type statement questions. The Pearson Chi-squared (χ^2) was used to understand correlations between respondents' socio-demographic characteristics and their perceptions about governance performance [33]. Null hypothesis (H_o) was that there is no significant difference between respondents' socio-demographic characteristics and their perceptions about governance performance. H_o was tested at 0.05 level of significance and was rejected if the p-value is lower than the significance level. However, when the p-value is higher than the significance level, then we accept H_o. We further used stack bar graphs to visualize the result of the survey. The stack bar graph plots responses with each row showing percentage breakdown at each step of the

Likert scale. An agreement with Likert-type statement is represented by a bar above zero [34]. The mode represents the response with the highest occurrence on a scale of 1 (strongly disagree) to 5 (strongly agree). SPSS version 23 and Statgraphics18 aided data analysis.

5 RESULTS AND DISCUSSION

5.1 Socio-demographic characteristics of respondents

As mentioned above, we obtained 238 valid responses in the survey, of which 148 were from Ghana and 90 were from Burkina Faso. The mean age of the Ghanaian respondents was 39.7 years old and that in Burkina Faso was 40.5 years old. In Ghana, 36% of the respondents were between 18 and 39 years old, and 57% was between 40 and 59 years old. On the contrary, in Burkina Faso, 53% belonged to 18–39 age bracket and 37% to 40–59 age bracket. These results imply that although the mean age in the two countries did not show much difference, more Ghanaian respondents tended to be older than those in Burkina Faso. Only 43% and 40% of the respondents in Ghana and Burkina Faso had some form of education (Table 2).

Our survey also revealed that 90% of the respondents in Ghana had lived close to the Volta River for 11 to 40 years whereas 63% did so in Burkina Faso (Table 2).

Table 2: Socio-demographic characteristics of respondents.

Socio-demographics		Burkina Faso	Ghana
Age	18–29	18 (20%)	9 (6%)
	30–39	30 (33%)	44 (30%)
	40–49	19 (21%)	53 (36%)
	50–59	14 (16%)	32 (21%)
	60+	9 (10%)	10 (7%)
Gender	Male	45 (50%)	97 (66%)
	Female	45 (50%)	51 (34%)
Education	No formal education	55 (60%)	84 (57%)
	Primary	18 (20%)	24 (16%)
	Junior high school	9 (10%)	18 (12%)
	Senior high school	4 (5%)	17 (11%)
	Tertiary education	4 (5%)	5 (4%)
Occupation	Artisan	2 (2%)	0 (0%)
	Farmer	63 (70%)	146 (98%)
	Student	3 (3%)	1 (1%)
	Teacher	6 (7%)	0 (0%)
	Trader	16 (18%)	1 (1%)
Years of residency in the study area	5–10	9 (10%)	55 (37%)
	11–20	50 (55%)	55 (37%)
	21–30	24 (27%)	17 (11%)
	31–40	7 (8%)	14 (10%)
	41–50	0 (0%)	3 (2%)
	51–60+	0 (0%)	4 (3%)
Total		90 (100%)	148 (100%)

5.2 Governance performance

To understand respondents' level of public participation in the Volta River Basin, we asked the respondents to respond to three statements (Table 1). The first statement was "my community is sufficiently involved in decision-making". The second statement was "customs, religion and norms of my community are recognized". The third statement was "my community has an avenue for receiving complaints or raising suggestions about the basin".

In response to the first statement, 24% of the respondents in Ghana agreed whereas 49% disagreed. In Burkina Faso, the results are somewhat different as 42% agreed and only 17% disagreed. About the second statement, 78% of the respondents in Ghana disagreed whereas 34% of the Burkina Faso respondents did so. Another 34% agreed with the statement. About the third statement, 52% in Ghana disagreed whereas 29% did so in Burkina Faso (Fig. 2). This means that the Ghanaian respondents tended to feel alienated from decision-making processes in connection to Volta River governance. Chi-squared analyses on these results revealed that although the Burkina Faso respondents tended to have higher participation than those in Ghana, gender disparity was observed. In Burkina Faso, more men (55%) agreed with the first statement. Regarding the third statement, more women (54%) agreed.

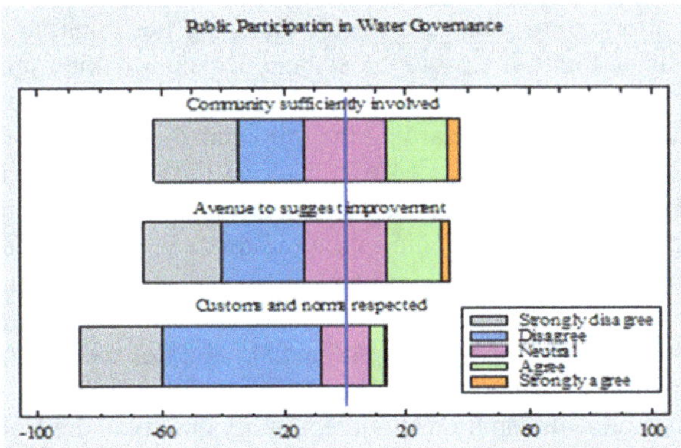

Figure 2: Public participation, Ghana.

To understand respondents' opinion about gender equity in the Volta River Basin, we posed two statements (Table 1): (1) "women and men have equal opportunity in decision making" and (2) "women and men have equal opportunity in sharing benefits". In response to the first statement, 72% of the respondents in Ghana disagreed whereas 78% in Burkina Faso did so. About the second statement, 54% in Ghana and 82% in Burkina Faso disagreed (Fig. 3). Chi-squared analyses on these results, however, did not show significant difference among women and men. These results suggest that the respondents in both countries largely acknowledged gender inequity in water governance.

To understand respondents' opinion about governance transparency, we asked three questions (Table 1): (1) "how easy was it to access information about spillage of Bagre dam water?"; (2) "how is it easy to understand decision-making processes about water governance?"; and (3) "how will you rate efforts made by water managers to make stakeholders understand the decision-making process?" In response to the first question, 53%

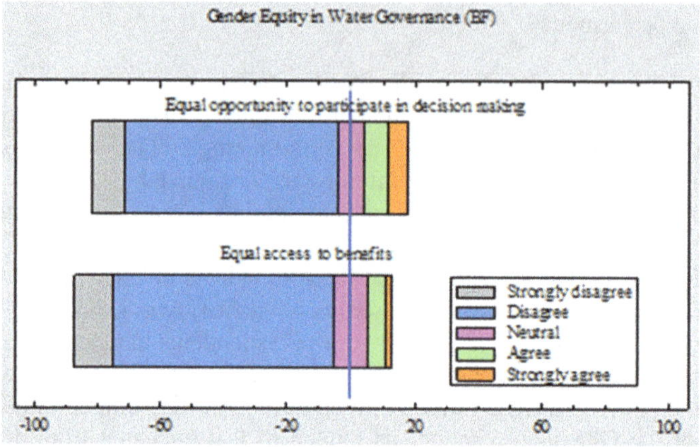

Figure 3: Gender equity, Burkina Faso.

of the respondents in Ghana found it easy to obtain information about spillage while 35% of them felt that it was difficult. In Burkina Faso, 59% of the respondents thought that it was difficult to access information whereas 15% of them did not think so. About the second question, 55% in Ghana found it easy to understand decision-making processes whereas In Burkina Faso, only 12% of the respondents found decision making process easy to understand while 63% of them was not sure. Regarding the third statement, 49% in Ghana highly rated water managers' performance whereas 64% in Burkina Faso rated low. These results show different perceptions about water governance between Ghana and Burkina Faso. Although the Ghanaian respondents tended to indicate the existence of governance transparency, our Chi-squared analysis revealed gender disparity about the second and the third statements among the respondents. We found that more men agreed with the second (72%) and third (68%) statements. This may be due to the patriarchal nature of the riparian communities of the Volta Basin where men dominate in public engagements.

To understand respondents' opinion about regulatory quality in the Volta River Basin, we asked two questions (Table 1): (1) "who manages the Volta River Basin?" and (2) "how will you rate the regulatory performance of water managers?" For the first question, six multiple choice answers were presented for the respondents to identify relevant institutions. These institutions are the Volta Basin Authority, Water Resource Commission in Ghana, Ministry of Water and Sanitation in Burkina Faso, Department of Water and Sanitation in Ghana, and the Forestry and Agriculture departments in both countries. The second question used a five-point Likert-scale ranging from "very dissatisfied" to "very satisfied".

In response to the first question, 57% of the respondents identified the Department of Agriculture. Also, about 45% chose the water resource commission. In Burkina Faso, the Ministry of Water was chosen by 40% of the respondents. Another 12% chose the Forestry Department. In stark contrast to the Ghanaian counterpart, only 9% chose the Department of Agriculture. Agricultural departments in the study areas inform farmers about good water management practices, but they are not responsible for managing water resources of the Volta River. In Ghana, the water resource commission, which was established by an Act of Parliament (Act 522 of 1996), is responsible for managing Ghana's water resources [35]. In Burkina Faso, the Ministry of Water and Sanitation is responsible for water governance [36].

Regarding the second question about rating regulatory performance, 58% of the respondents in Ghana were satisfied, whereas 36% of them thought otherwise. In Burkina

Faso, 54% of them were dissatisfied while 23% were satisfied with regulatory quality provided by water managers in the basin.

6 CONCLUSIONS

This paper examined riparian communities' water governance perceptions with particular emphasis on possible sources of rising tensions over the use of Volta River water. By using the questionnaire survey, it attempted to understand the perceptions about degrees of their public participation, gender equity, transparency, and regulatory quality.

Regarding public participation, our results revealed that the respondents generally had experienced limited levels of participation in water governance though the Burkina Faso respondents tended to show higher satisfaction level compared to those in Ghana. A large percentage of the Ghanaian respondents particularly had problems with discrimination against religion and norm in decision-making processes as they were minority Muslim people where Christianity tends to dominate Ghanaian society in general. As a result, 52% of the Ghanaian respondents felt being left out from governance whereas 39% in Burkina Faso, especially among males, felt involved in improving governance. The respondents generally acknowledged the existence of gender inequity in both countries.

With respect to transparency, the Ghanaian respondents tended to find that they had somewhat sufficient information about water governance. Regarding their access to information about Volta River water use, easiness to understand water governance processes, and water managers' performance, 45–55% of the respondents in Ghana answered favorably. In Burkina Faso, however, 59–65% of the respondents had negative perceptions about these transparency related issues.

This result, however, must be placed in a context of how the respondents understood water governance responsibilities among authorities. In Ghana 57% had a wrong perception and thought that the Agricultural Department was responsible for the Volta River Basin management. Only 45% knew correctly that the water resource commission was responsible. In Burkina Faso, 40% of the respondents knew that the Ministry of Water was responsible for managing water in the Basin.

These results, overall, suggest a further need to communicate and equitably engage with local communities about water governance. As most of the respondents and local communities in the Volta River Basin in general depend on agriculture for livelihoods, it may not be easy for them to distinguish water governance from agricultural activities. Flooding from Bagre dam spillage is another concerning issue for those who engage in farming along the Volta River, where they tend to expect agricultural authorities to mitigate. Although our survey was not designed to find cross-border communication efficiency, this is another topic to be explored with detail in the future in order to alleviate rising international tensions over water management in western Africa.

ACKNOWLEDGEMENT

We express profound gratitude to the 238 riparian community members in Ghana and in Burkina Faso who cooperated with us and provided information for this research during our field survey.

REFERENCES

[1] FAO, Global framework on water scarcity in agriculture. Report of Committee on Agriculture, COAG/2018/9, Twenty-sixth Session, Rome, 2018.

[2] Judeh, T., Haddad, M. & Özerol, G., Assessment of water governance in the West Bank, Palestine. *International Journal of Global Environmental Issues*, **16**(1/2/3), pp. 119–134, 2017.

[3] UNDP, *User's Guide on Assessing Water Governance*. Governance Centre: Oslo, 2013.

[4] Woodhouse, P. & Muller, M., Water Governance: An historical perspective on current debates. *World Development*, **92**, pp. 225–241, 2017. DOI: 10.1016/j.worlddev.2016.11.014.

[5] World Water Assessment Program, *Water for People Water for Life*. UNESCO-WWAP, pp. 4–6, 2003. https://unesdoc.unesco.org/ark:/48223/pf0000129556. Accessed on: 11 Feb. 2021.

[6] FAO, Water governance for agriculture and food security. Report on Committee on Agriculture, COAG/2014/6, Twenty-fourth Session, Rome, 2014.

[7] Global Water Partnership, Effective water governance: Learning from the dialogues. GWP Secretariat, Stockholm, 2003.

[8] Rogers, B.P. & Hall, A.W., Effective water governance. GWP Technical Committee (TEC) Background Paper, No. 7, 2003.

[9] Adhikari, B. & Tarkowski, J., Examining water governance: A new institutional approach. *Journal of Geography and Natural Disaster*, **3**(2), pp. 1–10, 2013.

[10] Batchelor, C., Water governance literature assessment. International Institute for Environment and Development (IIED), 2523, 2007.

[11] Ganoulis, J. & Fried, J., *Transboundary Hydro-Governance: From Conflict to Shared Management*. Springer, 2018.

[12] Lautze, J., De Silva, S., Giordano, M. & Sanford, L., Putting the cart before the horse: Water governance and IWRM. *Natural Resources Forum*, **35**(1), pp. 1–8. 2011.

[13] Kaufmann, D., Kraay, A., & Zoido, P., Governance matters. World Bank policy research working paper, No. 2196, 1999. https://papers.ssrn.com/sol3/papers.cfm?abstract_id=188568. Accessed on: 5 May 2020.

[14] Hanasz, P., A little less conversation? Track II dialogue and transboundary water governance. *Asia and the Pacific Policy Studies*, **4**(2), pp. 296–309, 2017. DOI: 10.1002/app5.183.

[15] Ampomah, B.Y., Adjei, B.A. & Youkhana, E., The transboundary water resources management regime of the Volta Basin. ZEF Working Paper Series, No. 28, 2008.

[16] Gao, Y. & Margolies, A., Transboundary water governance in the Volta River Basin, 2009. https://wikis.uit.tufts.edu/confluence/display/aquapedia/Transboundary+Water+Governance+in+the+Volta+River+Basin. Accessed on: 2 May 2020.

[17] World Bank, Project appraisal document on proposed grants to the Volta Basin Authority from the cooperation in international waters in Africa for a Volta River Basin strategic action programme implementation project. Global Water Practice (GWADR), Africa Region, 2015.

[18] Yankey, I. K., Ghana's water relations with Burkina Faso: Hydropolitical standpoint. Master's thesis, Sosyal Bilimler Enstitüsü, 2019. http://www.openaccess.hacettepe.edu.tr:8080/xmlui/handle/11655/6050. Accessed on: 2 May 2020.

[19] Schulte-Wülwer-Leidig, A., Gangi, L., Stötter, T., Braun, M. & Schmid-Breton, A., Transboundary cooperation and sustainable development in the Rhine Basin. *Achievements and Challenges of Integrated River Basin Management*, Intechopen, pp. 123–147, 2018.

[20] Paul, F.J.E., Francesc, R., Windekind, C.B., Mark, H., Troy, D.G. & McCutcheon, B., Good governance in protected areas: an evaluation of stakeholders' perceptions in British Columbia and Ontario Provincial Parks. *Journal of Sustainable Tourism,* **21**(1), pp. 60–79, 2013. DOI: 10.1080/09669582.2012.671331.

[21] Mahon, R., Fanning, L. & McConney, P., Assessing governance performance in transboundary water systems. *Environmental Development,* **24**(Nov. 2017), pp. 146–155, 2017.

[22] Jacobson, M., Fiona, M., Ingvild, O., Paavani, R. & Håkan, T., User's guide on assessing water governance. United Nations Development Programme (UNDP), Phoenix Design Aid, Denmark, 2013.

[23] Ghana Statistical Service – Bawku Municipality, District analytical report – Bawku municipality. Population and Housing Census, pp. 1–71, 2014. http://www.statsghana.gov.gh/docfiles/2010_District_Report/UpperEast/BawkuMunicipality.pdf.

[24] Ghana Statistical Service – Binduri, District analytical report – Binduri district. Population and Housing Census, pp. 1–79, 2014. https://new-ndpc-static1.s3.amazonaws.com/CACHES/PUBLICATIONS/2016/06/06/Binduri+2010PHC.pdf.

[25] Ghana Statistical Service – Garu, District analytical report – Garu Tempane district. Population and Housing Census, pp. 1–79, 2014. http://www2.statsghana.gov.gh/docfiles/2010_District_Report/Upper%20East/GARU%20TEMPANE.pdf.

[26] Ministry of Food and Agriculture, Facts and figures, 2015. http://agricinghana.com/wp-content/uploads/2017/07/AGRICULTURE-IN-GHANA-Facts-and-Figures-2015.pdf. Accessed on: 11 Dec. 2019.

[27] WikiZER, Bagre dam, 2020. https://www.wikizero.com/en//Bagre_Dam. Accessed on: 14 Jul. 2020.

[28] Global Water Partnership, Capitalizing the development process of the action plan for IWRM and its implementation in Burkina Faso. GWP/WA: Ouagadougou, Burkina Faso, pp. 3–4, 2009. https://www.gwp.org/contentassets/c9eeff4decf445d3b92bbf52711af009/capitalizing-burkina-iwrm-plan.pdf. Accessed on: 5 Nov. 2019.

[29] City population, 2020. https://citypopulation.de/php/burkinafaso-communes-admin.php?adm2id–BF480101. Accessed on: 15 Apr. 2020.

[30] IUCN, Improving water governance in the Volta Basin-Phase 2, PAGEV (2), 2009. https://www.iucn.org/sites/dev/files/import/downloads/pagev_ii.pdf. Accessed on: 9 Aug. 2020.

[31] Israel, G.D., Determining sample size. Fact sheet PEOD-6, University of Florida, pp. 1–5, 1992.

[32] Boone, H.N. & Boone, D.A., Analyzing Likert data. *Journal of Extension*, **50**(2), pp. 1–5, 2012.

[33] Kent State University Libraries, SPSS tutorials: Chi-square test of independence. 2020. https://libguides.library.kent.edu/SPSS/ChiSquare. Accessed on: 26 Apr. 2020.

[34] Statgraphics18, Likert plot, 2017. http://cdn2.hubspot.net/hubfs/402067/PDFs/Likert_Plot.pdf. Accessed on: 13 Aug. 2020.

[35] Water Resource Commission, About WRC, 2020. http://www.wrc-gh.org/. Accessed on: 14 Jul. 2020.

[36] International Water and Sanitation Centre (IRC), Burkina Faso creates new ministry for water and sanitation, 2013. https://www.ircwash.org/news/burkina-faso-creates-new-ministry-water-and-sanitation-1. Accessed on: 14 Jul. 2020.

RIVER RESTORATION FOR THE REPLACEMENT OF LOST SPAWNING GROUNDS DUE TO DAM CONSTRUCTION

REGINA SANTOS[1,2], LUÍS SANCHES FERNANDES[1], FERNANDO PACHECO[2], MARIZA MONTEIRO[1] & JOAQUIM JESUS[1]
[1]Centre for the Research and Technology of Agro-Environment and Biological Sciences, University of Trás-os-Montes and Alto Douro, Portugal
[2]Chemistry Research Centre, University of Trás-os-Montes and Alto Douro, Portugal

ABSTRACT
The construction of Baixo Sabor Hydroelectric Scheme (BSHS) in the Sabor River (located in northeast Portugal) without fishways has blocked fish migration. However, the Vilariça River, a tributary of the Sabor River, located downstream of BSHS represents an alternative habitat for fish, especially for specie more abundant in this river, the potamodromous cyprinid Iberian barbel (*Luciobarbus bocagei*). This way, to replace the spawning grounds lost in Sabor River, several compensation measures were implemented in a 2 km long river stretch of the Vilariça River. The measures included restoration activities like bank reinforcement, boulder placement, river water profile increase, and placement of submerged weirs with fish ramps to increase the lotic/lentic sequence. The river water profile increase was done with an adductor system constructed to transfer water from the BSHS to upstream the 2 km long river stretch during spawning. In the river stretch, a River Connectivity Index (RCI) was used and sampling sites of the Iberian barbel were set on to assess the effect of the compensation measures. The RCI and the wetted perimeter method showed that the water velocity of 1.5 m^3/s provides good longitudinal connectivity. The results showed that barbel spawns preferentially in a coarse substrate with shallow water and moderate water velocity. The monitoring of the effectiveness of restoration measures during the first 6 years of exploration (2016–2020) showed that the major actions were well achieved, such as the importance of adductor system for attracting the breeding barbels to the river, the stabilization of the riverbank and the increase in the habitat heterogeneity. However, some structures must be intervened, like the replacement of the stone weir destroyed, replantation of native species, and the removal of exotic plants.
Keywords: river restoration, spawning grounds, dam construction, adductor system, Iberian barbel.

1 INTRODUCTION

One of the main impacts associated with the construction of the Baixo Sabor Hydroelectric Scheme (BSHS) is the barrier effect that prevents the migration of fish species such as the Iberian barbel (*Luciobarbus bocagei*) potamodromous cyprinid, which carry out significant migrations from the Douro River to the Sabor River. The establishment of a compensation measure for its conservation is justified by the great migratory movement for spawning that occurred before the construction of the BSHS. The fact that the Vilariça River is a tributary of the Sabor River and is located downstream of BSHS has the requirements for implementing compensation measures for lost habitat. This way, a river stretch of the 7 km long in the Vilariça River was restored to create adequate conditions for the reproduction of the fish species affected by the construction of the BSHS in the Sabor River. However, the interventions occurred with a greater incidence in the last 2 km, and for this reason the present work describes the interventions carried out in this river stretch.

In the Vilariça River, the native species more abundant is the Iberian barbel, an excellent indicator of ecological quality and environmental degradation due to their abundance in well-preserved Iberian rivers [1]. Due to their sensitivity to habitat changes, cyprinids are among the most endangered species in Europe as a result of increased river degradation. In this way,

WIT Transactions on Ecology and the Environment, Vol 250, © 2021 WIT Press
www.witpress.com, ISSN 1743-3541 (on-line)
doi:10.2495/WRM210041

information on the behaviour, migration and spawning of these species is essential for the success of the requalification actions [2]. As well as the assessment of fluvial connectivity to improve the migratory movement to the spawning areas. According to Solà et al. [3] the River Connectivity Index (RCI) can be applied to any river hydraulic infrastructure that constitutes an obstacle to the longitudinal movements of the Ichthyofauna and it can be used as a complementary element in hydromorphological condition assessment.

In addition, the study area is important for the conservation of the environment because the Sabor River is part of the Natura 2000 Network, within the Special Protection Area (SPA) and the Sites of Community Importance (SCI), which are classified under European Directives 79/409/EEC and 92/43/EEC, respectively.

The main objective of this work is to improve the habitat quality in the Vilariça River for supporting the barbel migration and reproduction. For that, it is necessary (i) to assess fluvial connectivity of the restoration measures that have been implemented; (ii) to determine the optimal discharge from the adductor system to improve the migration routes and conserve spawning grounds; and (iii) to identify the environmental factors that support spawning activity.

2 MATERIAL AND METHODS

2.1 Study area

The Vilariça River (located in northeast Portugal) is a medium-sized and third-order stream [4] located on the right bank of the Sabor River, whose mouth is located downstream of BSHS. The section under the study of 2 km long in the Vilariça River is located immediately before at the confluence with the River Sabor (Fig. 1). The Vilariça River basin covers an area of 324 km^2 with an altitude that ranged between 1,186 m in the mountains and 94 m in the convergence with the Sabor River [5]. In lower altitude the basin is occupied by agricultural areas (42% of the basin area), artificial areas (2%) and water bodies (1%) and in higher elevation the land use dominant is semi-natural areas (31%) and forest (25%), according to Land Use Land Cover of 2015 [5].

The riparian gallery, limited to a small strip on the banks, is dominated by *Populus nigra*, clusters of *Salix salviifolia* and *Fraxinus angustifolia*, and the presence of the non-native giant cane (*Arundo donax*) that invades some areas [6]. The climate is characterised by warm and dry summers and cold and wet winters. The annual precipitation is an average of 730 mm/year and occurred mostly in winter [7]–[9]. The average water temperature between 1993 and 2017 was 18.8°C and 22.3°C in spring and summer, respectively [9], [10]. The BSHS consists of two hydroelectric dams, the Baixo Sabor Dam and the Feiticeiro Dam located at 12.6 km and 3 km upstream of the confluence of the Douro river, respectively. The total energy produced with the contribution of pumping is 444 GWh [11].

2.2 River restoration measures

To improve the reproduction habitat of fish species several restoration measures were implemented in the 2 km long river stretch of the Vilariça River (Fig. 1(c)). The restoration measures were carried out throughout 2014 and consisted of constructing an adductor system and improving channel hydromorphology and riverbank. The adductor system, called Sabor-Vilariça, was constructed for transferring water from the BSHS to upstream of the 2 km long river stretch of the Vilariça River. From the adductor system was release the experimental flows of 1.0 m^3/s, 1.25 m^3/s, 1.5 m^3/s and 2.0 m^3/s. The adductor is composed of a reinforced

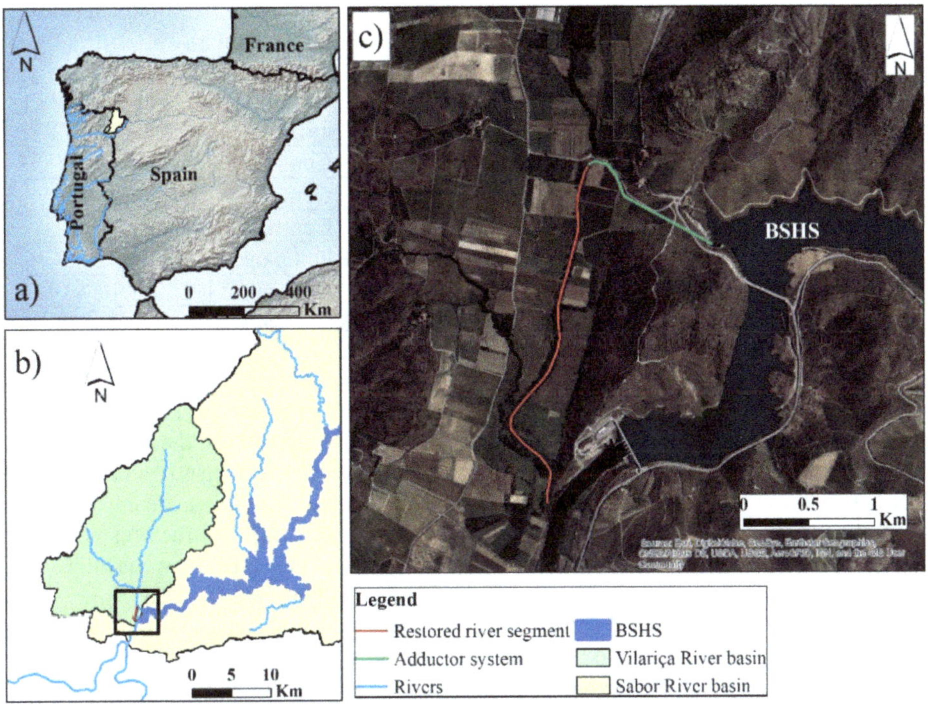

Figure 1: Location of the study area. (a) The Iberian Peninsula and the Sabor River in northern Portugal; (b) The Vilariça River and the Baixo Sabor Hydroelectric Scheme in the Sabor River; and (c) The restored river segment of 2 km long in the Vilariça River.

concrete structure, which is about 1000 m long, with safety gates and flow control. The flow passes in the adductor, in free surface and, in the terminal phase, in a transition structure with macro roughness for energy dissipation and finally, the flow enters in the Vilariça River by a restitution structure [12].

The channel hydromorphology interventions included the removal of fines in the riverbed, placement of blocks in the river and the construction of six transversal structures (weirs), six refuges structures with woody material (lunkers) and two islands planted with willow. The six transversal structures consist of two block stone weirs with rocky ramps and four large woods formed by trees trunks with central notches (Table 1). Due to the destruction of one of the large wood, it was later substituted by a deflector. The length of the five transversal structures varies between 7.1 and 17.9 m and the spillway between 1.25 and 3 m, being that in large woods the location of the spillway is central, and in stone weirs are in the right bank. And finally, the interventions on the riverbanks consisted of planting native species, removing the non-native giant cane and constructing vegetated riprap and vegetated gabion with live willow cuttings for bank protection [12]. The implemented restoration measures were continuously monitored from 2016 to 2020 [13].

To evaluate the level of transposability of the transversal structures was used the River Connectivity Index (RCI) applied for the natural flow (0.57 m^3/s) and the experimental flows

Table 1: Characteristics of the transversal structures placed in the 2 km long river stretch of the Vilariça River to improve the quality of the fish habitat.

Transversal structure	Length (m)	Spillway (m)	Location of spillway	Type of obstacle RCI
Large wood E1	7.1	1.25	Central	A
Deflector E2	9	n.a.	Central	n.a.
Large wood E3	15.7	1.3	Central	A
Stone weir E4	13.5	2.5	Right bank	C
Large wood E5	13.8	1.5	Central	A
Stone weir E6	17.9	3	Right bank	C

Note: n.a. = not applicable.

released from the adductor system. In structures in which water passes over the obstacle (weir type) creating a small waterfall (type A), the parameters measured are the following: Tw – width of spillway crest (m); Tz – water depth under spillway crest (m); h – the difference level between the spillway crest and the water level immediately downstream of the weir (m); and z – water depth immediately downstream of the weir (m). In structures with a very low slope, where water passes over the obstacle and does not generate any waterfall (type C) the longitudinal slope (%) and the average runoff speed (m/s) were evaluated [3].

The final RCI score obtained varies between 0 and 110 and obstacles are classified into 5 quality classes: ≥95 correspond to very good quality (all the present groups of fish can pass); 75–94 good (the majority of the present fish groups can pass); 50–74 moderate (the majority or some of the present fish groups can pass); 25–49 poor (only one or few species of the present groups of fish can pass) and <25 bad (no species of the present groups of fish or only some in very exceptional hydrological situations can pass) [3]. In the large wood, the RCI index calculation was used based on the type A obstacle and in the stone weir the type C obstacle (Table 1).

2.3 Wetted Perimeter Method

The Wetted Perimeter Method was used to obtain submerged habitat (and theoretically available for the fishes) given by the Wetted Perimeter and the flow. The fieldwork was developed in three representative sections on barbel habitat during the reproduction period (between April and July of 2015). This method is based on a comparative analysis between the flows in a section of a watercourse and the corresponding wetted perimeters, assuming that there is a relationship between both. The method requires that for the section of a watercourse, the curve that relates the flows with the respective wetted perimeters is obtained to determine the hydraulic conditions in which the spawners are found. Therefore, the wetted perimeters were obtained for the natural flow (0.57 m^3/s) and the flow released from the adductor system (1.0, 1.25, 1.5 and 2.0 m^3/s).

2.4 Spawning grounds identification

In the 2 km long river stretch, between April and July 2015, the spawning grounds of barbel were identified to know the most suitable habitat for the eggs deposition. Therefore, an evaluation of the hydraulic parameters as water depth, current velocity, granulometry of the substrate and cover were carried out. The water depth and the current velocity were measured with an electromagnetic reel and the percentage of substrate and cover were determined visually. The substrate was assessed using a modified Wentworth scale: 0 – organic material detritus; 1 – silt, clay, loam (1–2 mm); 2 – sand (<2 mm); 3 – fine gravel (2–6 mm); 4 – medium gravel (6–20 mm); 5 – large gravel (2–6 cm); 6 – small stones (6–12 cm); 7 – large stones (12–20 cm); 8 – boulders (>20 cm); and 9 – rock [14].

3 RESULTS AND DISCUSSION

3.1 River restoration measures

The habitat restoration was implemented in the 2 km long river stretch of the Vilariça River for increasing habitat heterogeneity and improving fish migration and spawning beds. The restoration measures included the increase of the water volume and current speed in the river segment of 2 km long, through the construction of the adductor system between the BSHS and the Vilariça River to transfer water during the reproduction period of the native ichthyofauna (from 15 March to 31 July) (Fig. 2(a)).

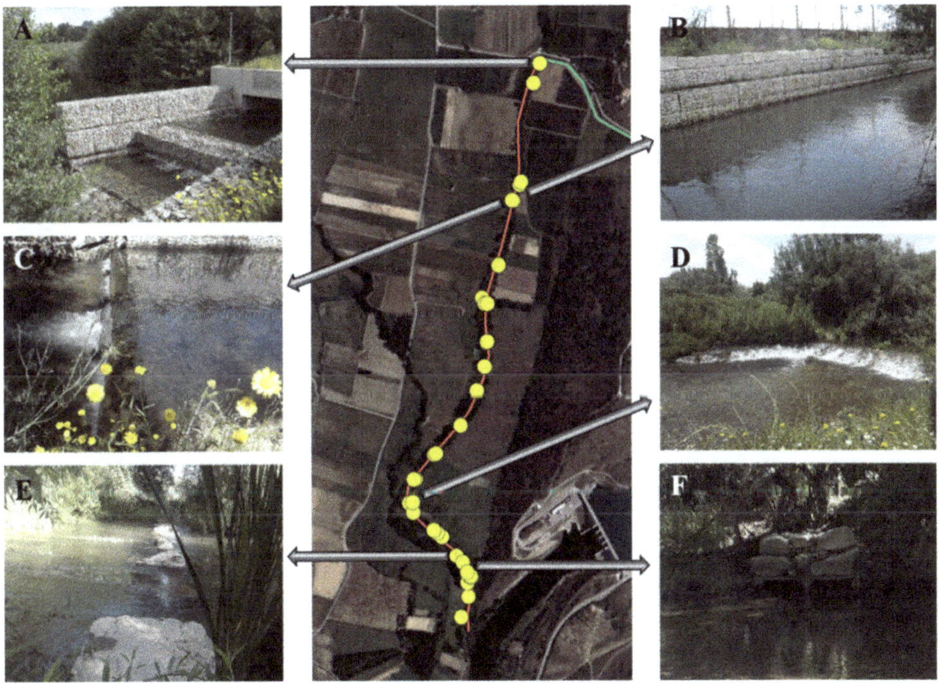

Figure 2: Examples of interventions implemented in the 2 km long river stretch of the Vilariça River to improve fish habitat. (a) Adductor system; (b) Reinforcement of the riverbanks with vegetated gabions; (c) Weir of the large wood; (d) Stone weir; (e) Placement of blocks in the riverbed; and (f) Lunkers.

In the riverbank, the restoration actions included the plantation of the native species (with live stakes willow), the removal of non-native giant cane followed by the reinforcement by vegetated riprap, and vegetated gabions with live stakes of willow (Fig. 2(b)). The reinforcement of the riverbank was applied for the consolidation of embankments in places with a marked erosion process.

The interventions in the channel hydromorphology consisted in the construction of three weirs of the large wood in slope areas not exceeding 5% to promote a sequence of riffles and pools for improving the water quality by decreasing the concentration in total suspended solids and consequently promoting the sedimentation of fines and the heterogeneity of the microhabitat (e.g. forming lotic/lentic sequence and with different substrate granulometry); (Fig. 2(c)); the construction of two stone weirs for creating a habitat that allows housing native fish populations, breeding adults and juveniles, especially during the dry season (Fig. 2(d)); the placement of blocks in the riverbed to create hydraulic irregularity and substrate heterogeneity, current speed and cover in the bed of the watercourse, promoting new zones of refuge for fish (Fig. 2(e)), the construction of six lunkers with woody material for refuge for fish fauna (Fig. 2(f)), as well as the construction of two islands planted with willow and the removal of fines in the riverbed.

The level of transposability of the transversal structures determined by the River Connectivity Index for the natural flow and the flow released from the adductor system are presented in Table 2. The results showed that all flows presented a good transposability, exception the flow of 0.57 m^3/s (natural flow), in which the flow takes place in a small part of the transversal structures and may cause difficulties in the downstream migration.

Table 2: Classification of the River Connectivity Index of transversal structures for natural flow (0.57 m^3/s) and the flows released from adductor system (1.0, 1.25, 1.5 and 2.0 m^3/s). The RCI score between 75–94 represent good quality and between 50–74 moderate quality.

Flow (m^3/s)	Large wood E1	Large wood E3	Weir E4	Large wood E5	Weir E6
0.57	60	85	60	85	60
1	85	85	85	85	85
1.25	85	85	90	85	90
1.5	85	85	90	85	90
2	85	85	90	85	90

3.2 Relationship between flow and wetted perimeter

The exponential curves adjusted for the three sections analysed show an asymptote of approximately 1.5 m^3/s (Fig. 3). Thus, it can be considered that flows higher than 1.5 m^3/s do not represent significant increases in available habitat. This situation shows a river channel physically well delimited by relatively steep banks, as is typical of low-order rivers that run along plains.

3.3 Spawning grounds

Table 3 shows the characterization of the spawning grounds identified in the 2 km long river stretch as water depth, flow velocity, type of substrate and cover. In general, the barbel spawns essentially in places with low depth, moderate current, a substrate formed mainly by gravel and sand (with granulometry between 2 and 64 mm) and preferably without any type of coverage.

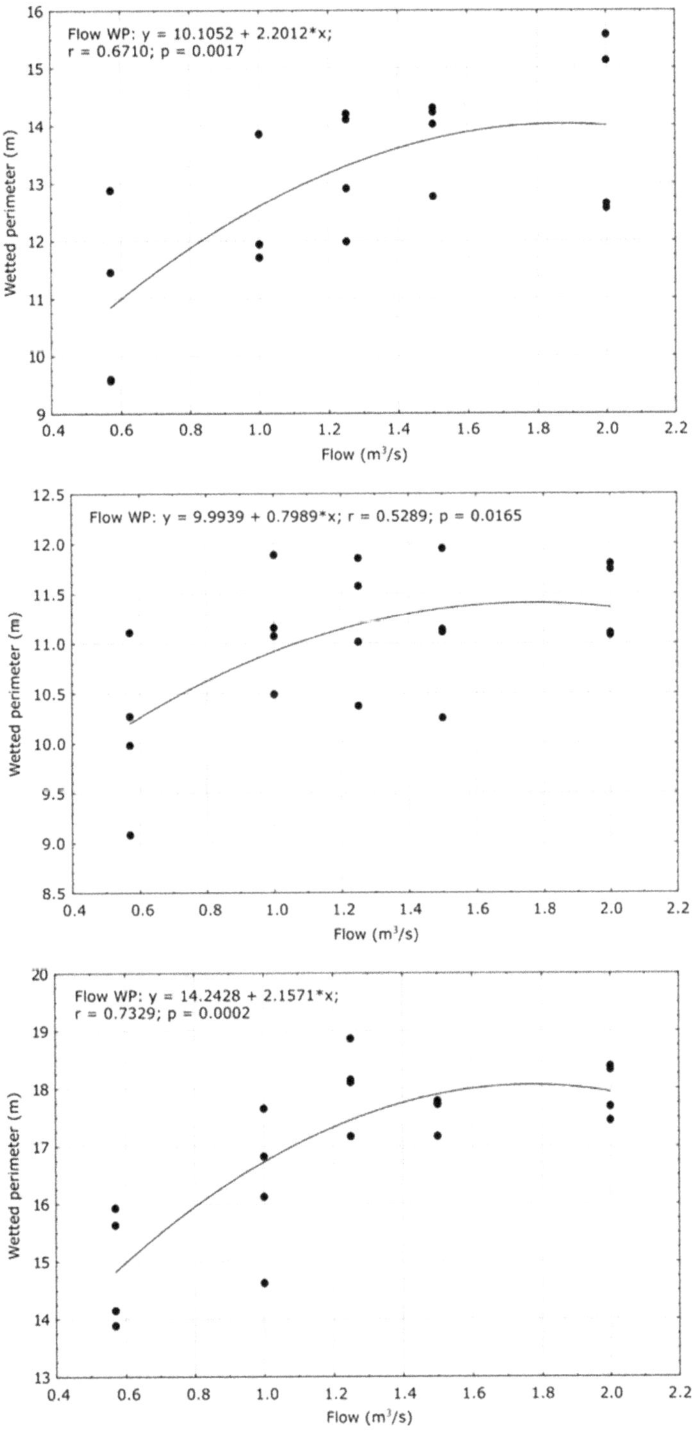

Figure 3: Relationship between wetted perimeter and flow with successive increments of about 0.5 m^3/s.

Table 3: Characteristics of the spawning grounds in run and riffle habitat identified in the 2 km long river stretch of the Vilariça River.

Type of habitat	Number of spawning grounds	Depth (cm) Mean (min–max)	Flow velocity (m/s) Mean (min–max)	Substrate (%)	Cover (%)
Run	11	40.73 (25–63)	0.427 (0.047–0.66)	82% gravel, 14% sand and 4% stones	64% riparian vegetation and 36% no vegetation
Riffle	7	31.14 (18–48)	0.678 (0.453–0.924)	50% sand, 43% gravel and 7% stones	86% no vegetation and 14% riparian vegetation
Total	18	37 (18–63)	0.525 (0.047–0.924)	67% gravel, 28% sand and 0.06% stones	56% no vegetation and 44% riparian vegetation

3.4 Monitoring of the effectiveness of restoration measures

The monitoring of the effectiveness of restoration measures was developed during 5 years (2016–2020) after the implementation measures in 2014. To evaluated the performance of the adductor system was measured the water quality in five sampling points, three located upstream and two downstream of the flow release from the adductor system. The parameters measured were: temperature, dissolved oxygen, rate of oxygen saturation, biochemical oxygen demand (CBO5), electrical conductivity at 200°C (average), pH, nitrates, ammoniacal nitrogen, and phosphates. In all sampling points, the water quality was good and reasonable according to the classification of the Water Framework Directive. The CPUE (catch per unit effort) of barbel, carried out using electric fishing at the reproduction period, showed that the barbel population remained without substantial variations with values between 1.5 and 2 individuals/minute in the 6 years of monitoring (2015–2020). Except in 2018 that there was a considerable increase with values greater than eight individuals/minute. These results show that the flow released by the adductor system of the 1.5 m^3/s, during the breeding season, has proved to be sufficient to attract the barbel to the Vilariça River.

In general, the built structures (transversal structures, blocks, and lunkers) in the 2 km long river stretch of the Vilariça River kept the main characteristics unchanged during the monitoring period. However, in 2020 some structures have shown to have their operationality affected concerning the function initially envisaged, namely the absence of the stone weir that was located near the mouth of the Vilariça River (E6). The destruction of this weir (due to floods) led to a decrease in water depth and consequently the emergence of some shelter structures for fish fauna (lunkers), which in the normal situation should have been submerged. However, the reconstruction of the stone weir E6 is already authorized by the authority and should be rebuilt in September this year (2021).

The enhancement measures of the riparian gallery as exotic control actions and planting native species have not been successful. The control of exotic species had an immediate effect of decreasing its cover, in 2014, but it was consistently increasing in following years since there was no removal of roots, followed by native plantations. The actions for planting native species showed, as of 2016, a mortality rate of around 100%. The cleaning actions also proved to be inefficient, due to the high dynamics of the watercourse, which carries a lot of solid elements every year. Nevertheless, the slope stabilization actions were well achieved, and the riparian gallery suffers much fewer consequences in the event of higher and potentially destructive flows, keeping the margins stabilized. In general, riparian communities since 2014 have been acquiring greater structural complexity, justified by the increase in tree cover, but it presents some degradation with low richness and an increased area of non-native giant cane.

4 CONCLUSION

The habitat restoration was implemented in the 2 km long river stretch of the Vilariça River for increasing habitat heterogeneity and improving fish migration and spawning beds. The work of restoration implied the removal of exotic plants, plantation of native species, and the construction of vegetated gabions and vegetated riprap to the consolidation of the riverbank. As well as the construction of an adductor system to increase the flow to attract the breeding barbels to the river and the construction of lunkers, boulder placement, and transversal structures to increase habitat heterogeneity. The evaluation of transposability of the transversal structures determined by the River Connectivity Index and the Wetted Perimeter Method showed that 1.5 m^3/s is the ideal flow for attracting the breeding barbels to the river. In fact, in the breeding period of the barbel, the preferential habitats are characterized by moderate currents, substrate composed of gravel and sand, greater luminosity, shallow waters, and beds more homogeneous.

The monitoring of the effectiveness of restoration measures developed during 5 years (2016–2020) showed good and reasonable water quality and that the release of 1.5 m^3/s by the adductor system is sufficient to attract the barbel to the Vilariça River. Relatively to build structures near the mouth of the Vilariça River became unusable, namely, the destruction of the stone weir and the inefficiency of some lunkers due to the decrease in water depth caused by the destruction of the weir. Relatively to the enhancement measures of the riparian gallery were verified the mortality of planting native species, expansion of exotics, however, the actions of stabilization of the riverbank were well achieved.

ACKNOWLEDGEMENTS

This research received financial support from Portuguese Foundation for Science and Technology (FCT), Ministry of Science, Technology, and Higher Education (MCTES), European Social Fund (FSE) through NORTE 2020 (North Regional Operational Program 2014/2020) and European Union (EU) to Regina Santos (Grant: SFRH/BD/141503/2018) for article writing. The authors integrated in the CITAB Research Centre was funded by the National Funds of FCT under the project UID/AGR/04033/2020. The authors integrated in the CITAB research centre are also integrated in the Inov4Agro – Institute for Innovation, Capacity Building and Sustainability of Agri-food Production. The Inov4Agro is an Associate Laboratory composed of two R&D units (CITAB & GreenUPorto). For the author integrated in the CQVR was funded by National Funds of FCT under the projects UIDB/QUI/00616/2020 and UIDP/00616/2020. And finally, we want to thank a Gestão da Produção de Energia, S.A. (EDP) for providing the data for publication. EDP is an energy producing company that operates the BSHS and financed all the restoration measures carried out in Vilariça River. We would also like to express our thanks to the whole team of the River Ecology Laboratory from the University of Trás-os-Montes and Alto Douro for the fieldwork carried out.

REFERENCES

[1] Martínez-Capel, F., Jalón, D.G., Werenitzky, D., Baeza, D. & Rodilla-Alamá, M., Microhabitat use by three endemic Iberian cyprinids in Mediterranean rivers (Tagus River Basin , Spain). *Fish. Manag. Ecol.*, **16**, pp. 52–60, 2009.

[2] Ovidio, B.M. & Philippart, J.C., Movement patterns and spawning activity of individual nase Chondrostoma nasus (L.) in flow-regulated and weir-fragmented rivers. *J. Appl. Ichthyol.*, **24**, pp. 256–262, 2008.

[3] Solà, C. et al., Longitudinal connectivity in hydromorphological quality assessments of rivers. The ICF index : A river connectivity index and its application to Catalan rivers. *Limnetica*, **30**, pp. 273–292, 2011.

[4] Strahler, A.N., Quantitative analysis of watershed geomorphology. *Trans. Am. Geophys. Union*, **38**, pp. 913–920, 1957.

[5] DGT Directorate-General of Territory Online. http://www.dgterritorio.pt/. Accessed on: 6 Mar. 2021.

[6] Boavida, I., Jesus, J.B., Pereira, V., Santos, C., Lopes, M. & Cortes, R.M., Fulfilling spawning flow requirements for potamodromous cyprinids in a restored river segment. *Sci. Total Environ.*, **635**, pp. 567–575, 2018.

[7] Santos, R.M.B., Sanches Fernandes, L.F., Cortes, V.M.R. & Pacheco, F.A.L., Hydrologic impacts of land use changes in the Sabor River Basin: A historical view and future perspectives. *Water (Switzerland)*, **11**, 1464, 2019.

[8] Santos, R.M.B., Sanches Fernandes, L.F., Cortes, V.M.R. & Pacheco, F.A.L., Development of a hydrologic and water allocation model to assess water availability in the Sabor River Basin (Portugal). *Int. J. Environ. Res. Public Health*, **16**, 2419, 2019.

[9] SNIRH National Information System for Water Resources Online. https://snirh.apambiente.pt/. Accessed on: 6 Mar. 2021.

[10] Santos, R.M.B., Monteiro, S.M., Cortes, R.M.V., Pacheco, F.A.L. & Fernandes, L.F.S., Seasonal effect of land use management on gill histopathology of Barbel and Douro Nase in a Portuguese watershed. *Sci. Total Environ.*, **764**, 142869, 2021.

[11] AHBS Aproveitamento hidroelétrico do Baixo Sabor (Projecto), Apresentação à Comissão Executiva. EDP – Energias de Portugal. Torre de Moncorvo, p. 22, 2012.

[12] AHBS Aproveitamento hidroelétrico do Baixo Sabor. Medida compensatório MC1 – Habitat de compensação da Vilariça. Laboratório de Ecologia Fluvial, Universidade de Trás-os-Montes e Alto Douro, Vila Real, p. 55, 2016.

[13] AHBS Aproveitamento hidroelétrico do Baixo Sabor, Relatório de monitorização da MC1 – Habitat de compensação da Vilariça. Fase de exploração (ano 2020). Laboratório de Ecologia Fluvial, Universidade de Trás-os-Montes e Alto Douro, Vila Real, p. 106, 2021.

[14] Bovee, K.D., Lamb, B.L., Bartholow, J.M., Stalnaker, C.B., Taylor, J. & Henriksen J., Stream habitat analysis using the instream flow incremental methodology. U.S. Geologicial Survey, Biological Resources Division Information and Technology Report USGS/BRD-1998-0004, p. 143, 1998.

SECTION 2
ECOLOGICAL AND ENVIRONMENTAL IMPACT

RECONDITIONING THE SEWERAGE AND DRINKING WATER SYSTEM WITHIN THE FACULTY OF MATHEMATICS, UNIVERSITY OF GUAYAQUIL, ECUADOR

ARMANDO SALTOS, PATRICIA VILLA, JACINTO ROJAS & MANUEL GOMEZ DE LA TORRE
Faculty of Math and Physical Sciences, University of Guayaquil, Ecuador

ABSTRACT
An evaluation of the sewer and rainwater systems, the same ones that currently discharge from the University of Guayaquil's main network, was carried out. These networks have previously exceeded their operational capacity, causing flooding in the rainy season. In addition, the drinking water supply system was evaluated. The evaluation identified interconnection problems between the rainwater and sewage networks. This research proposes improvements to the sewage systems with the design of new networks: to identify the flow of rainwater and sewers, avoid contamination of wastewater and minimize the impact of rainy season flooding. Installed capacity is currently insufficient, because the university population grew exponentially. The existing drinking water reservoirs made of asbestos cement are in bad condition; the pipes and accessories of the supply networks are corroded, as galvanized iron was used for their construction (materials widely used at that time). We established new potable water conduction networks, new sanitary units for the university's Faculty of Math and Physical Sciences building, and made the reconditioned sewage and rainwater networks work separately.
Keywords: drinking water, sewer systems, drainage, plumbing remodelling, rainwater, storage tank, building information modelling, Revit modelling, University of Guayaquil.

1 INTRODUCTION

The information used as a starting point for the development of this research included topographic surveys; records of sanitary units or public toilets; records of the storage system, such as tanks and cisterns and their pumping units; and the functionality of the plumbing systems, drinking water supply and sewers. The student population during 2018–2019 was also taken into account [1].

Before being evacuated, wastewater must meet certain quality parameters that meet the requirements for drinking water proposed by the World Health Organization (WHO) [2]. WHO and United Nations (UNICEF) have suggested action plans to ensure the continuous supply of safe water, sanitation and hygiene. In education centres, handwashing units with water and soap dispensers should be installed [3]. As technology advances, the efficiency of systems is improved. Based on this, many authors propose the installation of intelligent systems for water dispensers, to minimize the consumption of resources [4].

When designing the sanitary infrastructure, the elements must have dimensions according to the number of users [5]. The *Norma Ecuatoriana de la Construcción* (Ecuadorian Construction Standard) (NEC), establishes the minimum parameters to be met in all designs, before construction of hydro-sanitary installations. This, under normal conditions, ensures good functioning under quantity and quality expectations, in space and time inside the property, house, or building.

According to studies carried out in Japan, the toilet ventilation is recommended to be installed at the bottom of the walls, or at the sink level, to increase elimination of bad

WIT Transactions on Ecology and the Environment, Vol 250, © 2021 WIT Press
www.witpress.com, ISSN 1743-3541 (on-line)
doi:10.2495/WRM210051

odours [6]. This approach was considered for this project's proposed ventilation system design.

Ottawa University has several publications where the sizing of the distribution network and the drinking water supply system are analysed. This methodology was applied to evaluate the actual conditions of the drinking water and sewer networks in the University of Guayaquil, Ecuador [7].

When designing sanitary infrastructure, the necessary equipment must be provided according to the number of users and their type, whether the population is steady or fluctuates [5].

There is simulation software that helps to determine the degree of uncertainty for rainwater collection. The uncertainty can be estimated considering different demand scenarios for drinking water, with hourly and daily variations in a normal distribution [8]. It is not common to store rainwater in Ecuador.

The most severe problems found in buildings are usually generated by the sanitary infrastructure, especially caused by pipe failures, due to corrosion, ruptures, deflections, etc.: These cause transport deficiencies in the drinking water system, as well as in wastewater discharge [9].

The functionality of sanitary installations should be evaluated before any redesign is performed. Surveys are a useful tool and should be conducted to evaluate user's perceptions of the operation of the sewer system, bad odours, and the ease of discharge of the effluent. The functionality of sanitary installations should be valued before any redesign is performed: This result should be contrasted with subsequent inspections.

The construction industry has taken a technological leap in project design with the implementation of computer tools such as Building Information Modelling (BIM), which facilitates the design, planning and reduction of project times and costs: This tool was used in our research.

For water pipeline installation modelling with BIM, it is necessary to acquire the operating parameters in advance, so the software can automatically generate the pipe diameters to be used and the pipeline routing without considering the sizing of the pipes. The software has some parameters that are required for the network layout, so it is necessary to identify these values [10].

BIM software has three different modelling suites, the Construction Model (CM), which works user's data; the BIM Design Model, used for evaluation; Evaluation Model, which uses data details [11]. This research project is based on determining solutions from the analysis of results obtained in Phase 1 of this research, where problems were found in the sanitary and drinking water networks in the Faculty of Mathematical and Physical Sciences (FMPS) building facilities in the University of Guayaquil.

In the proposal to be presented for the reconditioning of the drinking water and sewage systems, we considered: storage tanks, adaptations in the ventilation systems in sanitary pipelines, increases in the number of toilets, making the rainwater and sewage systems independent, a proposal for the construction of a cistern, and implementation of ditches for rainwater evacuation.

2 METHODOLOGY AND MATERIALS

Fig. 1 illustrates the methodology to be developed in the implementation of the proposed solution for the sewage, rainwater and drinking water networks of the FMPS building, based on the results obtained in the evaluation of the first phase of the project, where we determined there was a lack of toilets, deterioration of the drinking water tanks with their connections, and bad interconnections between the sanitary and storm sewer network.

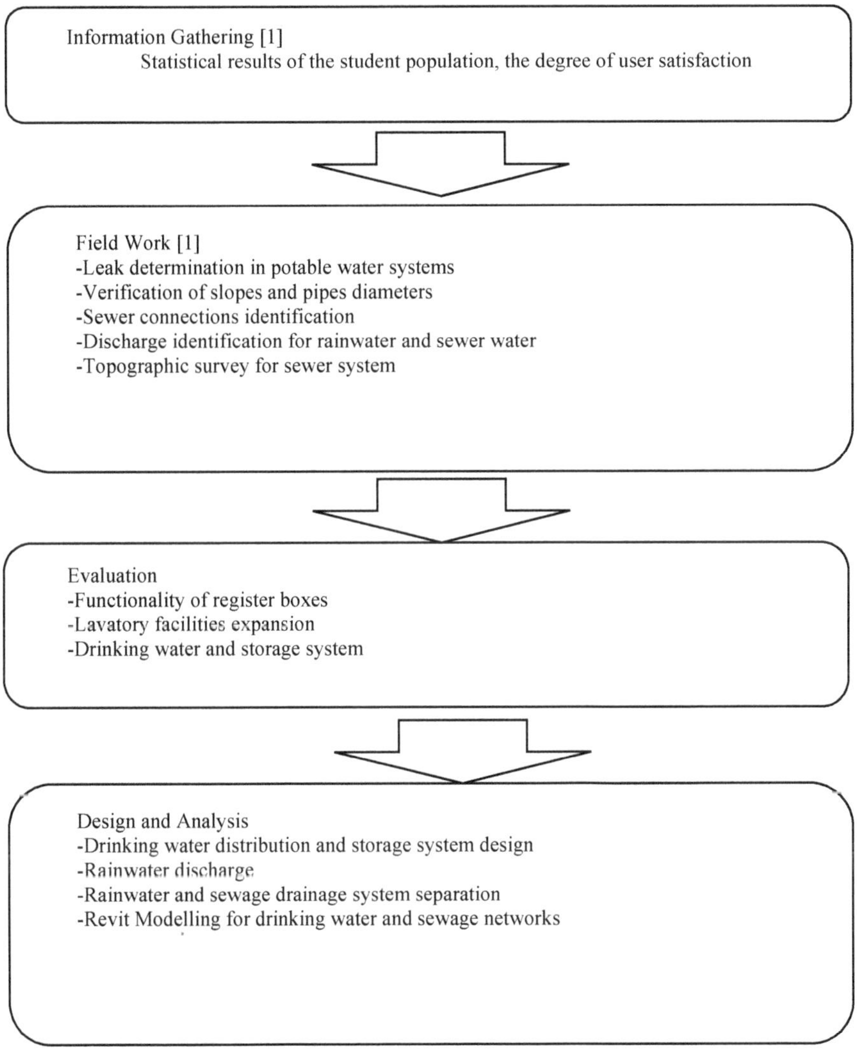

Figure 1: Applied methodology for the sewage and drinking water systems for the Faculty of Mathematical and Physical Sciences building in the University of Guayaquil.

In the evaluation carried out in the first phase of the project, surveys and the evaluation of the sewage and drinking water systems were included, they considered the existing student population during the period 2018–2019 and historical data from the last 20 years, the flows of two lines flowing in the FMPS building and its records.

After tabulating the surveys, generating the statistics for modelling, estimating the demand for water supply in the FMPS building, and the level of user satisfaction, our research continues with the final phase of the project, where solutions are proposed for the three different areas concerning sanitation.

This methodology generates the proposed solution for the reconditioning of healthy drinking water systems of the FMPS building, once Phase 1 of the research project [1] has been completed, where information was collected through fieldwork to determine leaks, the definition of slopes in the network, state of the toilets, diameters of the collectors, topography and general evaluation of the functionality of the sanitary devices. The design and modelling was implemented using the diagrams generated in BIM.

3 SANITARY SEWER SYSTEM

According to the information collected from the sanitary sewer network located in the library and the physics laboratory area, the number of existing record boxes needed to be evaluated [1].

In the library area, there are seven wastewater registration boxes: Their discharges are concentrated in (under) the women's restrooms and the dining room on the ground floor. The men's restrooms on the first floor are interconnected with the main system in the postgraduate offices and computer lab area. The student's wellness bathroom and the dining room drainage are also connected into this system [1].

The rainwater discharge runs towards box #4, as can be seen in Fig. 2. This box will be released and later connected directly to an evacuation channel, as shown in the proposed sewer solutions plan. The sewage water system located under the Physics laboratory is comprised of 11 interconnected boxes coming from eight restrooms and three from rainwater drainage.

The discharge system coming from the restrooms has very poor ventilation, as do the laboratory and library areas. Only the men's restrooms on the ground floor have good ventilation.

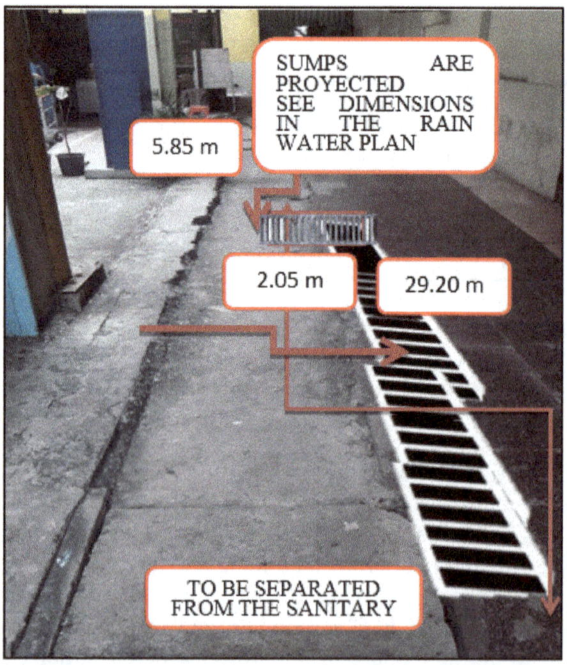

Figure 2: Rainwater evacuation channels and interconnections of the shelter with the Physics laboratory evacuation channels.

4 DRINKING WATER SYSTEM

The flow rate must be measured to estimate the deliverability of the system under actual demand conditions. In one of the registration boxes, it was not possible to measure the flow rate, because it entered directly into a storage vessel. To measure the flow rate, the water feeding line had to be disconnected to document the flow rate.

Flow measurements were made in May, June, and July of the year 2019; registering an average flow value of 1.7 m^3/h, considering the two pipelines that supply each water storage vessel present [1].

The working hours of the FMPS building are considered in regular business days, starting at 7:00 and ending at 22:00, so to calculate water demand a time-lapse of 14 h was considered. As there are two supply lines of 1.7 m^3/h each, simple math gives a total of about 82 m^3 of water demand, throughout the day. Therefore, a security factor of about 20% over the calculated demand was considered, which gives a total estimate for need of 100 m^3 [1].

According to the survey applied, as proposed in the methodology, a population of 2,500 permanent users was estimated for the FMPS building's facilities, this includes: teachers, students, administrative staff and maintenance staff. Consumption of 40 litres per inhabitant/day is required to meet demand. This represents a total of 100 m^3 per day.

Furthermore, during the evaluation performed on the first stage of this project, two underground water storage vessels were found to be the main suppliers of drinking water for the faculty. Each one has a capacity of 10 m^3: The first one is in the library area, and the second one at a garden nearby the FMPS building. Finally, an elevated water storage vessel is also present, having a total volume of 12 m^3 [12].

According to the results, drinking water demand is estimated to be 100 m^3/d. This supports that the FMPS building demand had increased; therefore, the water storage system needs to be expanded to meet the current demand [1].

The following investigation proposes to build up an extra 50 m^3 capacity cistern, located in the garden adjacent to the FMPS building, with the dimensions of approximately 15 m × 3 m × 1.40 m. With the construction of this new cistern, we would reach storage of 70 m^3 in underground vessels. To improve the supply system, elevated tanks will be required to be installed: 16 with a capacity of 1,000 litres each. They will be distributed according to a detailed plan (Fig. 3). The installation of these elevated tanks leads to the elimination of the concrete tank that was originally installed with a capacity of 2 m^3, weighing 2 tons. These 16 tanks will be distributed in the upper sector of the administrative premises [1].

Due to bad condition or deterioration suffered, eight water storage tanks will be replaced and installed on a terrace in front of the administrative premises.

With this new distribution, the Faculty building would get a total storage volume of 95 m^3, satisfying 95% of the water demand for the FMPS building's population. These elevated tanks will be fed by a pumping unit with a capacity of 1 HP. The storage tanks will be interconnected and installed as shown in Fig. 3.

All pipelines used to interconnect the storage system will be visible; therefore, good quality materials are required, such as polyvinylchloride (PVC) and polypropylene (PP) pipes with ultraviolet (UV) protection, to prevent premature damage to the system.

The three-dimensional (3D) design of the FMPS building and its toilet/sanitary facilities were done with the Revit BIM Software, to simulate the connections of the drinking water network based on the catalogue from the Plastigama company website.

BIM is an architecture, engineering, construction and operation industry modelling software. The application of this software permits managing the design and modelling of

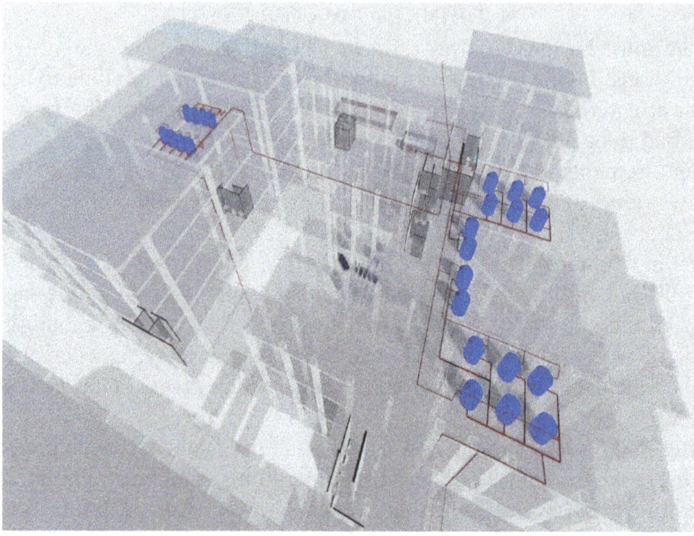

Figure 3: Proposed drinkable water storage system (royal blue) with pipes connected (red).

buildings, project data and facilities throughout their entire useful life. In addition, this software covers several fields, such as: Technology, Process, Policy, Interactions, CAD Format Models, Stages, Data Flows, and Project Life Cycle Phases [8].

To use the program, the building plans are required, then they must be imported into the software; after this, we generated the modelling of the drinking water supply system and the pipe installation scheme (Fig. 4). In Fig. 5, the whole drinking water system is projected in 3D. The BIM library was used, fulfilling the national Ecuadorian standards as required by the NEC.

5 DATA ACQUISITION

Consumption and pressure measurements were made in the FMPS building water and sewer network during May 2019. Using statistical calculations, the pressures used for this study's design and remodelling projects were estimated. Based on Table 1, an average pressure of 16.170 mwc was estimated. The flow rates measured provided us with an estimate of 1,700 m^3/h and 1,730 m^3/h for each of the original water storage tanks [1].

Figure 4: Pipeline modelling and installations generated by BIM.

www.witpress.com, ISSN 1743-3541 (on-line)

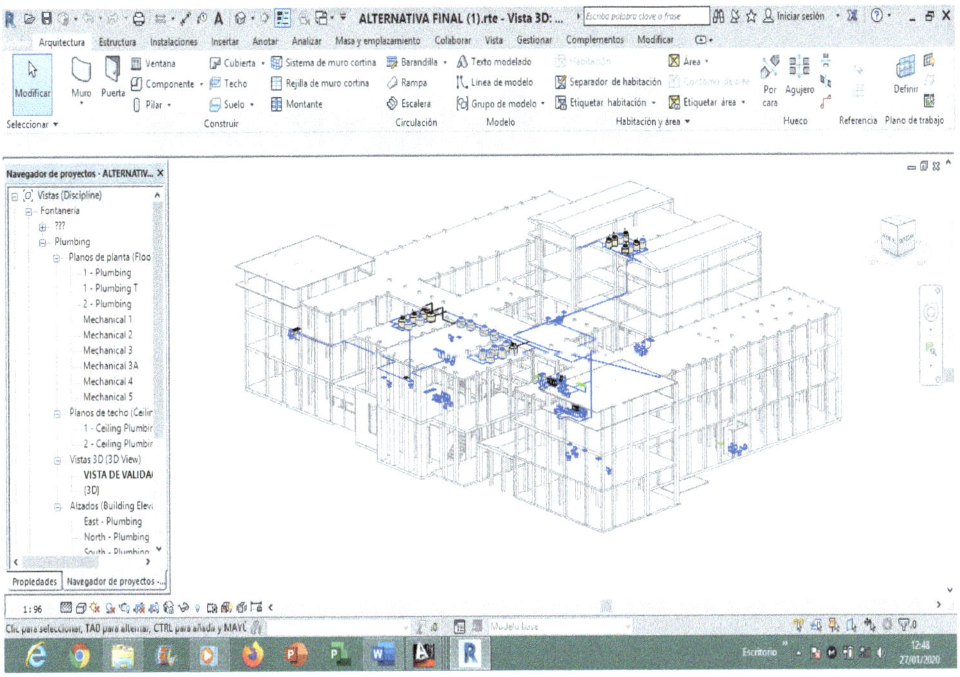

Figure 5: Design of the drinking water installations (blue) for the Faculty of Mathematical and Physical Sciences remodelling project.

Table 1: Pressure measurements obtained from the drinking water network.

Time	Pressures and dates (Unit: Meter of the water column (mwc))				
	05/29/19	05/30/19	05/31/19	06/03/19	06/04/19
7:00–8:00	18.304	19.008	19.712	18.304	19.008
8:00–9:00	17.600	18.304	18.304	18.304	16.896
9:00–10:00	16.896	15.488	16.896	17.600	15.488
10:00–11:00	17.600	16.192	17.600	15.488	15.488
11:00–12:00	16.192	13.376	16.192	16.192	14.784
12:00–13:00	14.080	12.672	15.488	16.896	14.784
13:00–14:00	14.080	13.376	14.784	16.896	14.080
14:00–15:00	13.376	13.376	14.784	16.896	13.376
15:00–16:00	14.784	14.080	14.080	17.600	14.080
16:00–17:00	14.784	15.488	15.488	17.600	15.488
17:00–18:00	16.192	15.488	16.896	17.600	16.192
18:00–19:00	16.896	16.192	16.896	18.304	16.192
19:00–20:00	16.896	16.896	17.600	18.304	16.896

6 WATER STORAGE SYSTEM PROPOSAL DEVELOPMENT

With all the data gathered regarding water demand, flow rates and the proposed distribution of water storage tanks, we performed a simulation to determine the feasibility of the proposed design (Fig. 6). Corrections were done in the software to guarantee water deliverability to the plumbing system and users.

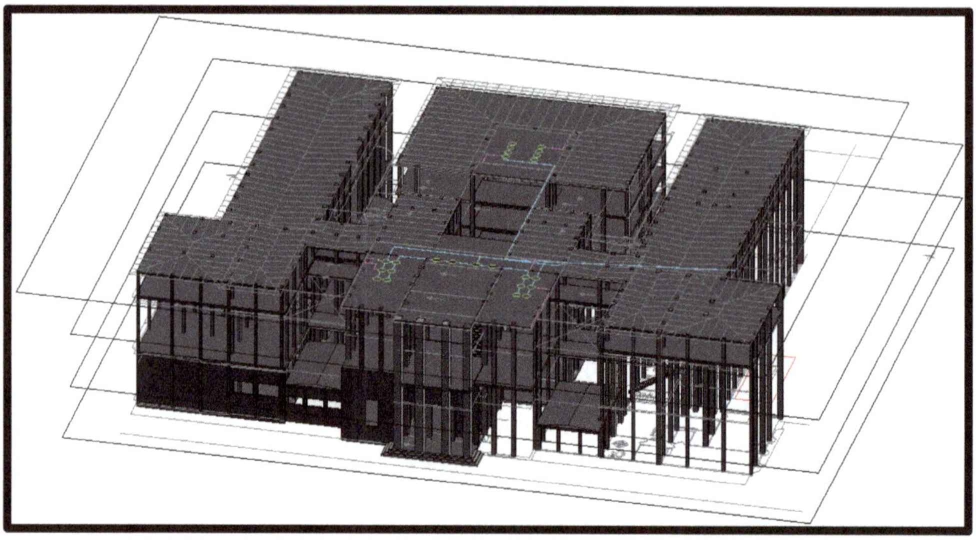

Figure 6: Design showing 3D view of the proposed system simulated by BIM.

Currently, the system has two flow meters installed and is located nearby the underground water storage tanks (cisterns) close to the FMPS complex. They record the drinking water flow rates to be stored in other cisterns located in the FMPS building. From the underground storage, the water is sent to the elevated storage tanks by pumping units and finally, the stream gets to the user by gravity. In this whole process, pressure is considered a critical parameter for water deliverability assurance.

Different modelling options were options considered in this proposal (Figs 7–12).

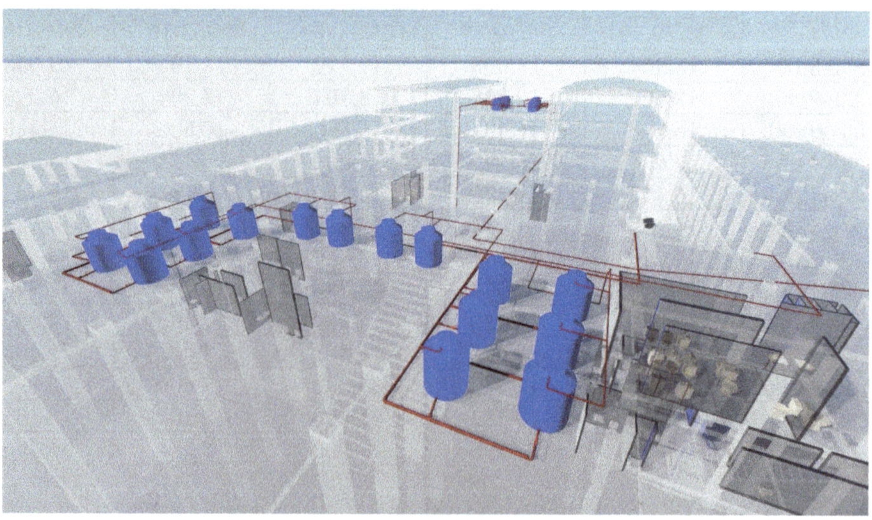

Figure 7: Water storage tank distribution (royal blue) in the main terrace of the administrative premises of the Faculty of Mathematical and Physical Sciences building.

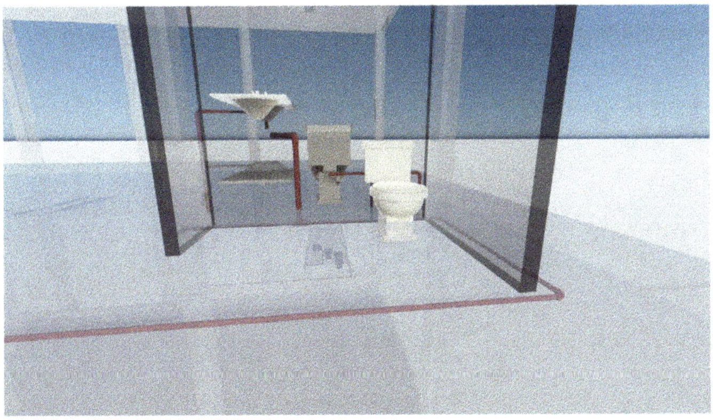

Figure 8: Model for improvement proposal regarding new pipes (red) and their accessories (cream) in a restroom/lavatory area.

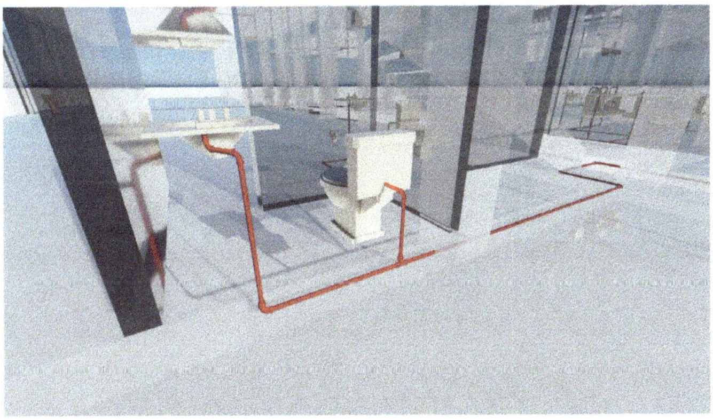

Figure 9: Drinking water (incoming, red) network design using the Plastigama catalogue.

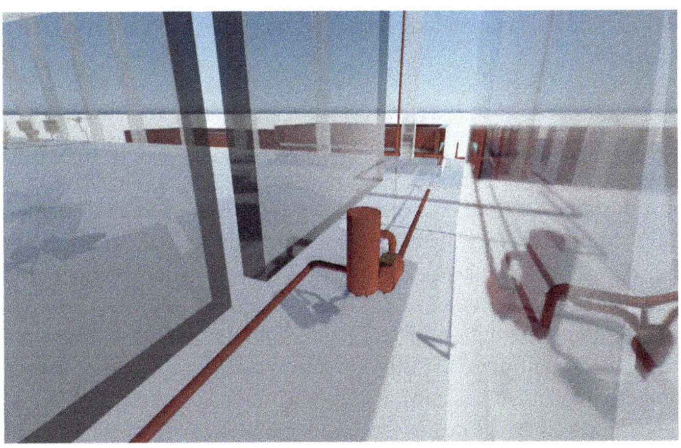

Figure 10: Pumping system units (red) needed to deliver water to the elevated tanks.

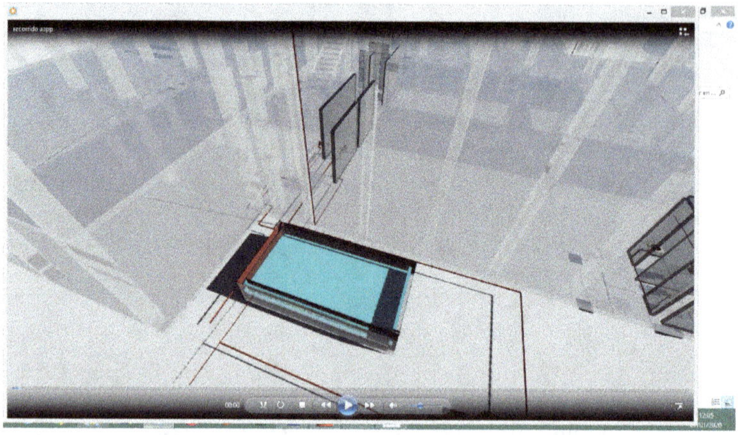

Figure 11: Cistern (turquoise) to be located in the Faculty of Mathematical and Physical Sciences building's garden.

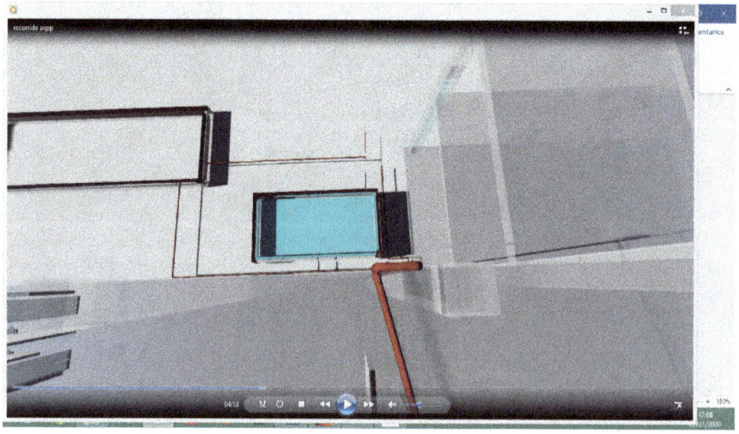

Figure 12: Location of the current working cistern (gray) and the proposed one (turquoise) to be built.

7 RAINWATER SEWAGE

In the first stage of the project, we discovered that the sewage and the rainwater diversion systems were interconnected. These interconnections were in the physics lab and library areas. Inspecting the area on foot, several slopes were found in the lateral and rear areas. This topographic condition was taken as an opportunity by the surroundings faculty buildings, which installed their drainage systems to lead to the sewage collector at the rear area of the Faculty of Mathematical and Physical Sciences building (Fig. 13).

We proposed the implementation of an evacuation channel for rainwater after a detailed topography study was completed. This channel will collect the rainwater from the surroundings and conduct them to a sink located beside a sidewalk. We used Revit software for modelling the proposal and Lumino software permitted a 3D video tour. These videos are available and access links are provided in the references [13], [14].

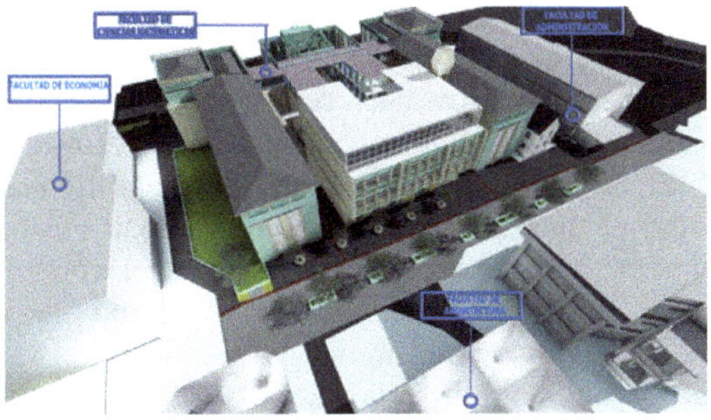

Figure 13: Front view of the Faculty of Mathematical and Physical Sciences building, showing drains in red, and surrounding buildings tagged in blue.

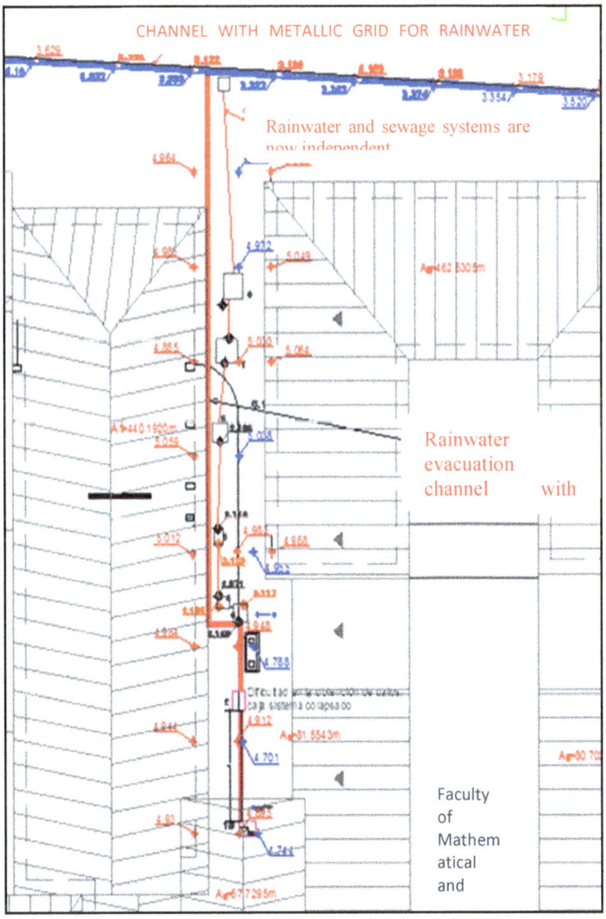

Figure 14: Rainwater evacuation plan (thick red line) inside the premises of the Faculty of Mathematical and Physical Sciences building towards the channel (thick blue) located at the rear of the building. Note that sewage system pipes near the faculty building (thin red) are now independent from the rainwater channel.

Rainwater evacuation channels are shown in Fig. 14, as well as the Physics laboratory and Library area where the rainwater boxes and the wastewater boxes are. The rainwater drainage and the sewer system have now been separated: there are two rainwater discharge systems, consisting of a duct channel with covered grids, also shown in Fig. 14.

In the rear area of the FMPS building, a gutter-type evacuation channel was proposed, which runs from the rear of the Faculty of Mathematical and Physical Sciences building adjacent to the Faculty of Architecture; then, this channel connects to the Faculty of Administration's channel, and later discharges laterally into the drain located in the sidewalk in front of the Faculty of Administration, as seen in Figs 15 and 16.

In the front area of the building, the rainwater discharge sewers become connected to the sidewalk drains that flush into the Faculty of Administration. All the rainwater drainpipes of the building will be connected to the new rainwater evacuation channel proposed in the FMPS building's rainwater system improvement project (Figs 15 and 16).

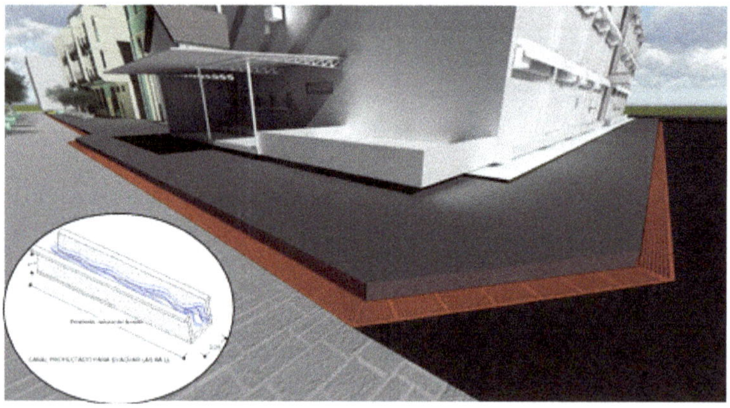

Figure 15: Proposed rainwater evacuation plan (red) for the Faculty of Mathematical and Physical Sciences' premises.

Figure 16: Rainwater drainpipes (red) leading to the evacuation channel proposed in our project.

8 LAVATORY FACILITIES INSTALLED AT FMPS BUILDING

According to the evaluation of the FCI-050 Project, "Evaluation of the Sanitary and Pluvial Sewer System of the Faculty of Mathematical and Physical Sciences of the University of Guayaquil" [1], the currently installed sanitary systems are insufficient for the existing overall population of the FMPS building complex. So at this first stage, the lavatory facilities were found insufficient.

The largest population being served by the lavatory facilities is the students.

Table 2: Sanitary facilities in the Faculty of Mathematical and Physical Sciences building at the University of Guayaquil, Ecuador, selected by user type and sex.

Lavatory facilities			
	Sinks	Lavatory	Urinal
Male students	9	4	37
Female students	4	4	0
Male lecturers	0	1	2
Female lecturers	0	0	0

9 CONCLUSIONS

The building complex for the Faculty of Mathematical and Physical Sciences in the University of Guayaquil, Ecuador, has drinking water networks that are in bad condition due to corrosion, as they have exceeded their useful life. Therefore, the University administration must change the entire drinking water system, the water storage tanks, and expand the volumes of water reserves to supply the current demand for water of the faculty's building based on its user population, as well as the implementation of new sanitary devices.

Sanitary facilities in the men's restrooms need to be expanded by nine toilets, four sinks and 36 urinals. On the other hand, facilities for women are sufficient [1].

To meet the water demand of the FMPS facilities users, 16 tanks of 1,000 L must be installed, distributed on the terrace of the main faculty building.

Our topographic survey revealed that the FMPS building's levels are lower at the rear of the building, which is the cause of flooding seen during the rainy season. Unfortunately, this tiered condition cannot be easily repaired; therefore, it is necessary to separate the rainwater discharge from the regular sewage system, to evacuate both more quickly, which is the best solution for rainfall-related flooding proposed in the present investigation.

10 RECOMMENDATIONS

We recommend implementation of a permanent maintenance plan to increase the new facilities' durability, and monitoring of the quality of water delivered and disposed of throughout the process. This plan should include leak monitoring.

To reduce flooding in the rear area of the faculty building, we recommend permanently evacuating garbage from the sewage system and evacuation channels.

To reduce water consumption, we believe it is necessary to install automatic water dispensers, flushometres, and tanks for toilets with a reservoir raised to 1.5 m of 4 L to reduce water consumption.

REFERENCES

[1] Saltos Sánchez, A., Rojas, J., Villa Ríos, P. & Tobar Barreno, G., Evaluación del sistema de alcantarillado sanitario y pluvial de la Facultad de Ciencias Matemáticas y Físicas de la Universidad de Guayaquil [Evaluation of the plumbing and rainwater gutter system at the University of Guayaquil's Faculty of Mathematical and Physical Sciences building]. *Revista Ciencia e Investigación [Journal of Science and Research]*, **3**(ICCE2018), pp. 16–27, 2018. https://revistas.utb.edu.ec/index.php/sr/article/view/586.

[2] Meride, Y. & Ayenew, B., Drinking water quality assessment and its effects on residents' health in Wondo genet campus, Ethiopia. *Environmental Systems Research*, **5**(1). DOI: 10.1186/s40068-016-0053-6.

[3] Huttinger, A. et al., Water, sanitation, and hygiene infrastructure and quality in rural healthcare facilities in Rwanda. DOI: 10.1186/s12913-017-2460-4.

[4] González-Amarillo, C.A., Cárdenas-García, C.L. & Mendoza-Moreno, M.A., M2M system for efficient water consumption in sanitary services, based on intelligent environment. *DYNA (Colombia)*, **85**(204), pp. 311–318, 2018.

[5] Blasco, E.J., Instalaciones sanitarias en edificaciones. Lima, Perú: Colegio de Ingenieros del Perú [Plumbing Installations in Buildings in Lima, Peru]. College of Engineers of Peru, 2005.

[6] Cheng, C.-L., Peng, J.-J., Ho, M.-C., Liao, W.-J. & Chern, S.-J., Evaluation of water efficiency in green building in Taiwan. *Water*, **8**(6), p. 236, 2016. DOI: 10.3390/w8060236.

[7] Salvo, P., Sparling, M. & Telmosse, R., Higher education – University of Ottawa investigates to identify underground infrastructure academic grade. *NASTT's 2018 No-Dig Show*, 2018.

[8] Silva, A.S. & Ghisi, E., Uncertainty analysis of daily potable water demand on the performance evaluation of rainwater harvesting systems in residential buildings. *Journal of Environmental Management*, pp. 82–93, 2016.

[9] Bartkiewicz, E. & Zimoch, I., Analysis of the risk of pipe breaks based on hydraulic model. *Safety and Reliability – Safe Societies in a Changing World*, pp. 1511–1516, 2018.

[10] Angulo, C.M., Aplicación de metodología BIM para la optimización en los diseños hidrosanitarios en las edificaciones [BIM usage methods for optimizing hydrosanitary designs for buildings]. Universidad Francisco De Paula Santander, 2020.

[11] Fadoul, A., Tizani, W. & Osorio-Sandoval, C.A., *A knowledge-based model for constructability assessment of buildings' design using BIM*, Lecture Notes in Civil Engineering, vol. 98, pp. 147–159, 2020.

[12] Carmona, R.P., Instalaciones hidrosanitarias, de gas y de aprovechamiento de aguas de lluvia [Hydrosanitary plumbing installation, gas installation, and use of rainwater], 2015.

[13] Saltos, A., 2021. Sewer system tour. https://www.youtube.com/watch?v=MBnHztphhaE&feature=youtu.be. Accessed on: 15 Feb. 2021.

[14] Saltos, A., 2021. Drinking water tour. https://www.youtube.com/watch?v=MBnHztphhaE&feature=youtu.be. Accessed on: 15 Feb. 2021.

HEAVY METALS IN RIVERS AFFECTED BY MINING ACTIVITIES IN ECUADOR: POLLUTION AND HUMAN HEALTH IMPLICATIONS

KENNY ESCOBAR-SEGOVIA[1,2], SAMANTHA JIMÉNEZ-OYOLA[1,3], DANIEL GARCÉS-LEÓN[1], DANIELA PAZ-BARZOLA[1], EDUARDO CHAVEZ NAVARRETE[1], PAOLA ROMERO-CRESPO[1] & BRYAN SALGADO[1]
[1]Escuela Superior Politécnica del Litoral, Ecuador
[2]Universidad Espíritu Santo, Ecuador
[3]Universidad Politécnica de Madrid, Spain

ABSTRACT

Artisanal and small scale gold mining operations have been an important source of toxic elements in the environment, with possibly negative consequences for ecosystems and human health. This study aims to assess the pollution degree and human health risk due to heavy metals exposure in mining communities. The data correspond to samples collected in rivers located in the vicinity of Ponce Enríquez, one of the most important gold mining sites in Ecuador. The concentrations of As, Cd, Cu, Pb, and Zn were analysed in surface water (n = 13) and sediment (n = 7) samples. Heavy metals contents were compared with (a) the corresponding backgrounds values (B_i) and (b) the permissible limits established by the Ecuadorian regulation. Furthermore, the Geo-accumulation Index (I_{geo}) and Metal Index (MI) were estimated for sediments and surface waters, respectively. Lastly, the potential adverse effects on human health due to the exposure to HMs in polluted rivers were carried out based on the hazard index (HI) and total cancer risk (TCR), using the traditional deterministic approach. The results showed that surface waters and sediments were enriched in heavy metals. The 57% of the sediment samples showed severe to extremely pollution ($I_{geo} > 3$) due to the high contents of As and Cd. On the other hand, surface water was in the strongly to seriously affected category (MI > 3) in 23% of the sample's locations, with As, Cu and Pb as the main contaminants. Likewise, residents exposed to polluted rivers showed HI and TCR values above the safe exposure threshold (HI = 1 and TCR = 10^5), being as the main contributor to the overall risk. Based on the preliminary results found in this research, future rivers monitoring and control efforts are needed to reduce the pollutions levels and the health hazard of the inhabitants of the mining areas.

Keywords: mining, pollution, heavy metals, health risk.

1 INTRODUCTION

Uncontrolled mining operations are causing serious environmental problems due to the large amount of pollutants generated. These can be introduced into the water system by surface runoff, causing water pollution not only in the mining area but in a larger region [1], [2]. Heavy metal(loid) contamination is known to be harmful for the environment. Presence of contaminants in soils, waters and air can be caused by natural events; however, the anthropogenic activities are the main contributors to the pollution [3]. Mining is one of the most polluting anthropogenic activities, especially in underdeveloped countries, where environmental controls are inadequate [4], [5]. The high concentrations of toxic elements around mining regions causes serious risk to ecosystems and human health [6].

Mining waste may include toxic elements such as As, Cd, Cr, Ni, and Pb, some of them recognized as human carcinogens by the International Agency for Research on Cancer [7], [8]. Exposure to these metals, even in small quantities, can cause significant human-health and ecosystem risks [9], [10]. In Ecuador, inadequate mining practices over the years have caused the environmental degradation of mining environments [11], [12]. Mining processing

WIT Transactions on Ecology and the Environment, Vol 250,
www.witpress.com, ISSN 1743-3541 (on-line)
doi:10.2495/WRM210061

operations have been conducted along rivers and streams [13], discharging mine tailings into streams that eventually flow into rivers [14], [15].

The human health risk depends on the pollutants concentrations in the environmental systems, their toxicity, and the exposure frequency [16]. Furthermore, in countries with tropical weather and very hot seasons like Ecuador, rivers are an important source of enjoyment for the community, especially for the children [17]. Given the elevated levels of exposure of residents have to heavy metal(loid)s and other toxic compounds in the mining communities, the aims of this research were to (a) evaluate the heavy metal(loid) pollution in surface waters and sediments in the Ponce Enríquez mining area, using the Geo-accumulation Index (I_{geo}) and Metal Index (MI) and (b) assess the human health risks due to exposure to multiple potentially toxic elements present in surface waters, using deterministic approach. The results of this work will provide an insight about the human health risk associated to the pollution in the mining area.

2 METHODOLOGY

2.1 Study area

Ponce Enríquez is in the southwest part of the country, and the western part of Azuay District (Fig. 1). This area has had massive mining activities since the 1980s [18]. The inadequate and illicit mining practices have caused significant environmental degradation in the area. Most of the mining contamination comes from Cerro Bella Rica and reaches the Siete River through illegal discharges from the treatment plants. Furthermore, mine tailings are stored in technically deficient dumps [15], where acid mine drainage have been lixiviated into soils and streams which eventually discharge into rivers. As a consequence, surface waters and sediments contain high concentration of heavy metals [14], [18]–[21].

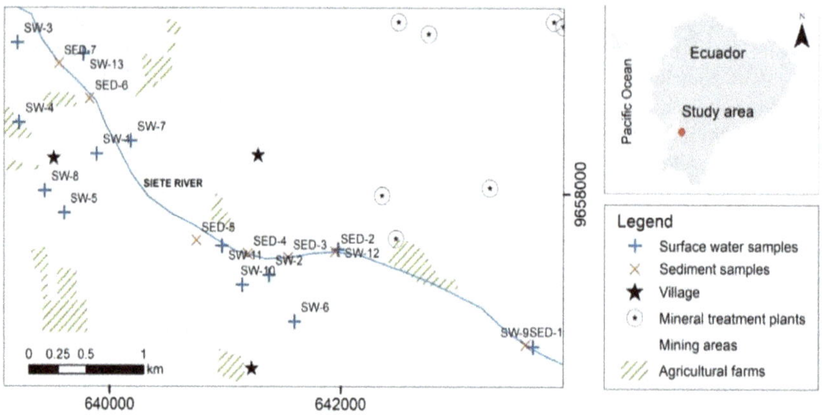

Figure 1: Location of the study area in a national scale, and location of the sampling sites.

2.2 Data collection and chemical analysis

The concentrations of As, Cd, Cu, Pb, and Zn were analysed for 13 surface water samples and 7 sediment samples collected in 2018. The samples came from the Siete River and some surrounding streams. Water samples were acidified with nitric acid and refrigerated until analysis. Sediment samples were taken with a Van Veen grab. The samples were sun dried for 48 h and classifyed to remove impurities. Then, samples were milled and passed through a 2 mm sieve, to be finally stored in a polyethylene bag prior to analysis.

Heavy metal(loid) concentrations in water samples were determined by Inductively Coupled Plasma Mass Spectroscopy (Agilent 7700x ICP-MS). Certified standard solution concentrations were used to ensure quality results with a standard deviation <10%. On the other hand, heavy metal concentrations in sediments were analysed by Inductively Coupled Plasma Optical Emission Spectrometry (ICP-OES). For quality control during the analysis, parallel process samples were established as reference material, obtaining relative standard deviations <5%.

2.3 Heavy metal(loid) pollution

2.3.1 Geo-accumulation index (I_{geo})

The pollution level of river sediments was quantified with the geo-accumulation index (I_{geo}) according to eqn (1). This index compares the analytical values of each heavy metal with the geochemical background values [22].

$$I_{geo} = \log_2 \left[\frac{C_i}{1.5 \text{ x } B_i}\right], \tag{1}$$

where C_i (mg kg^{-1}) is the element i concentration, and B_i (mg L^{-1}) is the element i geochemical background value. The constant 1.5 is used due to possible variations in reference data. The B_i values (mg kg^{-1}) used in this study were: As = 7.7, Cd = 0.17, Cu = 60, Pb = 5.4, Zn = 64. Those values correspond to sediment samples within the studied area, but in sites which are not affected by mining pollution [23]. The I_{geo} has seven quality classes: practically unpolluted ($I_{geo} \leq 0$), unpolluted to moderately polluted ($0 < I_{geo} < 1$); moderately polluted ($1 < I_{geo} < 2$), moderately to heavily polluted ($2 < I_{geo} < 3$), heavily polluted ($3 < I_{geo} < 4$), heavily to extremely polluted ($4 < I_{geo} < 5$); and extremely polluted ($I_{geo} > 5$) [24].

2.3.2 Metal index (MI)

The water surface quality was analysed using the metal index (MI) [25], [26]. The MI provides a general understanding of water quality. It has six water quality classifications: very pure (MI < 0.3), pure (0.3 < MI < 1); slightly affected (1 < MI < 2); moderately affected (2 < MI < 4); strongly affected (4 < MI < 6); and seriously affected (MI > 6). The MI was calculated with the following eqn (2):

$$MI = \sum_{i=1}^{N} \frac{C_i}{(MAC)_i}, \tag{2}$$

where C_i is the individual heavy metals concentration of each sample, and MAC is the maximum permissible concentration of the country's metal-based standards. According to the water quality guidelines established by the Ecuadorian legislation: 50, 1, 5, 1, and 30 (µg·L^{-1}) were the MAC values used for As, Cd, Cu, Pb and Zn, respectively [27].

2.4 Human health risk

The human health risks assessments were carried out using the Spatial Analysis and Decision Assistance (SADA) software, whose risk models follow the risk assessment guidelines proposed by the United States Environmental Protection Agency (USEPA) [28]. The traditional deterministic approach was used due to its simplicity and easy understanding [29]–[31]. This approach assigns a single representative value to each input parameter in the risk equation, leading to a single-value risk output.

For a deterministic approach, the 95% upper confidence limit (UCL) of the data mean was calculated with the Statistical Software ProUCL 5.1 developed by USEPA [32]. This study considered two kinds of receptors, adults and children; and two kinds of exposure routes: water incidental ingestion and water dermal contact during swimming activities. The received dose (ADD: $mg{\cdot}kg^{-1}{\cdot}day^{-1}$) of each exposure was calculated using eqns (3) and (4) according with USEPA recommendations [33].

$$ADD_{ingestion} = \frac{C \times EF \times ET \times IR \times ED}{BW \times AT}, \quad (3)$$

$$ADD_{dermal\ contact} = \frac{C \times EF \times ET \times SA \times Kp \times ED}{BW \times AT}, \quad (4)$$

where C value is the water heavy metal concentration ($\mu g{\cdot}L^{-1}$); EF is the annual exposure frequency ($days{\cdot}year^{-1}$); ET is the exposure time ($hour{\cdot}day^{-1}$); IR is the ingestion rate ($L{\cdot}day^{-1}$); ED is the life exposure duration (years); SA is the skin surface area exposed (cm^2); AF is the adherence factor ($mg{\cdot}cm^{-2}{\cdot}day^{-1}$); kp is the skin permeability constant ($cm{\cdot}hour^{-1}$); AT is the averaging time (days); and BW is the body weight (kg). The values used in this study were taken from the specific literature and are summarized in Table 1.

Table 1: Parameters used in the risk assessment.

Parameters		References values	
		Adults	Children
EF	Exposure frequency ($days{\cdot}year^{-1}$)	60[3]	60[3]
ET	Exposure time ($hour{\cdot}day^{-1}$)	2.6[3]	2.6[3]
IR	Incidental ingestion of water ($L{\cdot}day^{-1}$)	0.053[2]	0.090[2]
ED	Exposure duration (years)	30[1]	6[1]
SA	Skin surface area exposed (cm^2)	18,400[3]	6,800[3]
AF	Adherence factor ($mg{\cdot}cm^{-2}$)	0.07[1]	0.2[1]
kp	Constante de permeabilidad ($cm{\cdot}hour^{-1}$)	Zn = 0.0006; As, Cd, Cu, Pb = 0.001[4]	
BW	Body weight (kg)	72[3]	15.6[3]
AT	Averaging time		
	Carcinogens (days)	365 × ED[1]	365 × ED[1]
	Non-carcinogens (days)	365 × 70[1]	365 × 70[1]

[1]USEPA [34].
[2]USEPA [16].
[3]Spence and Walden [35].
[4]USDoE [37].

The potential risk of systemic effects was quantified in terms of hazard quotient (HQ) (eqn (5)), which results from dividing the ADD by the reference dose (RfD: $mg{\cdot}kg^{-1}{\cdot}day^{-1}$). To assess the risks of more than one route of exposure, the Hazard Index (HI) (eqn (6)) was introduced, which is the sum of HQs of all applicable routes of exposure. For HQ and HI above 1 the risk is considered unacceptable [33], [34].

$$HQ = \frac{ADD}{RfD}, \tag{5}$$

$$HI = \sum_{i=1}^{n} HQ_i. \tag{6}$$

To characterize the risk for exposure to carcinogenic compounds, the individual probability of develops cancer throughout his life (CR) is established according eqn (7). The CR is a function of the ADD and the slope factor (SF: $(mg \cdot kg^{-1} \cdot day^{-1})^{-1}$), which is the upper limit of the estimate of the increased risk of suffering cancer throughout life due to exposure to a unit dose of the carcinogenic agent. CR was determined for each exposure, and then CRs were summed and expressed as a Total Cancer Risk (TCR) in eqn (8). Finally, the TCR values were compared with the acceptable reference value of $10E^{-05}$, where the 1 person developing cancer probability per 100,000 inhabitants exposed is considered [36].

$$CR = ADD \text{ x } SF, \tag{7}$$

$$TCR = \sum_{i=1}^{n} CR_i. \tag{8}$$

Toxicity data (RfD and SF) were obtained from the Risk Assessment Information System (RAIS) database website [37]. The cancer risk was calculated for As and Pb, elements that have slope factors reported. The overall equations used to calculate the human health risk were implemented in R language.

3 RESULTS

3.1 Heavy metal(loid) pollution

The concentration of As, Cd, Cu, Pb and Zn in surface waters and sediments are presented in Tables 2 and 3, respectively. In surface waters, high values of As, Cu, Pb and Zn concentrations exceeding Ecuadorian standards were registered in 55% of the samples, where 64% of these occurred in the central part of the study area. Furthermore, SW-11 and SW-13 samples showed very high concentrations of As and Cu, compared to the other samples and the Ecuadorian regulation. Regarding sediments pollution, this study found elevated heavy metals sediments concentration, mainly Cd and As, with values up to 600 and 52 times higher than their background values (B_i). Cd was above the B_i in all the sampling locations. The results concur with the informed by other authors [18], [38], who reported elevated heavy metals sediments concentrations in the studied mining area.

3.1.1 Geo-accumulation index

The I_{geo} is an effective tool to characterize the sediment pollution levels by evaluating metal content above background values [22]. Based on I_{geo}, sediments from the Siete River presented high pollution levels. According to Table 4, 57% of the samples resulted in heavily to extremely polluted category. The highest I_{geo} values were reported for As and Cd. For Cu, the I_{geo} values denoted a moderate to extremely polluted degree. Also, I_{geo} values for Pb ranged from unpolluted to moderate polluted whereas Zn occurred in unpolluted levels in 42% of samples ($I_{geo} < 0$). The box plot in Fig. 2 showed a large data dispersion of I_{geo} values. The highest I_{geo} was reported to Cd, followed by As, Cu, Pb and Zn. The range of values were between −4 and 10, being As-I_{geo} = −2.94 the minimum value of the whole set of samples, and Cd-I_{geo} = 8.63 the maximum estimated value. The I_{geo} mean of Cu, Pb and Zn were below 2 (unpolluted to moderately polluted) while the values of As and Cd were above level 4; heavily to extremely polluted.

Table 2: Concentrations of heavy metal in surface water samples (SW: $\mu g \cdot L^{-1}$). Values in bold were above the Ecuadorian water quality guidelines.

Samples	As	Cd	Cu	Pb	Zn
SW-1	0.87	0.03	0.78	0.14	7.95
SW-2	9.33	0.04	2.43	0.96	9.06
SW-3	12.03	0.03	2.31	0.37	8.93
SW-4	10.85	0.03	2.36	0.20	7.13
SW-5	1.36	0.01	0.40	0.06	5.68
SW-6	0.87	0.02	0.67	0.17	11.30
SW-7	9.99	0.05	1.63	0.47	14.17
SW-8	0.90	0.03	0.68	0.09	8.65
SW-9	5.82	0.01	0.63	0.15	5.89
SW-10	11.97	0.08	3.83	**1.06**	13.54
SW-11	**62.04**	0.17	**27.26**	0.46	10.33
SW-12	44.60	0.29	**8.65**	0.69	19.62
SW-13	**263.46**	0.80	**293.95**	**154.32**	**96.55**

Table 3: Concentrations of heavy metal in sediment samples (SED: $mg \cdot kg^{-1}$). Values in bold were above the background values for sediments.

Samples	As	Cd	Cu	Pb	Zn
SED-1	1.50	**0.24**	21.59	3.97	49.86
SED-2	7.14	**1.68**	**101.19**	**8.32**	**70.37**
SED-3	4.31	**0.89**	42.67	2.79	45.34
SED-4	**240.97**	**60.67**	**308.52**	**14.08**	**121.14**
SED-5	**201.19**	**51.56**	**285.93**	**14.36**	**98.35**
SED-6	**402.67**	**101.15**	**1561.39**	**114.61**	**356.92**
SED-7	**232.50**	**57.25**	**1480.21**	**43.38**	**195.43**

Table 4: Values of Geo-accumulation Index. Values in bold show heavily to extremely pollution.

Samples	Geo-accumulation index				
	As	Cd	Cu	Pb	Zn
SED-1	−2.94	−0.09	−2.06	−1.03	−0.95
SED-2	−0.69	2.72	0.17	0.04	−0.45
SED-3	−1.42	1.80	−1.08	−1.54	−1.08
SED-4	**4.38**	**7.89**	1.78	0.80	0.34
SED-5	**4.12**	**7.66**	1.67	0.83	0.03
SED-6	**5.12**	**8.63**	**4.12**	3.82	1.89
SED-7	**4.33**	**7.81**	**4.04**	2.42	1.03

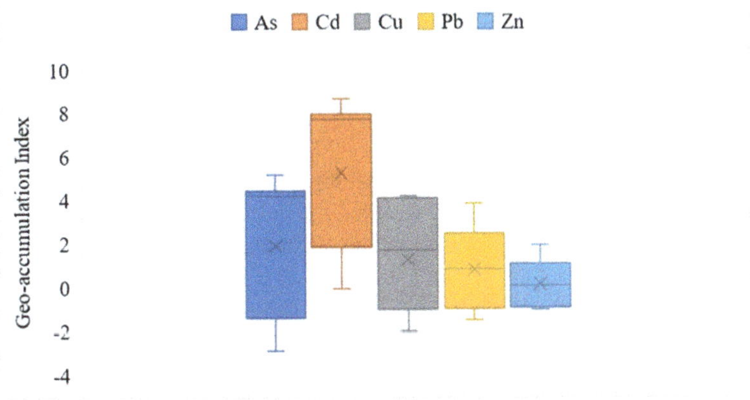

Figure 2: Box plot of I_{geo} values for sediment samples.

The I_{geo} values detected in this study were similar to those reported by Aguilar et al. [20]. Furthermore, the I_{geo} values in the study area were in the same range as observed in the Colombia, Brazil, Iran and China mining areas [39]–[42], with moderate to severe degree of pollution where As–Cd–Cu metals are the most disturbing contaminants. On the other hand, I_{geo} for Pb was significantly lower than the ones reported in other mining areas [43], [44] which could be explained by the geology. Furthermore, lowers I_{geo} were detected in mining regions in Brazil [45] and India [46], with I_{geo} values of 4 and 2, respectively.

3.1.2 Metal index (MI)

The categorization of water quality based on the metal index is shown in Fig. 3. As a result, 38% of the samples were pure, the remaining percentage were from slightly to seriously affected. The samples SW-11 and SW-13 were the ones identified as seriously affected. These results evidence the potential impact of mining activities. It is well known that mining wastes in the area usually contains: arsenopyrite, pyrite, pyrrhotite, and sphalerite [11], which may explain the pollution levels in surface waters.

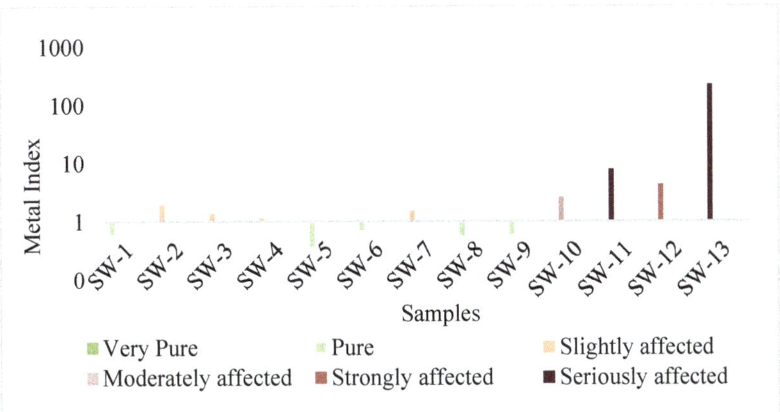

Figure 3: MI column diagram of surface water samples.

3.2 Human health implications

Table 5 shows the outcomes for the carcinogenic (TCR) and non-carcinogenic (HI) risk for adults and children receptors. Surface waters have revealed elevated concentration of heavy metal(loid)s. These elements, especially As and Pb, represents a significant health hazard for inhabitants. The cumulative non-carcinogenic risk (HI) for children was over the safe exposure limit level (HI > 1) for water incidental ingestion. In contrast, the HI for adults was below the recommended limit. While As resulted the pollutants of major concern, exposure to Cd, Cu, Pb and Zn was negligible (HQ values were in order of 10^{-01} and 10^{-05}) for both receptors (Fig. 4(a)).

Regarding the carcinogenic risk, the TCR through incidental ingestion of water was above the safe exposure threshold for both adults and children receptors, but it was below the recommended value ($10E^{-05}$) for dermal contact. In addition, As was the main contributor to the cancer risk level for almost 99.75% of TCR for both receptors (Fig. 4(b)).

Table 5: Deterministic HI and TCR in surface waters for both receptors.

	Ingestion		Dermal contact	
	Adults	Children	Adults	Children
HI	1.60E − 01	1.27E + 00	5.64E − 02	9.62E − 02
TCR	2.89E − 05	4.53E − 05	9.80E − 06	3.42E − 06

Children are the most vulnerable receptors; the HI and TCR values for children were almost two times higher than adults values. The exposure to heavy metal(loid)s through polluted waters incidental ingestion was the main cause to the non-carcinogenic and carcinogenic risk for children.

The outcomes in this study were consistent with the informed by other authors in mining areas around the word [3], [39], [42], [47], [48]. Therefore, exposure reduction policies should be adopted to reduce the risk levels to which the population around mining locations is exposed. The frequency of EF exposure used in this study was 60 events·year^{-1}. However, under tropical weather like Ecuador, where rivers are used for recreational purposes, EF can increase considerably, mainly for children. Therefore, population-specific parameters should be determined locally to obtain more reliable results.

4 CONCLUSIONS

This study reveals the elevated heavy metal concentration in surface waters and sediment samples from the Ponce Enríquez mining area. The 57% of the sediment samples showed severe to extremely pollution, mainly with the As and Cd elements. Furthermore, 23% of the water samples were from strongly to seriously affected category, with As, Cu and Pb elements as the main pollutants. The risk outcomes by deterministic methods showed that exposure in local rivers are unsafe for human health. The carcinogenic and non-carcinogenic risk was above the safe exposure limit showing that residents were exposed to an unacceptable risk level, mainly through incidental ingestion of water. The main contributor to the overall risk was As.

This study provides important knowledge about the risk to human health to which the inhabitants of mining communities are exposed. To obtain more realistic results, population-specific parameters should be determined locally. In addition, a continuous environmental monitoring could help to reduce the pollution in the mining area.

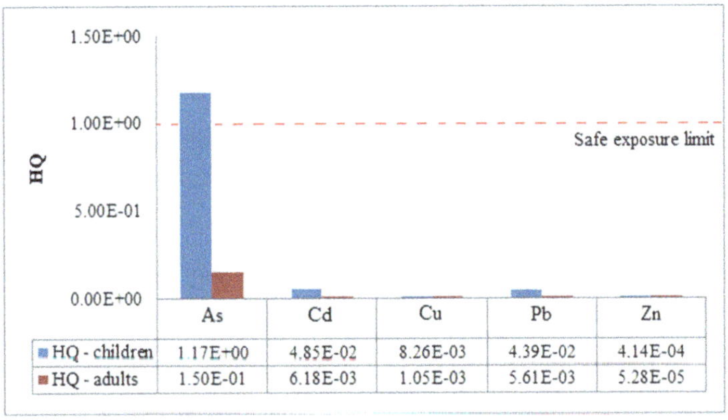

(a)

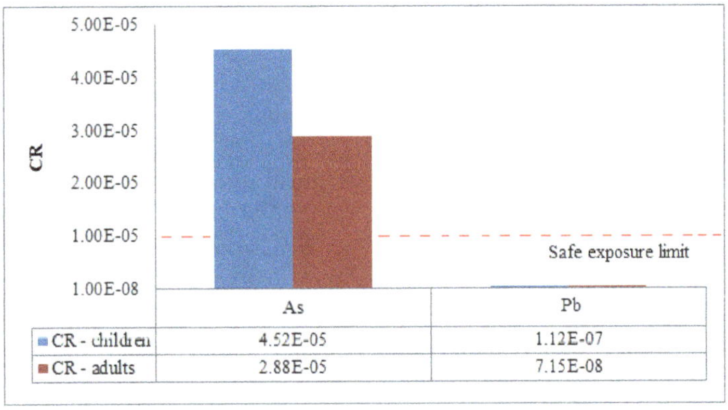

(b)

Figure 4: Heavy metal(loid)s in the human health risk outcomes. (a) Hazard quotient; and (b) Cancer risk.

REFERENCES

[1] García-Sánchez, A., Alonso-Rojo, P. & Santos-Francés, F., Distribution and mobility of arsenic in soils of a mining area (Western Spain). *Science of the Total Environment*, **408**(19), pp. 4194–4201, 2010. DOI: 10.1016/j.scitotenv.2010.05.032.

[2] Wu, Y. et al., Migration and transformation of arsenic: Contamination control and remediation in realgar mining areas. *Applied Geochemistry*, **77**, pp. 44–51, 2017. DOI: 10.1016/j.apgeochem.2016.05.012.

[3] Ngole-Jeme V.M. & Fantke P., Ecological and human health risks associated with abandoned gold mine tailings contaminated soil. *PLoS One*, **12**(2), pp. 1–14, 2018. DOI: 10.1371/journal.pone.0172517.

[4] Bundschuh, J. et al., One century of arsenic exposure in Latin America: A review of history and occurrence from 14 countries. *Science of the Total Environment*, **429**, pp. 2–35, 2012. DOI: 10.1016/j.scitotenv.2011.06.024.

[5] Vangsnes, G.F., The extractive industries and society the meanings of mining: A perspective on the regulation of artisanal and small-scale gold mining in southern *Ecuador. Extractive Industries and Society*, **5**(2), pp. 317–326, 2018. DOI: 10.1016/j.exis.2018.01.003.

[6] Rapant, S., Dietzová, Z. & Cicmanová, S., Environmental and health risk assessment in abandoned mining area, Zlata Idka, Slovakia. *Environtal Geology*, **51**(3), pp. 387–397, 2006. DOI: 10.1007/s00254-006-0334-x.

[7] IARC, Evaluation of carcinogenic risk for humans. *Lyon*, 1987.

[8] Mulware, S.J., Trace elements and carcinogenicity: a subject in review. *3 Biotech*, **3**(2), pp. 85–96, 2013. DOI: 10.1007/s13205-012-0072-6.

[9] WHO, *Guidelines for Drinking-Water Quality*, 4th edn, 2011.

[10] USEPA, *2018 Edition of the Drinking Water Standards and Health Advisories Tables*, Washington, DC, 2018. DOI: EPA 822-S-12-001.

[11] Peña-Carpio, E. & Menéndez-Aguado, J.M., Environmental study of gold mining tailings in the Ponce Enriquez mining area (Ecuador). *Dyna*, **83**(195), pp. 237–245, 2016. DOI: 10.15446/dyna.v83n195.41745.

[12] Velásquez-López, P.C., Veiga, M.M. & Hall, K., Mercury balance in amalgamation in artisanal and small-scale gold mining: Identifying strategies for reducing environmental pollution in Portovelo-Zaruma, Ecuador. *Journal of Cleaner Production*, **18**(3), pp. 226–232, 2010. DOI: 10.1016/j.jclepro.2009.10.010.

[13] Williams, T.M. et al., Regional geochemical reconnaissance of the Cordillera Occidental of Ecuador: Economic and environmental applications. *Applied Geochemistry*, **15**(4), pp. 531–550, 2000. DOI: 10.1016/S0883-2927(99)00059-1.

[14] Carling, G.T. et al., Particulate and dissolved trace element concentrations in three southern Ecuador rivers impacted by artisanal gold mining. *Water, Air, and Soil Pollution*, **224**(2), 2013. DOI: 10.1007/s11270-012-1415-y.

[15] Jordá, R., Romero, P., Peña, E., Jiménez, S., Garcés, D. & Chang R., Análisis preliminar de la estabilidad de escombreras y balsa de relaves en el Distrito Minero Ponce Enríquez, Ecuador, pp. 19–21, 2017.

[16] USEPA, *Exposure Factors Handbook: 2011 Edition*, Washington, DC, 2011.

[17] MAE-PRAS, Ministerio de Ambiente de Ecuador. Programa de Reparación Integral de la zona de estudio Tenguel – Camilo Ponce Enríquez. *Quito*, 2015.

[18] Appleton, J.D., Williams, T.M., Orbea, H. & Carrasco, M., Fluvial contamination associated with artisanal gold mining in the Ponce Enríquez, Portovelo-Zaruma and Nambija areas, Ecuador. *Water, Air, and Soil Pollution*, **131**(1–4), pp. 19–39, 2001. DOI: 10.1023/A:1011965430757.

[19] Appleton, J., Carrasco, M., Maldonado, R. & Orbea, H., Assessment of mercury contamination in the Ponce Enriquez artisanal gold mining area, Ecuador. *British Geological Survey*, 1996.

[20] Aguilar, A., Vitvar, T. & Peña-Carpio, E., A Multi-index analysis approach to heavy metal pollution in river sediments in the ponce enríquez Area, Ecuador. *Water*, **11**(590), 2019. DOI: 10.3390/w11030590.

[21] Tarras-Wahlberg, N.H., Flachier, A., Lane, S.N. & Sangfors, O., Environmental impacts and metal exposure of aquatic ecosystems in rivers contaminated by small scale gold mining: The Puyango River basin, southern Ecuador. *Science of the Total Environment*, **278**(1–3), pp. 239–261, 2001.

[22] Muller, G., Index of geoaccumulation in sediments of the Rhine river. *GeoJournal*, **2**, pp. 108–118, 1969.

[23] PRODEMINCA, Monitoreo Ambiental de las areas mineras en el sur del Ecuador. Ministerio de Energía y Minas, 1998.
[24] Wang, Y. et al., Assessment of multiple exposure to chemical elements and health risks among residents near Huodehong lead-zinc mining area in Yunnan, Southwest China. *Chemosphere*, **174**, pp. 613–627, 2017. DOI: 10.1016/j.chemosphere.2017.01.055.
[25] Withanachchi, S.S. et al., Water quality in surface water: A preliminary assessment of heavy metal contamination of the Mashavera river, Georgia. *International Journal of Environmental Research and Public Health*, **15**(4), pp. 1–25, 2018. DOI: 10.3390/ijerph15040621.
[26] Tamasi, G. & Cini, R., Heavy metals in drinking waters from Mount Amiata (Tuscany, Italy). Possible risks from arsenic for public health in the Province of Siena. *Science of the Total Environment*, **327**(1–3), pp. 41–51, 2004. DOI: 10.1016/j.scitotenv.2003.10.011.
[27] MAE-TULSMA, Texto Unificado de Legislación Secundaria Medio Ambiental. Ministerio de Ambiente de Ecuador. *Quito*, 2015.
[28] University of Tennessee, SADA. Spatial analysis and decision assistance. *Tennessee*, 2005.
[29] Jiménez-Oyola, S. et al., Multi-pathway human exposure risk assessment using Bayesian modeling at the historically largest mercury mining district. *Ecotoxicology and Environmental Safety*, **201**(March), pp. 110833, 2020. DOI: 10.1016/j.ecoenv.2020.110833.
[30] Saha, N., Rahman, M.S., Ahmed, M.B., Zhou, J.L., Ngo, H.H. & Guo, W., Industrial metal pollution in water and probabilistic assessment of human health risk. *Journal of Environtal Management*, **185**, pp. 70–78, 2017. DOI: 10.1016/j.jenvman.2016.10.023.
[31] Jiménez-Oyola, S. et al., Análisis espacial del riesgo para la salud por exposición a metales pesados en el campo minero Ponce Enríquez. pp. 24–26, 2019. DOI: 10.18687/laccei2019.1.1.117.
[32] EPA, ProUCL statistical support software for site investigation and evaluation. Environmental Protection Agency, p. 1, 2016.
[33] USEPA, Risk assessment guidance for superfund volume I: Human health evaluation manual (Part F). Supplemental Guidance for Inhalation Risk Assessment, Washington, DC, 2009.
[34] USEPA, Risk assessment guidance for superfund volume I: Human health evaluation manual (Part E). Supplemental Guidance for Dermal Risk Assessment, Washington, DC, 2004.
[35] Spence, L.R. & Walden, T., *RISC User's Manual Version 4.0*. p. 463, 2001.
[36] USEPA, Guidelines for carcinogen risk assessment, Washington, DC, 2005.
[37] USDoE, U.S. Department of Energy, RAIS: Risk Assessment Information System. 2020.
[38] Tarras-Wahlberg, N., Flachier, A., Fredriksson, G., Lane, S., Lundberg, B. & Sangfors, O., Environmental impact of small-scale and artisanal gold mining in southern Ecuador. *AMBIO Journal of Human Environment*, **29**(August), pp. 484–491, 2000. DOI: 10.1639/0044-7447(2000)029.
[39] Alonso, D.L., Pérez, R., Okio, C.K.Y.A. & Castillo, E., Assessment of mining activity on arsenic contamination in surface water and sediments in southwestern area of Santurbán paramo, Colombia. *Journal of Environmental Management*, **264**(February), 2020. DOI: 10.1016/j.jenvman.2020.110478.

[40] Yasami, N., Ghaderi, M. & Taghilou, B., Heavy metal assessment in stream sediments from the rivers passing through the mining area. *International Journal of Environmental Science and Technology*, **16**(0123456789), pp. 1–20, 2018. DOI: 10.1007/s13762-018-1840-6.

[41] Zhang, Z., Lu, Y., Li, H., Tu, Y., Liu, B. & Yang, Z., Assessment of heavy metal contamination, distribution and source identification in the sediments from the Zijiang River, China. *Science of the Total Environment*, **645**, pp. 235–243, 2018. DOI: 10.1016/j.scitotenv.2018.07.026.

[42] Santos de Souza, E. et al., Assessment of risk to human health from simultaneous exposure to multiple contaminants in an artisanal gold mine in Serra Pelada, Pará, Brazil. *Science of the Total Environment*, **576**, pp. 683–695, 2017. DOI: 10.1016/j.scitotenv.2016.10.133.

[43] El Azhari, A., Rhoujjati, A., El Hachimi, M.L. & Paul Ambrosi, J., Pollution and ecological risk assessment of heavy metals in the soil-plant system and the sediment-water column around a former Pb/Zn-mining area in NE Morocco. *Ecotoxicology and Environmental Safety*, **144**(June), pp. 464–474, 2017. DOI: 10.1016/j.ecoenv.2017.06.051.

[44] Kusin, F.M., Azani, N.N.M., Hasan, S.N.M.S. & Sulong, N.A., Distribution of heavy metals and metalloid in surface sediments of heavily-mined area for bauxite ore in Pengerang, Malaysia and associated risk assessment. *Catena*, **165**, pp. 454–464, 2018. DOI: 10.1016/j.catena.2018.02.029.

[45] Salomão, G.N. et al., Geochemical mapping and background concentrations of iron and potentially toxic elements in active stream sediments from Carajás, Brazil – Implication for risk assessment. *Journal of South American Earth Sciences*, **92**(February), pp. 151–166, 2019. DOI: 10.1016/j.jsames.2019.03.014.

[46] Raj, D., Chowdhury, A. & Maiti, S.K., Ecological risk assessment of mercury and other heavy metals in soils of coal mining area: A case study from the eastern part of a Jharia coal field, India. *Human and Ecological Risk Assessment*, **23**(4), pp. 767–787, 2017. DOI: 10.1080/10807039.2016.1278519.

[47] Chen, G., Wang, X., Wang, R. & Liu, G., Health risk assessment of potentially harmful elements in subsidence water bodies using a Monte Carlo approach: An example from the Huainan coal mining area, China. *Ecotoxicology and Environmental Safety*, **171**(December 2018), pp. 737–745, 2019. DOI: 10.1016/j.ecoenv.2018.12.101.

[48] Pavilonis, B., Grassman, J., Johnson, G., Diaz, Y. & Caravanos, J., Characterization and risk of exposure to elements from artisanal gold mining operations in the Bolivian Andes. *Environmental Research*, **154**(August 2016), pp. 1–9, 2017. DOI: 10.1016/j.envres.2016.12.010.

FILTRATION OF DISSOLVED ORGANIC NUTRIENTS FROM FISH FARM WASTEWATER USING A MACROALGAE BIOFILTER

JACK R. HALL* & GEORG MARTIN
Estonian Marine Institute, University of Tartu, Estonia

ABSTRACT

Intensive animal aquaculture damages the environment by releasing large quantities of nutrients which drives eutrophication in aquatic ecosystems. Macroalgae are efficient in uptaking nutrients as they grow with past studies suggesting their integration into aquaculture systems as a means to improve wastewater quality. This study was designed to assess the feasibility of using macroalgae as a biological filtration system for the removal of dissolved nutrients found in finfish farm wastewater. To test this, an experimental fish farm and mesocosm system was established on the northern coast of Saaremaa island, West Estonian archipelago. The green algae *Ulva intestinalis* was selected as a good candidate to assess the efficacy of a macroalgae biofiltration system to uptake nutrients. The results obtained show at best a 18–25% reduction in waste water nutrient concentrations for the nitrogenous compounds nitrite and nitrate for mesocosms stocked with macroalgae compared with the control. The system experienced an average 60% reduction in nitrogen and phosphorus concentrations in wastewater outflow compared to concentrations present within the finfish mesocosm. Additionally, the subsequent biomass gain of the incubated macroalgae species *Ulva intestinalis* is reported to be 4% per day at its maximum rate. The results obtained in this study indicate that *Ulva intestinalis* can be integrated into aquaculture systems as a nitrogen biofilter. In addition, the macroalgae biomass produced may offer aquaculture operations an additional income stream improving farm economics.

Keywords: aquaculture, biofilter, macroalgae, nutrients, bioremediation, ulva, nitrogen, IMTA.

1 INTRODUCTION

The decline of wild fish stocks has necessitated the need for aquaculture to offset the increasing global demand for fish protein [1]. However, as demand for fish protein grows the increasing intensification of finfish aquaculture requires ever greater inputs, in particular, that of fish feed [2], [3]. However, as no operation is 100% efficient, wastes in such systems are formed as either by-products or as unused inputs. As a consequence, Intensive finfish aquaculture is known to cause several adverse environmental impacts through the excessive output of nutrients, particularly nitrogen and phosphorous, into aquatic ecosystems [4]–[6]. For instance, the production of finfish in Japan has been found to generate on average approximately 0.8 kg of N and 0.1 kg of phosphorus irrespective of fish species cultivated [7]. Furthermore, the waste discharged by 63,000 tons of finfish or the total amount of cultivated fish in Japan for the year of 1999 is equivalent to that of 5 million people [7]. Consequently, coastal eutrophication has been identified as on major ecological impact and one that disproportionally effects enclosed and coastal systems associated with finfish aquaculture. This highlights the need for the development of strategies that limit the output of nutrients as a means to protect the environment and to ensure the future expansion and intensification of finfish aquaculture is conducted in a sustainable manner. Therefore, the challenge that is presented is to develop methods and technologies that minimize the negative environmental impacts of finfish aquaculture wastes.

* *ORCID: https://orcid.org/0000-0002-7446-855X*

WIT Transactions on Ecology and the Environment, Vol 250, © 2021 WIT Press
www.witpress.com, ISSN 1743-3541 (on-line)
doi:10.2495/WRM210071

Currently the most practical and economic solution for reducing the concentration of eutrophying compounds in finfish effluent is to treat it directly before it is discharged as waste into the environment. An alternative biological solution to this process is reported has been numerous studies [8]–[10]. By integrating macroalgae into the production process nutrients can be absorbed and concentrated from the effluent before discharged. In this sense the macroalgae acts as a biofilter, uptaking nutrients from the system. In addition to improving water quality, it has been suggested that the cultivation of macroalgae within finfish systems is a technology that can bring economic benefits to farm operations through additional revenue streams and the diversification of the organisms cultured [11], [12]. Several studies have reported the benefits of integrating macroalgae into the finfish culture process, however most of these studies have consisted primarily of small-scale laboratory experiments. As such, there is a present need for the scaling up of experimental systems in order to validate the feasibility of integrating macroalgae into real world farming operations.

In the present study we integrate the green macroalgae *Ulva intestinalis*, into a finfish aquaculture system with the aim to assess its ability to uptake waste nutrients and as a consequence its usefulness as a biofilter. As macroalgae has value outside of just its nutrient uptake ability, we simultaneously examined the production of macroalgal biomass and the safety of this biomass for use as feed material in terms of heavy metal concentrations in respect to European Union directives.

2 MATERIALS AND METHOD

The study on the growth and biofiltration capacity of the macroalgae *Ulva intestinalis* was conducted in open air conditions on the north coast of Saaremaa island, Estonia (latitude: 58.439559, longitude: 22.062178). The experiment was conducted over one vegetative season (2020) and comprised of three, 4–5 week growth trials. Wild grown *U. intestinalis* material for use in the experiment was collected from local coastal areas on Saaremaa island. *U. intestinalis* was selected as the biofilter because it is found to grow in large quantities locally. Additionally, it is characterized by its high nutrient uptake rates as reported by literature and confirmed by nutrient uptake trials conducted in the previous year that demonstrated it to have a better efficacy for nutrient uptake than that of other species found to occur naturally within the region. Upon collection the macroalgae material was cleaned of epiphytes and encrusting organisms before being placed in seawater filled containers and transported directly to the experiment site.

2.1 Experimental protocol

To test the biofiltration ability of *Ulva intestinalis*, a land based mesocosm experiment was constructed. The mesocosm experiment consisted of mesocosm chambers placed in four parallel series, 12 chambers in length. Each mesocosm chamber in the series was connected to the previous chamber with a pipe in such a manner that water could flowed through from the first mesocosm chamber into the subsequent chamber next in the series before outflowing as wastewater into the adjacent sea. The first mesocosm in each of the four series was connected with piping to a larger 6,000 l mesocosm chamber that was stocked with Rainbow trout (Oncorhynchus mykiss) (50 kg). Wastewater from the trout stocked mesocosm flowed at an equal rate into each of the four series. Seawater was pumped, aerated and filtered from the Baltic sea directly into the trout stocked mesocosm providing a continuous flow of new water to the system. *U. intestinalis* material was weighed and stocked (approximately 2,000 g each) into each of the mesocosms ranked fourth, fifth and sixth from the initial mesocosm in three of the four series. The fourth series contained no macroalgal material and was designated the control series. Vigorous aeration was supplied via an air pump and tubing system to the fourth, fifth and sixth mesocosm in each of the series. Automatic loggers were used to continuously record factors of temperature (°C), pH and dissolved oxygen (mg/L) for the trout mesocosm as well as the mesocosms tanks in series. Salinity (PSU) was measured for all mesocosms every 3.5 days using a portable sensor. The system was allowed to acclimatize for 3.5 days after the designated mesocosms were stocked before any sampling commenced. At the conclusion of each of the three trials the macroalgal material was removed and replaced with new material.

Water samples were taken in triplicate from the trout mesocosm as well as from the mesocosms ranked three, five, nine and twelve in each of the series and analyzed by an accredited water quality laboratory 3.5 days. Water samples were analysed for nutrient content (nitrite, nitrate, total nitrogen, phosphate and total phosphorus). Any macroalgae material found to be in poor condition was weighed and removed from the experiment. If the macroalgae material was found to be decaying the nutrient sample was discarded and sampling for that mesocosm deferred until the next sample period. As per government licensing requirements the nutrients were also monitored at the beginning and end of the experiment where water inflowed and outflowed from the experiment.

At the conclusion of each of the three trials the change in *U. intestinalis* weight from the initial stocked weight for each of the stocked mesocosms was recorded. Macroalgae material was retained from each of the macroalgae stocked mesocosm and was analyzed for trace metal concentrations.

2.1.1 Statistical analysis

In order to compare the nutrient concentrations between each of the three trials the recorded data was adjusted to the percentage change relative to the fish stocked mesocosm (mesocosm zero) for each of the sample mesocosms in the series. A students t-test was used to compare the means of macroalgae stocked mesocosms with the control. A significance level of 95% ($\alpha = 0.05$) was set for all tests. Macroalgae growth rate for each of the stocked mesocosms was calculated as percentage growth per day using the recorded initial and final waits and time.

3 RESULTS

3.1 Nutrient removal

The concentration of the nutrients tested was found to decrease throughout the mesocosm series. Both Nitrite and Nitrate were observed to be up taken by the mesocosms containing the macroalgae *Ulva intestinalis* when compared to the associated control. Under favorable growth conditions *U. intestinalis* demonstrated a significant increase in the uptake of both nitrate and nitrite resulting in a decrease of 18.4% and 25.2% of the nutrients respectively when compared to the control series (students t-test; $p < 0.05$) (Fig. 1). The phosphorous nutrient data was found to have a large degree of variability among the samples and due to this high variability, no significant difference between the control series and macroalgal stocked series for these nutrients was observed. Overall, the system demonstrated a high degree of nutrient removal efficiency, with up to 60% of both nitrate and nitrite removed from the system and 60% of phosphate and 30% of phosphorous also removed relative to concentration of nutrients measured in the trout mesocosm.

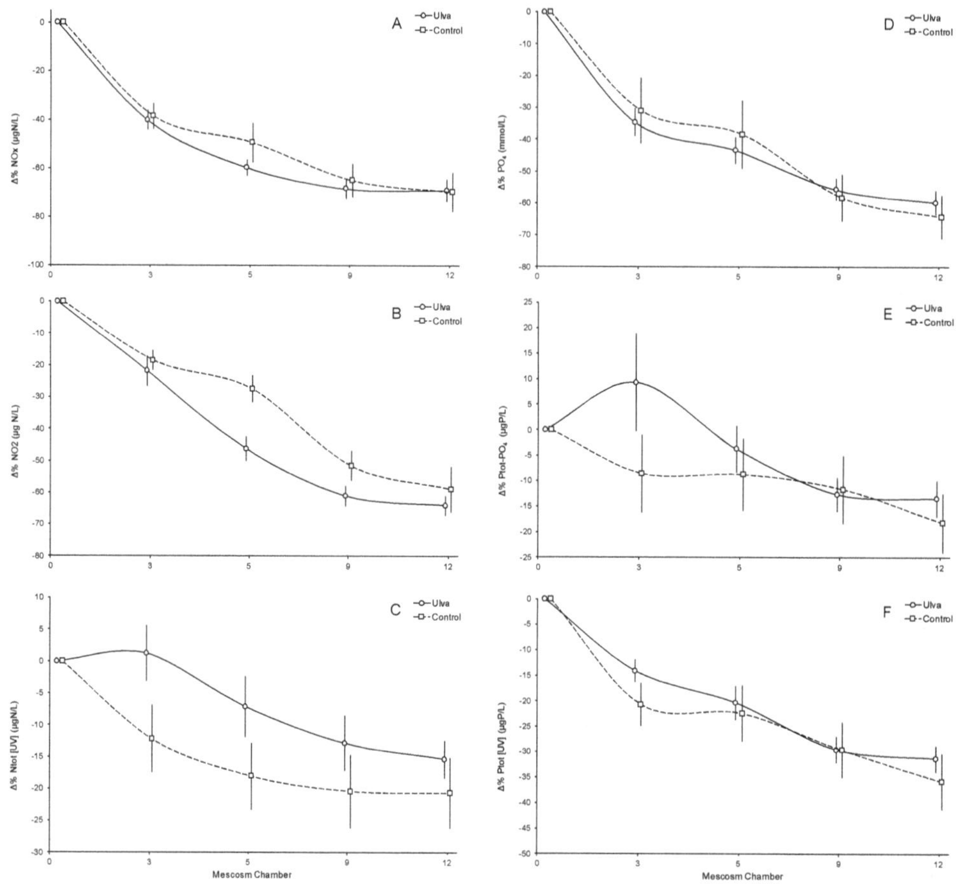

Figure 1: Change in the mean concentration of the nutrients (A = nitrite; B = nitrate; C = total nitrogen; D = phosphate; E = phosphorus; and F = total phosphorus) as a percentage relative to the initial trout stocked mesocosm across the mesocosm series. The mesocosms ranked four, five and six in the Ulva series were stocked with macroalgae. The control series contained no macroalgae. Error constructed as ±1 standard error.

3.2 Physical parameters and biomass growth

Water temperature was found to vary to a high degree throughout the experiment coinciding with air temperatures which followed typical diurnal and seasonal trends in temperature. The variation in temperature between the first and final tanks in the mesocosm series was recorded to be on average 3 degrees warmer than that of the final. The highest the water temperatures reached was 25 degrees in June. In contrast salinity remained highly stable throughout the experiment with little fluctuation and a mean PSU of 7.2 observed. Dissolved oxygen fluctuated following the expected trend of daytime photosynthesis and nighttime respiration. Mesocosm pH fluctuated to a similar degree in accordance with day and night cycles. When experiencing the appropriate abiotic factors for growth, *U. intestinalis* biomass was found to increase steadily over the course of the experiment. Mean biomass growth was recorded to be 1.9% per day (under growth conditions) with the highest growth observed occurring in the third growth trial at 4% per day.

Table 1: Mean (±SE) nutrient concentration across the mesocosm series.

	0	3	5	9	12
NOx (µg N/L)					
Ulva	10.39 ±1.29	5.95 ±0.75	3.85 ±0.44	2.91 ±0.3	2.66 ±0.28
Control	10.39 ±2.3	6.19 ±1.34	4.96 ±1.27	3.25 ±0.69	2.7 ±0.58
NO2 (µg N/L)					
Ulva	1.19 ±0.12	0.89 ±0.08	0.69 ±0.06	0.5 ±0.04	0.46 ±0.04
Control	1.19 ±0.21	0.97 ±0.16	0.83 ±0.14	0.53 ±0.09	0.45 ±0.07
Ntot [UV] (µg N/L)					
Ulva	262.78 ±11.17	257.15 ±11.13	241.86 ±9.68	228.46 ±7.48	223.85 ±6.65
Control	262.78 ±19.83	224.36 ±17.8	211.67 ±19.55	212.96 ±15.43	205.58 ±9.9
PO4 (µg P/L)					
Ulva	12.33 ±0.76	9.03 ±0.8	7.08 ±0.61	5.69 ±0.44	5.39 ±0.45
Control	12.33 ±1.34	7.94 ±1.25	6.82 ±1.23	5.62 ±0.82	4.75 ±0.77
Ptot-PO4 (µg P/L)					
Ulva	22.35 ±1.3	24.42 ±1.66	23.1 ±1.3	21.42 ±1.33	21.23 ±1.35
Control	22.35 ±2.3	21.01 ±2.42	20.94 ±2.4	21.46 ±2.3	19.65 ±1.99
Ptot [UV] (µg P/L)					
Ulva	35.53 ±1.93	33.45 ±2.09	30.18 ±1.47	27.11 ±1.44	26.62 ±1.41
Control	35.53 ±3.43	28.95 ±2.66	27.76 ±2.78	27.08 ±2.51	1.4 ±2.09

3.3 Heavy metals

The heavy metal analysis of *Ulva intestinalis* material for arsenic, lead, mercury and cadmium was conducted by an accredited laboratory and is reported in relation to the maximum accepted levels under the European Union directive for unwanted substances in animal feed (2002/32/EC). The concentration of arsenic was found to be near the limit of acceptable levels for animal feed (2 mg/kg (2002/32/EC)) but was much lower than the maximum level of seaweed meal and feed material made from seaweed (40 mg/kg (2012/744/EU)). Excluding one sample which was found to have concentrations several times higher than the allowable maximum, lead levels in the sample *U. intestinalis* material were found to be below the established limits of 10 mg/kg (2002/32/EC) for green fodder (2012/744/EU; 2013/1275/EU). The mercury content of all macroalgal samples analyzed was found to be below the maximum level of 0.1 mg/kg (2002/32/EC) for feed material. In all macroalgal samples analyzed, cadmium content was found to be considerably lower than that of the maximum allowable level of 1 mg/kg (2002/32/EC) for feed material of plant origin.

Table 2: Heavy metal concentration of *Ulva intestinalis* cultivated under enhanced nutrient conditions.

	Range	Mean ± SD
Arsenic (mg/kg)	4.8–2.2	3.27 ± 1.17
Cadmium (mg/kg)	0.09–0.06	0.07 ± 0.01
Lead (mg/kg)	12.0–3.6	7.58 ± 3.54
Mercury (mg/kg)	0.03–0.01	<0.02 ± 0.01

4 DISCUSSION

The effluent produced by finfish aquaculture poses numerous detrimental effects to the environment through excessive nutrient loading [4]–[6]. The cultivation of macroalgae in a manner that is integrated into finfish aquaculture operations is suggested as a means to offset these environmental impacts and improve water quality [8]–[10]. Within this context, the macroalgae species *Ulva intestinalis* used in this study demonstrates its potential to efficiently remove some of these eutrophying compounds. In particular *U. intestinalis* demonstrated useful potential in the efficient removal of nitrite and nitrate from the growth medium and as a consequence improved water quality. Past laboratory-based studies have indicated macroalgae species such as *Gracilaria* and *Ulva* constitute ideal candidates for use in integrated biofiltration systems within a wastewater and polyculture context. An example being Hernández et al. [9] which investigated *Ulva rotundata*, *Enteromorpha intestinalis*, *Gracilaria gracilis* for their biofiltration capacity of waste effluents produced by Sea bass (*Dicentrarchus labrax*) with small scale experiments. They demonstrated these species to be useful biofilters but concluded that the scaling up of biofiltration designs and integration into finfish aquaculture system was necessary in order to adequality evaluate a given species utility. The present study represents this scaling up and confirms *U. intestinalis* as species suitable for integration into current land-based aquaculture operations as a biofilter.

Although *U. intestinalis* is an efficient nitrogen filter with a relatively small amount of macroalgal material (i.e. ~2,000 g wet weight) removing significant concentrations of nitrogen based nutrients from the system relative to the density of the stocked fish, it absorbed little in the way of phosphorus based nutrients. Such a result is consistent with findings in the literature which states *Ulva* species as having a lower affinity for and slower uptake of phosphorus when compared with nitrogen [13], [14]. It is therefore suggested that finfish aquaculture operations may see a benefit in culturing macroalgae species with a strong affinity for phosphorus compounds in tandem with *U. intestinalis* as a means to uptake the full spectrum of eutrophying dissolved compounds produced by finfish effluent. Such a system could act as a highly efficient, low cost, biofiltration system.

The cultivation of macroalgae alongside onshore finfish facilities may be an attractive solution to improving water quality for farm operators. Nevertheless, for operations to be successful care must be taken in order to ensure the macroalgae is cultivated under the correct growth conditions. For instance, in outdoor operations such as the one experimented with in this study, temperatures spikes represented a serious issue, at times resulting in the die off of portions of the macroalgal material. Such die offs result in an inefficient system being less efficient and counter the filtration goal as decaying material releases additional nutrients into back into the system. Additionally, water flow through rates must be carefully calibrated to ensure adequate nutrients are supplied to the macroalgae with considerations made for the macroalgae stocked at the end of the flow through series where nutrients may have already been stripped from the water leading to starvation. In this regard, it may make sense for aquaculture operations to implement designs that incorporate a larger number of shorter flow through series as means to avoid macroalgae experiencing limited nutrients.

The nutrients stripped from the finfish wastewater mesocosm resulted in an increase in *Ulva intestinalis* biomass. Such growth represents a potential economic opportunity to finfish aquaculture operation as macroalgae biomass holds value for numerous industries and is typically highly difficult to obtain in industrial quantities. For instance, macroalgae and more specifically *Ulva* has numerous applications in industries ranging from Food, medicine, therapeutics feed and fertilizers [11]. At its peak we obtained a growth rate of 4% per day by wet weight with a system not specifically optimized for growth. Temperature appeared to be the major restricting factor with the open-air system experiencing midday temperature spikes resulting in acute stress for *U. intestinalis* in the mesocosms. With improved designs, which could be as simple as providing shades to block direct light for the mesocosms, operations can likely further improve growth rates. In this sense future onshore aquaculture operations may take advantage of developing alternative revenue streams through grown biomass having the overall effect of increasing both farm profitability and sustainability.

To ensure the safety and suitability of *U. intestinalis* enriched by finfish effluent for consumption by humans or as an additive to animal feed, the macroalgae material was analyzed for its heavy metal concentrations in relation to guidelines set out by the European Union (2002/32/EC). All samples bar one, which was found to contain high levels of lead (likely contamination by lead weights used to secure aeration hoses) and was excluded, were found to be within the maximum limit for safe consumption for the heavy metals arsenic, lead, mercury and cadmium. By nature, *U. intestinalis* is a fast-growing species and as such its high turnover is unlikely to accumulate appreciable amounts of heavy metals from its environment before it is harvested. As such, there is no indication from this study that a fast growing macroalgae produced from a biofiltration system would represent a substantial threat to either human or animal health due to high heavy metal concentrations.

Integrating macroalgae into aquaculture systems offers a biological solution to the issue of nutrient loading from wastes associated with intensive finfish cultivation. The present study demonstrates *Ulva intestinalis'* ability as species to prevent N loading from finfish effluent and to improve water quality. The results validate previous past laboratory-based studies that have suggested biofilters can be scaled for use in real world aquaculture systems. This study also further supports that the production of macroalgae biomass cultivated alongside finfish aquaculture can provide economic opportunities for farming operations.

ACKNOWLEDGEMENTS

This study was funded by European Maritime and Fisheries Fund through special project call organized by Estonian Ministry of Rural affairs (Project number 821017780003). We would like to thank Ösel Aquafarms OÜ and Ösel Harvest OÜ and especially Andro Ots and Henry Ots for good cooperation and providing with experimental site and needed fishfarming equipment. Special thanks go to Kaido Noormägi for maintaining the trout mesocosm. Thanks to Marin Teeveer and Teemar Püss for constructing the mesocosm experiment. We thank Hanna Eliisa Luts for assistance with macroalgae sample collection and all staff of the department of Marine Biology of Estonian Marine Institute who assisted us in the process.

REFERENCES

[1] Pauly, D., Watson, R. & Alder, J., Global trends in world fisheries: impacts on marine ecosystems and food security. *Philosophical Transactions of the Royal Society: Biological Sciences*. **360**(1453), pp. 5–12, 2005. https://doi.org/10.1098/rstb.2004.1574.

[2] Bostock, J. et al., Aquaculture: global status and trends. *Philosophical Transactions of the Royal Society B: Biological Sciences.* **365**(1554) pp. 2897–2912, 2010. https://doi.org/10.1098/rstb.2010.0170.

[3] Huntington, T.C. & Hasan, M.R., Fish as feed inputs for aquaculture practices, sustainability and implications: A global synthesis. FAO Fisheries and Aquaculture Technical Paper, 518, pp. 1–61, 2009.

[4] Wang, J., Beusen, A.H., Liu, X. & Bouwman, A.F., Aquaculture production is a large, spatially concentrated source of nutrients in Chinese freshwater and coastal seas. *Environmental Science and Technology*. **54**(3), pp. 1464–1474, 2019. https://doi.org/10.1021/acs.est.9b03340.

[5] Iwama, G.K., Interactions between aquaculture and the environment. *Critical Reviews in Environmental Science and Technology.* **21**(2), pp. 177–216, 1991. https://doi.org/10.1080/10643389109388413.
[6] Herbeck, L.S., Unger, D., Wu, Y. & Jennerjahn, T.C., Effluent, nutrient and organic matter export from shrimp and fish ponds causing eutrophication in coastal and back-reef waters of NE Hainan, tropical China. *Continental Shelf Research*, **57**, pp. 92–104, 2013. https://doi.org/10.1016/j.csr.2012.05.006.
[7] Suzuki, Y., Maruyama, T., Numata, H., Sato, H. & Asakawa, M., Performance of a closed recirculating system with foam separation, nitrification and denitrification units for intensive culture of eel: towards zero emission. *Aquacultural Engineering*, **3–4**, pp. 165–182, 2003. https://doi.org/10.1016/j.aquaeng.2003.08.001.
[8] Marinho-Soriano, E., Nunes, SO., Carneiro, M.A., Pereira, D.C., Nutrients' removal from aquaculture wastewater using the macroalgae *Gracilaria birdiae. Biomass and Bioenergy*, **33**(2), pp. 327–331 2009. https://doi.org/10.1016/j.biombioe.2008.07.002.
[9] Hernández, I., Martínez-Aragón, J.F., Tovar, A., Pérez-Lloréns, J.L. & Vergara, J.J., Biofiltering efficiency in removal of dissolved nutrients by three species of estuarine macroalgae cultivated with sea bass (*Dicentrarchus labrax*) waste waters 2. Ammonium. *Journal of Applied Phycology*, **5**, pp. 375–384, 2002. https://doi.org/10.1023/A:1022134701273.
[10] Hernández, I., Fernández-Engo, M.A., Pérez-Lloréns, J.L. & Vergara, J.J., Integrated outdoor culture of two estuarine macroalgae as biofilters for dissolved nutrients from Sparus auratus waste waters. *Journal of Applied Phycology,* **6**, pp. 557–567, 2005. https://doi.org/10.1007/s10811-005-9006-6.
[11] Dominguez, H. & Loret, E.P., *Ulva lactuca*, a source of troubles and potential riches. *Marine Drugs*, **6**, p. 357, 2019. https://doi.org/10.3390/md17060357.
[12] Chopin., T. et al., Integrating seaweeds into marine aquaculture systems: A key toward sustainability. *Journal of Phycology*, **37**(6), pp. 975–986. 2001. https://doi.org/10.1046/j.1529-8817.2001.01137.x.
[13] Li, S., Yu, K., Huo, Y., Zhang, J., Wu, H., Liu, Y., Shi, D. & He, P., Effects of nitrogen and phosphorus enrichment on growth and photosynthetic assimilation of carbon in a green tide-forming species (*Ulva prolifera*) in the Yellow Sea. *Hydrobiologia*, **776**(1), pp. 161–171, 2016. https://doi.org/10.1007/s10750-016-2749-z.
[14] Luo, M.B., Liu, F. & Xu, Z.L., Growth and nutrient uptake capacity of two co-occurring species, Ulva prolifera and Ulva linza. *Aquatic Botany*, **100**, pp. 18–24, 2012. https://doi.org/10.1016/j.aquabot.2012.03.006.

USE OF AGRO-INPUTS IN AGRICULTURAL PRODUCTION AND WATER RESOURCES POLLUTION: A CASE STUDY OF CHU-MANGO IN THE MEKONG RIVER DELTA, VIETNAM

TRUONG HONG VO TUAN KIET[1], PHAM THI NGUYEN[2] & NGUYEN THI KIM THOA[1]
[1]Mekong Delta Development Research Institute, Can Tho University, Vietnam
[2]School of Economics, Can Tho University, Vietnam

ABSTRACT

This paper employed the Environmental Impact Quotient (EIQ) model to determine the impact of chemical fertilizer and pesticide on farm workers, consumers and the ecology, as well as imply potential threats of Chu-mango cultivation on water resource pollution. The findings indicate that the number of root fertilizer (N–P–K) is 624.1, 542.5 and 443.5 kg/ha for seasons 1, 2 and 3, respectively. The nitrogen and phosphates misuse could leach into groundwater or enter via surface runoff into waterways. The ecology components in Chu-mango farming is the most vulnerable objective among three seasons. The field use EIQ in season 1 is the highest (1,058.56) in all of three seasons, then season 2 (747.06) and season 3 (592.34). Pesticides can move into water through direct application, runoff, and atmospheric deposition to contaminate water resources. One suggestion, therefore, is that the government could encourage mango growers to adopt good agricultural practices by training programs that raise their awareness of ecological protection. This policy would help to reject banned agrochemicals in mango production, which protects the water quality. Additionally, traceability code is needed, which would help farmers change the behavior of pesticide overuse, and is to carry out individual responsibilities for their products to communities. It not only reduces loadings of agricultural pollutants to water resources but also takes the potential trade benefit.

Keywords: Chu-mango, ground water, surface water, pollution.

1 INTRODUCTION

The global increasing demand for agricultural commodities, farmers are looking to increase productivity through the intensive use of inputs such as pesticides and chemical fertilizers as well as agricultural land expansion, irrigation system progression. This has transferred agricultural pollution to water bodies. Water pollution from unsustainable agricultural practices threatens human health and ecosystems. According to [1], agriculture accounts for 70% of total water consumption worldwide and is the single-largest contributor of non-point-source pollution to surface water and groundwater. Agriculture intensification is often accompanied by increased soil erosion, salinity and sediment loads in water and by the excessive use of agricultural inputs (pesticides and fertilizers) to rise productivity.

Residues of chemical fertilizers and pesticides may wash off the field into rivers or leach through the soil into ground water. Irrigation can move salt and other dissolved minerals to surface water. That is reason why main pollution source of rivers and streams in the United States of America comes from agricultural farming [2]. Surface-water and groundwater pollution in China stem from nitrogen in agricultural production [3]. Insecticides, herbicides and fungicides are applied intensively in agriculture in many countries [4] especially in developing countries where is weak rule enforcement, and limited knowledge and awareness among farmers on the use of hazardous chemicals pose enormous challenges to the safe and sustainable management of pesticides. This can cause poison fish and wildlife, contaminate food sources, and destroy the habitat that animals use for protective cover.

WIT Transactions on Ecology and the Environment, Vol 250, © 2021 WIT Press
www.witpress.com, ISSN 1743-3541 (on-line)
doi:10.2495/WRM210081

The study aims to identify the current status of chemical fertilizers and pesticides (insecticides, herbicides and fungicides) use in Chu-mango production as well as measure negative impacts of agro-inputs on health and ecology, especially is water resources. This helps more understanding regarding the causes and effects of agricultural water pollution in order to control and mitigate pollution loads from Chu-mango cultivation activities.

2 MATERIALS AND METHODS

2.1 Sampling techniques

Based on the secondary data of statistical yearbook in agricultural production, the study chose the three biggest production area of Chu-mango including: An Giang, Dong Thap and Vinh Long provinces in the Mekong River Delta (Fig. 1). Information on primary data was conducted several steps. Firstly, study group interviewed directly with agricultural extension workers in province and district level to identify big mango villages. Secondly, there were six discussion groups (4 people per group) in six Chu-mango production villages to determine essential elements before designing questionnaire. Thirdly, the research carried out trial survey 36 sampling observations (12 observations in each province). Finally, simple random technique was employed to select 522 sampling observations (212, 171, and 139 observations for seasons 1, 2, and 3, respectively) for computing the EIQ of the human health and ecological impacts.

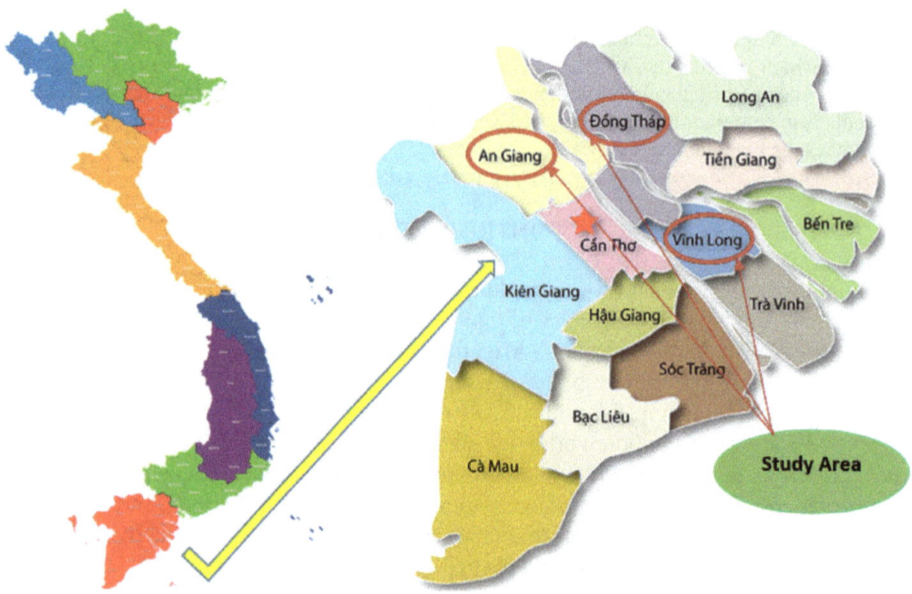

Figure 1: Study area in the Mekong Delta, Vietnam.

2.2 Environment impact quotient model

There are other measurements of ecological risk of agrochemical use [5]. However, assessments of pesticide applications still use Environment Impact Quotient (EIQ) for scientific and policy purposes [6]–[10]. The use of EIQ has been debated in the literature

[11]–[13]. The EIQ model was developed by Kovach et al. [14] at Cornell University to quantify the effects of various crop pests and disease-control strategies on humans and the ecology.

Table 1: Definition for symbols and ratings for each toxicity category [14].

Variables	Symbol	Score 1	Score 3	Score 5
Long-term health effects (chronic)	c	Little-none	Possible	Definite
Dermal toxicity (Rat LD50)	dt	>2,000 mg/kg	200–2,000 mg/kg	0–200 mg/kg
Bird toxicity (8 day LC50)	d	>1,000 ppm	100–1,000 ppm	1–100 ppm
Bee toxicity	z	Non-toxic	Moderately toxic	Highly toxic
Beneficial arthropod toxicity	b	Low impact	Moderate	Severe impact
Fish toxicity (96 h LC50)	f	>10 ppm	1–10 ppm	<1 ppm
Plant surface half-life	s	1–2 weeks	2–4 weeks	>4 weeks
Soil residue half-life (TI/2)	p	<30 days	30–100 days	>100 days
Mode of action	sy	Non-system	Systemic	
Leaching potential	l	Small	Medium	Large
Surface runoff potential	r	Small	Medium	Large

The variables in Table 1 are used to computer eight ecological impact indicators (applicator, harvester, exposure, groundwater, fish, birds, bees, other beneficial insects). These scores are then further aggregated to represent the ecological impact on each of the three major profiles: farmer, consumer and ecology. Eventually, the synthesized EIQ score is the average of the three scores and it is calculated for each pesticide active ingredient. The EIQ formula can be divided into influences on the farm worker (applicator and harvester), consumer (exposure and groundwater effects), and ecology (fish, birds, bees, beneficial arthropod). The EIQ is specific to individual active ingredients that is calculated by eqn (1). The field use EIQ is specific to individual pesticide formulations, which may have multiple active ingredients. The field use EIQ for pesticide formulation is the sum of the EIQ's of the individual active ingredients weighted by both the proportion of the formulation each active ingredient comprises and the application rate of the formulation (eqn (2)). The higher values of the field use EIQ indicate greater relative risk. The EIQ formula is defined:
Total EIQ = farm worker + consumer + ecology:

$$\left\{\left[c*(dt*5)+(dt*p)\right]+\left[\left(c*(s+p)/2*sy\right)+(l)\right]+\left[(f*r)+\left(d*(s+p)/2*3\right)+(z*p*3)+(b*p*5)\right]\right\}/3 \quad (1)$$

where:

c = chronic toxicity,
dt = dermal toxicity,
p = plant surface residue half-life,
s = soil residue half-life,
sy = systematicity,
l = leaching potential,
f = fish toxicity,
r = surface loss,
d = bird toxicity,
z = bee toxicity, and B = beneficial arthropod activity.

The Field Use EIQ = EIQ theorical values * % active ingredient * rate/ha (2)

Name list of active ingredients in the Table 2 is result of mango households' investigation in study area. Values of EI components (farm worker, consumer and ecology) and average EIQ are calculated based on eqn (1). In fact, these values are calculated by Cornell University (List of pesticide active ingredient EIQ values: A method to measure the environmental impact of pesticides, Table 2: list of pesticides (last updated May 2020)). They are considered as impact parameters (EIQ theoretical values) and is used in eqn (2) to calculate the field use EIQ.

Table 2: Theoretical values of the EIQ from Cornell University [15].

Active ingredient	EI component value			Average EIQ value
	Farm worker	Consumer	Ecology	
Paclobutrazol	21.30	6.55	51.45	26.43
Herbicide				
Glyphosate	8.00	3.00	35.00	15.33
Paraquat	31.95	6.33	35.92	24.73
2,4-D	24.00	7.00	31.00	20.67
Pesticide				
Cypermethrin	13.80	5.90	89.35	36.35
Chlorpyrifos	6.00	2.00	72.55	26.85
Emamectin benzoate	9.00	4.00	65.85	26.28
Abamectin	13.80	3.90	86.35	34.68
Imidacloprid	6.90	10.35	92.88	36.71
Permethrin	12.00	5.00	71.00	29.33
Fungicide				
Mancozeb	20.25	8.13	48.79	25.72
Propiconazole	12.00	19.00	63.90	31.63
Ziram	24.00	9.00	42.45	25.15
Carbendazim	25.00	40.50	86.00	50.50
Difenoconazole	15.00	23.50	86.00	41.50
Tebuconazole	20.00	31.00	70.00	40.33
Azoxystrobin	8.10	6.05	66.62	26.92
Metalaxyl	8.10	12.15	36.95	19.07
Trifloxystrobin	12.15	10.23	66.95	29.78

3 RESULTS AND DISCUSSION

3.1 The situation of chemical fertilizer use in Chu-mango production

Season 1 (*Off-season*): flowering from May to June, harvesting from mid-August to October. Season 2 (*Late season*): flowering from late August to October, harvesting from late

November to February of the next year (this is known as the festival season because the harvesting time is focused on important festivals such as those that take place in mid-October and mid-January according to the lunar calendar (Buddhist days), Christmas, New Year, and Lunar New Year. Season 3 is natural season (flowering from January to February, harvesting from mid-April to late June) or early season (flowering from November to December, harvesting from mid-February to April). The natural and early season of sunny season are collectively known as season 3, which takes place in favorable climate conditions. Hence, the agro-inputs during this season differs from those of season 1 and season 2.

Chu-mango growers provide nutrients to their mango orchards by applying the phosphorus, nitrogen, and potassium in the form of chemical fertilizers to increase productivity. The role of these elements in mango production has been mentioned in previous studies such as vegetative growth, flowering stimulation, fruit quality, and yield increase. Besides, they are not only influence directly on plants tissues but also effect on synthesis of other nutrients [16]. The nitrogen fertilizer is positive relationship with mango productivity in Florida [17]. Moreover, a proper combination of the phosphorus, nitrogen, and potassium contributes to improve mango productivity significantly. The phosphorus can stimulate progression of root, branch, fruit, and absorption of water and nutrients [18].

Information about chemical fertilizer use in Chu-mango farming is presented in the Table 3. Overall, there is difference in chemical fertilizer use among seasons. The study result shows that the number of root fertilizer (N–P–K) in season 1 is the highest (624.1 kg/ha), follows by season 2 (542.5 kg/ha) and in season 3 (443.5 kg/ha). By comparison with the result of [19], the number of chemical fertilizer use in countries is demonstrated 100.4 in Turkey, 665.5 in the Netherlands, 624.8 in Egypt, 373.2 in Japan, 301.5 in China, 287.5 in Britain, 205.4 in Germany; 180.1 in France, 160.8 in the USA, 126.4 in Italy, 121.4 in India, 115.4 in Greece and 106.9 in Indonesia kg/ha respectively. From the result shows that Chu-mango production in the Mekong River Delta uses chemical fertilizer highly. It is noticeable that the nitrogen fertilizer overuse can cause reduction of yellow skin in mature mango fruits, the anthracnose rise [20]. Furthermore, the nitrogen and phosphates overuse can leach into groundwater or move via surface runoff into waterways. Phosphate is not as soluble as nitrate and ammonia and tends to get adsorbed onto soil particles and enter water bodies thorough soil erosion.

Table 3: The number of chemical fertilizer in Chu-mango cultivation. *(Source: Field Survey Data, 2018.)*

Items	Season 1 (n = 212)	Season 2 (n = 171)	Season 3 (n = 139)
Root fertilizer			
N: nitrogen (kg/ha)	268.8 ± 523.2	232.0 ± 298.6	181.0 ± 211.7
P: phosphorus (kg/ha)	207.6 ± 575.2	150.3 ± 209.0	144.3 ± 180.1
K: potassium (kg/ha)	147.7 ± 276.8	160.2 ± 291.4	118.2 ± 156.1
Microelements (gr/ha)	0.103 ± 0.973	0.011 ± 0.146	0.088 ± 1.049
Leaf fertilizer (Liquid) for flowering stimulation			
N: nitrogen (kg/ha)	10.4 ± 12.4	11.5 ± 24.0	7.7 ± 8.6
P: phosphorus (kg/ha)	1.6 ± 8.6	1.1 ± 6.6	1.2 ± 7.0
K: potassium (kg/ha)	21.0 ± 28.3	21.3 ± 33.5	15.9 ± 22
Microelements (gr/ha)	21.8.5 ± 86.1	33.6 ± 150.8	60.5 ± 542.0

In Chu-mango cultivation, growers usually use the liquid fertilizer (N–P–K) to spray on mango leaves for flowering stimulation. The liquid fertilizer consists of a large amount of microelements (Mn, Zn, Cu, Mo, S and B) compared to the root fertilizer. In general, two key factors in the liquid fertilizer are the nitrogen and potassium. The role of potassium nitrate is flowering stimulation for mango orchards in off-season as well as increase flower induction, fruit set and fruit retention [21].

In summary, chemical fertilizer use in mango production need a carefully management and control. If it is misused, it will take place negative impacts on human health (farm worker, consumer) and ecology (soil, water, air, terrestrial and aquatic ecosystems). The nitrogenous fertilizer is considered one of the most important inputs in agricultural production. Its overuse causes air pollution by nitrogen oxides (NO, N_2O, NO_2) emissions. The nitrogenous fertilizer overuse in mango cultivation only absorb a part to soil, and the rest of the nitrogenous fertilizer is lost evaporation, and react organic compounds in the clay soil and the remaining interfere surface and ground water. It can cause algae blooms, and kill fish by removing oxygen from the water.

3.2 Human health and ecology impacts in Chu-mango production

The study of [22] showed that EIQ classification values for all the pesticide used in mango production is 30%, 25% and 45% of those pesticides have been rated as low (EIQ = 0 to 20), moderate (EIQ = 21 to 40) and high (EIQ $\geq$ 41) respectively. Among three seasons, the findings of the study shows that none of active ingredients belong to moderate risk group among three seasons. There are four active ingredients EIQ more than 41 (high risk group) including: the paclobutrazol, mancozeb, propineb and ziram. The rest of active ingredients is low risk group.

In season 1 (Table 4), the greatest percentage of pesticide is used by the fungicide, at 50.82%. Less than a small number, namely 40.86%, is applied from the paclobutrazol (a growth control for flowering stimulation). The insecticide accounts for 4.88% of pesticide usage, leaving the herbicide at only 0.43. Noticeably, the paclobutrazol, and mancozeb, propineb and ziram of the fungicide make up 91.3% of total pesticide use in Chu-mango farming.

The most outstanding feature of the Table 5 is that the vast majority of the fungicide use in Chu-mango cultivation (64.74%) contrasting with a minority of the herbicide (0.65%) using in season 2. Just over a quarter, namely 27.89%, go to the paclobutrazol, follows by the insecticide at 6.72%. Particularly, four active ingredients play vital role in Chu-mango farming including: the paclobutrazol, and mancozeb, propineb and ziram. These active ingredients are responsible for 87.53% in total of pesticide usage.

In season 3 (Table 6), the fungicide still occupies the highest proportion of the total pesticide, at 76.22%, whereas the opposite is true of herbicide (0.45%). The paclobutrazol ranks second in terms of popularity, at 16.93%, follows by the insecticide at 6.39%. The season 3 resembles the season 2; the paclobutrazol, and mancozeb, propineb and ziram are popular active ingredients in Chu-mango production with approximately 86.76% of the total pesticide use. Thus, reduction of pesticide use in Chu-mango production can conducted by decreasing these active ingredients.

Table 4: The practical values of the health and ecology impacts (EIQ) in season 1. *(Source: Field Survey Data, 2018.)*

Active ingredients	Quantity	EIQ component			EIQ average	Percent
		Farmer	Consumer	Ecology		
Paclobutrazol	**17.57**	**374.20**	**115.07**	**903.87**	**464.32**	**43.86**
Herbicide	**0.24**	**4.37**	**1.13**	**8.30**	**4.60**	**0.43**
Glyphosate	0.12	0.99	0.37	4.33	1.90	0.18
Paraquat	0.08	2.40	0.48	2.70	1.86	0.18
2,4-D dimethyl	0.04	0.98	0.29	1.27	0.84	0.08
Insecticide	**1.65**	**15.53**	**7.70**	**131.75**	**51.66**	**4.88**
Cypermethrin	0.53	7.30	3.12	47.29	19.24	1.82
Chlorpyrifos	0.59	3.51	1.17	42.49	15.72	1.49
Emamectin	0.28	2.50	1.11	18.27	7.29	0.69
Abamectin	0.06	0.80	0.23	5.02	2.02	0.19
Imidacloprid	0.20	1.37	2.05	18.44	7.29	0.69
Permethrin	0.00	0.04	0.02	0.25	0.10	0.01
Fungicide	**19.32**	**357.85**	**233.09**	**1,023.16**	**537.98**	**50.82**
Mancozeb	8.89	179.99	72.26	433.66	228.61	21.60
Propineb	5.38	64.51	102.14	343.52	170.04	16.06
Ziram	4.12	98.79	37.05	174.73	103.52	9.78
Carbendazim	0.08	2.10	3.40	7.22	4.24	0.40
Difenoconazole	0.40	5.93	9.28	33.97	16.39	1.55
Tebuconazole	0.23	4.66	7.22	16.31	9.40	0.89
Azoxystrobin	0.17	1.41	1.05	11.59	4.68	0.44
Metalaxyl	0.05	0.43	0.64	1.96	1.01	0.10
Trifloxystrobin	0.00	0.03	0.03	0.18	0.08	0.01
Field use EIQ	**38.78**	**751.95**	**356.99**	**2,067.08**	**1,058.56**	**100.00**

Unit: kg/ha.

Table 5: The practical values of the health and ecology impacts (EIQ) in season 2. *(Source: Field Survey Data, 2018.)*

Active ingredients	Quantity	EIQ component			EIQ average	Percent
		Farmer	Consumer	Ecology		
Paclobutrazol	**7.88**	**167.92**	**51.64**	**405.61**	**208.36**	**27.89**
Herbicide	**0.26**	**4.23**	**1.17**	**9.13**	**4.85**	**0.65**
Glyphosate	0.16	1.28	0.48	5.62	2.46	0.33
Paraquat	0.06	1.79	0.35	2.01	1.38	0.18
2,4-D dimethyl	0.05	1.16	0.34	1.50	1.00	0.13
Insecticide	**1.60**	**15.27**	**7.21**	**128.09**	**50.19**	**6.72**
Cypermethrin	0.57	7.92	3.39	51.30	20.87	2.79
Chlorpyrifos	0.61	3.67	1.22	44.34	16.41	2.20
Emamectin	0.19	1.74	0.77	12.72	5.08	0.68
Abamectin	0.06	0.89	0.25	5.55	2.23	0.30
Imidacloprid	0.15	1.05	1.58	14.19	5.61	0.75
Permethrin	–	–	–	–	–	–
Fungicide	**17.08**	**308.42**	**218.76**	**923.94**	**483.66**	**64.74**
Mancozeb	7.01	142.01	57.01	342.15	180.37	24.14
Propineb	5.63	67.59	107.02	359.93	178.16	23.85
Ziram	3.46	83.04	31.14	146.87	87.02	11.65
Carbendazim	0.16	3.98	6.45	13.71	8.05	1.08
Difenoconazole	0.37	5.59	8.76	32.07	15.47	2.07
Tebuconazole	0.22	4.45	6.90	15.58	8.98	1.20
Azoxystrobin	0.19	1.53	1.15	12.62	5.10	0.68
Metalaxyl	0.03	0.21	0.31	0.94	0.48	0.06
Trifloxystrobin	0.00	0.01	0.01	0.08	0.04	0.01
Field use EIQ	**26.82**	**495.84**	**278.78**	**1,466.77**	**747.06**	**100.00**

Unit: kg/ha.

Table 6: The practical values of the health and ecology impacts (EIQ) in season 3. *(Source: Field Survey Data, 2018.)*

Active ingredients	Quantity	EIQ component			EIQ Average	Percent
		Farmer	Consumer	Ecology		
Paclobutrazol	**3.79**	**80.83**	**24.86**	**195.24**	**100.29**	**16.93**
Herbicide	**0.15**	**2.23**	**0.63**	**5.12**	**2.66**	**0.45**
Glyphosate	0.10	0.78	0.29	3.41	1.49	0.25
Paraquat	0.03	0.93	0.18	1.05	0.72	0.12
2,4-D dimethyl	0.02	0.51	0.15	0.66	0.44	0.07
Insecticide	**1.21**	**11.17**	**5.94**	**96.52**	**37.88**	**6.39**
Cypermethrin	0.34	4.64	1.99	30.07	12.23	2.06
Chlorpyrifos	0.39	2.33	0.78	28.22	10.44	1.76
Emamectin	0.25	2.23	0.99	16.28	6.50	1.10
Abamectin	0.04	0.61	0.17	3.85	1.54	0.26
Imidacloprid	0.19	1.34	2.01	18.00	7.12	1.20
Permethrin	0.00	0.02	0.01	0.10	0.04	0.01
Fungicide	**16.13**	**296.20**	**197.10**	**861.37**	**451.51**	**76.22**
Mancozeb	7.84	158.71	63.72	382.39	201.58	34.03
Propineb	4.53	54.41	86.14	289.71	143.40	24.21
Ziram	2.86	68.54	25.70	121.24	71.83	12.13
Carbendazim	0.10	2.58	4.18	8.88	5.21	0.88
Difenoconazole	0.29	4.38	6.86	25.09	12.11	2.04
Tebuconazole	0.29	5.79	8.98	20.28	11.68	1.97
Azoxystrobin	0.19	1.51	1.13	12.45	5.03	0.85
Metalaxyl	0.03	0.22	0.34	1.02	0.53	0.09
Trifloxystrobin	0.00	0.06	0.05	0.32	0.14	0.02
Field use EIQ	**21.29**	**390.43**	**228.52**	**1,158.25**	**592.34**	**100.00**

Unit: kg/ha.

It is apparent that the ecology is the most vulnerable objective by pesticide use because its EIQ components is the highest in three EIQ components (farm worker, consumer) among three seasons. In detail, EIQ component of ecology in season 1 is 2,067.08. It is 2.7 times that of farm worker and 5.8 times that of consumer. In season 2, the ecology EIQ is 1,466.77, about 3 times farm worker EIQ and 5.3 times consumer EIQ. The number in season 3 is 1,158.25, approximately 3 times farm worker EIQ and 5.1 times consumer EIQ. Especially, Chu- mango cultivation in season 1 is the most negative impact on ecology in all of year. It is more than 1.4 and 1.8 times for seasons 2 and 3. Similarly, the field use EIQ in season 1 is the highest (1,058.56) in all of three seasons, then seasons 2 (747.06) and season 3 (592.34). Chu-mango production in season 3 is natural crop with favorable climate condition. Thus, the amount of pesticide use much less than compared to season 1 and 2. There is a dearth of empirical studies on EIQ on fruit in Vietnam, especially is mango fruit. This is difficult to compare the research result to previous studies in Vietnam. Therefore, this study provides empirical data on EIQ on mango fruit production as a reference source for next studies. The finding indicates that the field use EIQ of Chu-mango production in three seasons is lower than that of mango production in Kenya. This figure in Kenya is 10,006.73 kg/ha (44,049.67 kg/acre) [23], and greater than 9.5, 13.4, and 16.9 times in seasons 1, 2, and 3 respectively in Chu-mango production in the Mekong River Delta, Vietnam.

Main cause of chemical fertilizer and pesticide overuse of seasons 1 and 2 is to control insect pests, fungus, and disease as well supply nutrients mango orchards in unfavorable climate condition of off-season in order to ensure mango productivity. Agrochemicals can move to water through direct application, runoff, and atmospheric deposition. If it does not control well, it will contaminate water resources. This become more and more dangerous pose risks to human health by probable human carcinogens of pesticide, especially is in developing countries that farmers directly use highly hazardous pesticide. Besides, pesticide residues harmful impact on freshwater and marine organisms as impact of eutrophication on aquatic ecosystem [24].

In the circumstance, the government could encourage mango growers to adopt good agricultural practices by training programs that raise their awareness on ecological protection. This policy would help to reject banned agrochemicals in mango production, which protect water quality. Additionally, traceability is needed, which would help of farmer change behavior of pesticide overuse, is to carry out individual responsibilities for their products to communities. It not only reduces loadings of agricultural pollutants to water resources but also take the potential trade benefit. The study is conducted in the Mekong River Delta where is famous for the cultivation of mango as well as is considered central mango production in Vietnam, as it accounts for 62.8% of the mango production volume and for 46.3% of the mango production area in Vietnam [25]. Thus, the result is typical as a study case in Vietnam.

4 CONCLUSIONS

The N–P–K application in Chu-mango production in the Mekong River Delta is relatively high compared to other countries. The number of root fertilizer (N–P–K) in season 1 is the highest (624.1 kg/ha), follows by season 2 (542.5 kg/ha) and in season 3 (443.5 kg/ha). Chemical fertilizer use in mango production need a carefully management and control. If it is misused, it will take place negative impacts on human health (farm worker, consumer) and ecology (soil, water, air, terrestrial and aquatic ecosystems). Importantly, the nitrogen and phosphates overuse can leach into groundwater or move via surface runoff into waterways.

Phosphate is not as soluble as nitrate and ammonia and tends to get adsorbed onto soil particles and enter water bodies thorough soil erosion.

Chu-mango cultivation in season 1 is the most negative impact on ecology in all of year. It is more than 1.4 and 1.8 times for seasons 2 and 3. Similarly, the field use EIQ in season 1 is the highest (1,058.56) in all of three seasons, then seasons 2 (747.06) and season 3 (592.34). Chu-mango production in season 3 is natural crop with favorable climate condition. Thus, the amount of pesticide use much less than compared to season 1 and 2 (off-season). Agrochemicals can move to water through direct application, runoff, and atmospheric deposition. If it does not control well, it will contaminate water resources.

ACKNOWLEDGEMENTS

The study used data source from project "Value chain development of Vietnamese mango fulfilling requirement for domestic and international markets" (2017–2020), code: KHCN-TNB.ĐT/14-19/C14). The authors would like to thank the financial support from program "Technology and science program for sustainable development in Mekong Delta region" from Ministry of Technology and Science in Vietnam.

REFERENCES

[1] F.A.O., *Water Pollution from Agriculture: A Global Review*, Rome, 35 pp., 2017.

[2] U.S. E.P.A., *Water Quality Assessment and TMDL Information*, United States Environmental Protection Agency (US EPA): Washington, DC, 2016. https://ofmpub.epa.gov/waters10/attains_index.home.

[3] F.A.O., Guidelines to control water pollution from agriculture in China. Water Report 40, Rome, 197 pp., 2013.

[4] Schreinemachers, P. & Tipraqsa, P., Agricultural pesticides and land use intensification in high, middle and low income countries. *Food Policy*, **37**, pp. 616–626, 2012.

[5] Peterson, R.K.D., Comparing ecological risks of pesticides: The utility of a risk quotient ranking approach across refinements of exposure. *Pest Management Science*, **62**, pp. 46–56, 2006.

[6] Bahlai, C.A., Xue, Y., McCreary, C.M., Schaafsma, A.W. & Hallett, R.H., Choosing organic pesticides over synthetic pesticides may not effectively mitigate environmental risk in soybeans. *PLoS One*, **5**, p. e11250, 2010.

[7] Cross, P. & Edwards-Jones, G., Variation in pesticide hazard from arable crop production in Great Britain from 1992 to 2002: Pesticide risk indices and policy analysis. Crop Protection, **25**, pp. 1101–1108, 2006.

[8] Peck, G.M., Merwin, I.A., Brown, M.G. & Agnello, A.M., Integrated and organic fruit production systems for 'Liberty' Apple in the Northeast United States: A systems-based evaluation. *Hortscience*, **45**, pp. 1038–1048, 2010.

[9] Soltani, N., Stewart, C.L., Nurse, R.E., Van Eerd, L.L., Vyn, R.J. & Sikkema, P.H., Weed control, environmental impact and profitability of weed management strategies in glyphosate-tolerant corn. *American Journal of Plant Sciences*, **3**, pp. 1594–1607, 2012.

[10] Stewart, C.L., Nurse, R.E., Van Eerd, L.L., Vyn, R.J. & Sikkema, P.H., Weed control, environmental impact and profitability of weed management strategies in glyphosate-resistant soybean. *Weed Technology*, **25**, pp. 535–541, 2011.

[11] Kniss, A.R. & Coburn, C.W., Quantitative evaluation of the environmental impact quotient (EIQ) for comparing herbicides. *PLoS One*, **10**(6), 2015.

[12] Peterson, R.K.D. & Schleier, J.J., A probabilistic analysis reveals fundamental limitations with the environmental impact quotient and similar systems for rating pesticide risks. *PeerJ*, **2**, 2014.

[13] Gallivan, G.J., Beaton, D., McGee, B. & Berges, H., Survey of pesticide use and evaluation of the changes in pesticide risk on agricultural crops in Ontario in 2008. New Directions Report, 48 p., 2010.

[14] Kovach, J., Petzoldt, C., Degni, J. & Tette, J., A method to measure the environmental impact of pesticides. *New York's Food and Life Sciences Bulletin*, **139**, pp. 1–8, 1992.

[15] Cornell, C.A.L.S., New York State integrated pest management: List of pesticide active ingredient EIQ values: A method to measure the environmental impact of pesticides, Table 2: list of pesticides (last updated May 2020). College of Agriculture and Life Sciences, Cornell University, 2020. https://nysipm.cornell.edu/eiq/list-pesticide-active-ingredient-eiq-values/.

[16] Sen, P.K., Roy, P.K. & De, B.N., Hunger signs on mango. *Indian Journal of Horticulture*, **5**, pp. 35–44, 1947.

[17] Young, T.W., Koo, R.C.J. & Miner, J.T., Effects of nitrogen, potassium and calcium fertilization on Kent mangos on deep, acid, sandy soil. *Florida State Horticultural Society*, pp. 364–371, 1962.

[18] Samra, J.S. & Arora, Y.K., Mineral nutrition. *The Mango, Botany, Production and Uses*, ed. R.E. Litz, CAB International: Wallingford, UK, pp. 175–201, 1997.

[19] FAO, *Resource STAT-Fertilizer*, Food and Agriculture Organization of the United Nations, 2009. http://faostat.fao.org/site/575/DesktopDefault.aspx?PageID=575#ancor. Accessed on: 12 Mar. 2009.

[20] Bally, I.S.E., Hofman, P.J., Irving, D.E., Coates, L.M. & Dann, E.K., The effects of nitrogen on postharvest disease in mango (*Mangifera Indica* L. 'Keitt'). *Acta Horticulturae*, **820**, pp. 365–370, 2009.

[21] Shinde, A.K., Dabke, D.J., Jadhav, B.B., Landalkar, M.P. & Burondkar, M.M., Effect of dose and source of potassium on yield and quality of Alphonso mango (*Mangifera indica* L.). *Indian Journal of Agricultural Science*, **76**, pp. 213–217, 2006.

[22] Mazlan, N. & Mumford, J., Insecticide use in cabbage pest management in the Cameron Highlands, Malaysia. *Crop Protection*, **24**(1), pp. 31–39, 2005.

[23] Mwungu, C.M., Muriithi, B., Affognon, H., Githiomi, C., Diiro, G. & Ekesi, S., Health and environmental effects of adopting an integrated fruit fly management strategy among mango farmers in Kenya. *African Journal of Agricultural and Resource Economics*, **15**(1), pp. 14–26, 2020.

[24] Pait, A., DeSouza, A. & Farrow, D., Agricultural pesticide use in coastal areas: A national summary. U.S. Department of Commerce, National Oceanic and Atmospheric Administration, Sep. 1992.

[25] G.S.O. (General Statistic Office of Vietnam), *Statistical Year Book 2019*, Statistical Publishing House: Hanoi City, Vietnam, 2020.

SECTION 3
WATER AND THE COMMUNITY

VALUATION OF A COMMUNITY COMPANY AND ITS IMPACT ON DEVELOPMENT STRATEGIES

BETHY MERCHAN[1,2,3], ROMINA TORRES[1,2], JANNELY BENALCAZAR[1,2], PAÚL CARRIÓN-MERO[1,2], GRICELDA HERRERA-FRANCO[3,4], EDGAR BERREZUETA[5] & MARÍA JAYA-MONTALVO[6]
[1]Faculty of Engineering in Earth Sciences (FICT), Escuela Superior Politécnica del Litoral (ESPOL), Ecuador
[2]Center for Research and Projects Applied to Earth Sciences (CIPAT), Escuela Superior Politécnica del Litoral, Ecuador
[3]Geo-Resources and Applications (GIGA), Escuela Superior Politécnica del Litoral, Ecuador
[4]Faculty of Engineering Sciences, Santa Elena Peninsula State University (UPSE), Ecuador
[5]Departamento de Infraestructura Geocientífica y Servicios, Instituto Geológico y Minero de España (IGME), Spain
[6]Junta Administradora de Agua Potable Regional de Manglaralto (JAAPMAN), Ecuador

ABSTRACT
Water and sanitation are basic and essential rights for the adequate development of a population. In the context of the rural area studied in this work (Manglaralto, Ecuador), at the beginning of 1979 the Manglaralto Potable Water Board (JAAPMAN) was registered with the objective of managing the water supply in the area. This board is administered by local users, under the principles of joint work, local democracy, and self-management. Currently, the service provided reaches 26,362 people. JAAPMAN, during the 40 years that it has been operating non-profit, has generated significant assets in its operating infrastructure. With the creation of the Ministry of Environment and Water in Ecuador, there is the possibility that the community company will become managed by a public water company. This situation could cause conflicts and mistrust in society since the community has self-managed and solved its water supply for 40 years. This article aims to develop an inventory of physical infrastructure, equipment, machinery, plus operation and maintenance expenses, of the investment that has been made to date, to quantify its value and, more generally, estimate the company's expenses. In this context, the method considers the following: (i) preparation of inventory tables; (ii) estimate of the value of assets, other income and general expenses; and (iii) proposal for development strategies. The most relevant result is that it indicates that the accumulated equity of the community enterprise is approximately two million dollars, which shows the good management carried out to date and, in addition, the possibility of obtaining international financing and support. Finally, it was revealed that joint work in the company is constantly evolving and is executed efficiently and effectively.
Keywords: community enterprise, assets, business efficiency, development strategies.

1 INTRODUCTION

The value of a company could be defined as a measure of the evaluation of the activity of that company. This value is a key characteristic and its analysis is essential because the ability to evaluate the effects of the company's operation in the market is the fundamental strength of the company [1]. In addition, the value of a company is one of the important topics highlighted by stakeholders, such as investors, managers, employees, customers or suppliers. However, it is important to comment that to this day, discussions continue on the concept of business value or measurement of value [2]–[4]. According to [5], traditional business valuation methods are based on the balance sheet, the income statement or the cash flow statement; however, intellectual capital is also an asset. As indicated [6], it is an asset market and the book value could not be explained because the intellectual capital is zero. According to [7], intellectual capital is intangible and is made up of all the non-financial assets of a company, which are not reflected in the balance sheet.

In addition to traditional companies, so-called community companies can and should also be valued. Community companies are created based on the needs of the population to manage

WIT Transactions on Ecology and the Environment, Vol 250, © 2021 WIT Press
www.witpress.com, ISSN 1743-3541 (on-line)
doi:10.2495/WRM210091

services that are not provided in a timely manner by the state or a private company. The study of community management has been the subject of various theoretical analyzes in the last five decades [8]. In the 1970s, the observation of community management as a rural phenomenon began. In the 1980s, under the social capital approach (an initiative sponsored by the World Bank), community management was promoted as a scenario for strengthening the social fabric [9]. This trend continues to be applied in current empowerment approaches. In the 90s, starting from the theory of common goods, the archetype of the classical economy based on the interaction of the individual-society-state in a competitive environment was questioned, in order to present new alternatives of governance supported by the recognition of the community as collaborative collectivity and the possibility of human interaction under rules of cooperation [10]. Since the mid-1990s, collaborative management or co-management recovers the value of the interaction between actors or stakeholders in which individuals (as an expression of individual initiative), communities (as an expression of collective action), local, regional and national governments and other civil society organizations take part in the management of resources and services. In general, it can be said that the community management has been a traditional form of governance of natural resources in peasant and indigenous communities [11], currently being part of the society and culture of rural and urban-marginal areas.

In Ecuador's case, the 2008 Constitution of the Republic of Ecuador establishes two types of management: community and public. In addition, it establishes that public-community alliances should be made to strengthen community management [12]. According to the 2010 Population Census, there are about 14 million inhabitants of that total, 5 million live in rural areas, of which approximately 3.5 million have access to water through community management [13]. In the Manglaralto parish, approximately 26,362 people receive water resources from groundwater wells. The Manglaralto Regional Drinking Water Administration Board (JAAPMAN) makes an effort to manage drinking water in six communes, Libertador Bolívar, Cadeate, Rio Chico, Manglaralto, San Antonio and Montañita, through the drilling of wells, construction of dikes, distribution networks and reservoirs [14], [15].

The need for the creation of this Water Board is due to the lack of vital liquid in the area, which is characterized by being a semi-arid region [16], [17], adding to the fact that it presents other problems such as the growing demand for water, being a site with a high influx of tourists and the phenomenon of saline intrusion [18]. In this context, in search of sustainability [19], new studies have been developed, such as the creation of green filter systems in the Montañita commune [20], hydrochemical studies [18], flow model [21], watershed integral management [22] and even geotourism potential [23] as it is a socio-ecological system [24] with rescue practices of ancestral knowledge [22].

With the creation of the Ministry of the Environment and Water in Ecuador, there is a risk and the possibility that JAAPMAN will be managed by a central government company. This situation could cause conflicts and mistrust in society since the community has self-managed and solved its water supply for 40 years, therefore, it is necessary to quantify its accumulated patrimony as a reflection of the effort, dedication and joint work developed.

The main aim of this research is, to prepare the inventory of the assets of a community company and, mainly, to give an economic valuation to the movable and immovable property of the Manglaralto Board. In addition, it is proposed to address a general estimate of expenses. With the global valuation data, the configuration of management strategies and future development of JAAPMAN will be presented.

2 METHODOLOGY

In this work, three main phases have been considered for the development of research (Fig. 1).

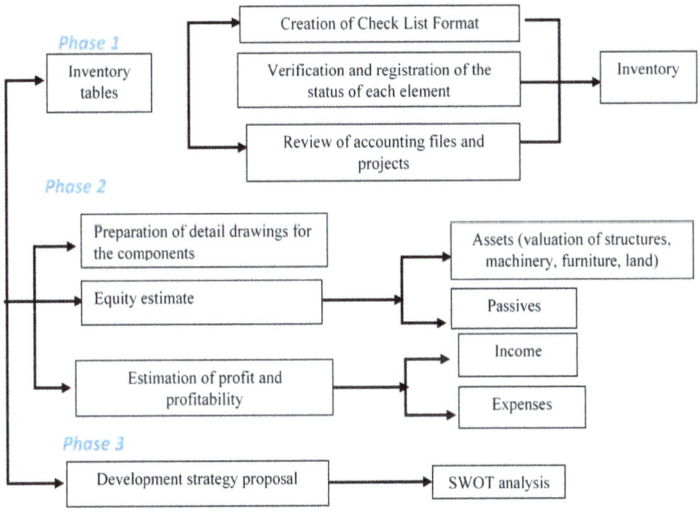

Figure 1: Scheme of steps to follow during the analysis of community business assets.

The first phase focused in detail on the creation of inventory tables: check list, verification and registration of the status of each element by component and review of accounting files and project information. In addition, the physical formats, inventory tables for wells, dams and reservoirs were created in the office, which contain all the detailed compilation of: (a) year of construction, (b) locations referenced with GPS, (c) records of the names of the elements of each component with their respective quantities, brands and states of functionality. All this in order to have an information base of the elements of each infrastructure, equipment, machinery and building, to be able to quantify, verify, value and reference each of its components managed by JAAPMAN. With the form ready, the field investigation was programmed, where the tours of the different components were carried out, and the types of investment infrastructures of the JAAPMAN could be physically verified. In each one of them a technical sheet was configured. Additionally, the files with technical specifications of the equipment and its maintenance were reviewed. The form fields were filled out, verifying and searching for the pertinent information. The files were reviewed in the JAAPMAN offices where documents of purchases and acquisitions of materials and equipment were verified. Equipment and materials without physical documentation support were valued with input from the different projects. The implementation of improvements and maintenance to the infrastructures through the self-management of JAAPMAN was also considered.

The second phase was initially focused on the development of detail plans and calculation of current economic value from the field survey where the information was collected for the preparation of the plans for each component belonging to JAAPMAN. Next, the detailed plans were made that contain, (a) plan of the type of well; (b) wells and dikes location plan; (c) general scheme of distribution networks; where you can view the references of dikes, wells and reservoirs. With this information, the referential budget for each well, dam, reservoir, distribution networks and administrative office was configured. Quantities were calculated working for each item with its respective analysis of unit price (APU). Wear was considered due to the elapsed time of each element, an inspection was carried out to determine the status, functioning and remaining useful life, as well as to define the economic value that would correspond to it due to depreciation. Once the depreciation, maintenance and improvement concepts have been obtained, the total investment of the JAAPMAN is calculated. It is important to note that once the referential budget for the structures had been obtained, it was necessary to carry out a depreciation using the Ross Heidecke method, which uses the Fitto and Corvini tables [25]. This type of method was necessary since it applies depreciation factors that relate the useful life and the state of conservation of the structure. For machinery and furniture, a depreciation was carried out using the decreasing method by sum of digits. In this phase, the estimation of income from collections in water supply, savings, expenses (salaries and maintenance of the infrastructures) and overdue portfolios by consulting the accounting books was also addressed. All this to be able to make a general balance of the company to obtain the current economic value.

The last and third phase focused on developing a proposal for development strategies, with all the information collected, and once the patrimony had been quantified, the proposal that JAAPMAN should implement in the short, medium and long term was drawn up. To obtain the strategies, a workshop was implemented with representatives of JAAPMAN, community members, and technicians linked to JAAPMAN activities, with which the analysis of strengths, opportunities, weaknesses and threats (SWOT) was carried out.

3 RESULTS

3.1 Creation of check list formats and general inventory

These formats allowed the gathering of information for the inventory of the elements of the company (e.g. structures). They were made in digital format to facilitate their use and future storage. The information generated was delivered to the company for continuous and future use in economic and technical matters.

3.2 Preparation of detailed drawings for components

The elaboration of the detailed plans of each component (Fig. 2) allowed to define the location of the infrastructure and, mainly, its state. From this information your budget estimate was made. Information in format digital of these planes It allows its continuous updating and the incorporation of new components if necessary. In addition to the plans, a sheet was generated with the basic information for its use in component inventory.

The following steps were followed to estimate the equity:

a) Structure assessment: The inventory carried out allowed the identification of 27 infrastructures divided into 16 wells, 6 reservoirs, 3 dams, 1 distribution network and 1 company headquarters building. The information on these structures is presented in Table 1, where it is observed, year of construction, useful life and its state of conservation. In Table 2, the general assessment of the available structures is presented.
b) Machinery valuation: As machinery you have a truck and a motorcycle. To find the real present value of the machinery, it was necessary to use the decreasing depreciation method by sum of digits, which consists of applying a higher depreciation in the first

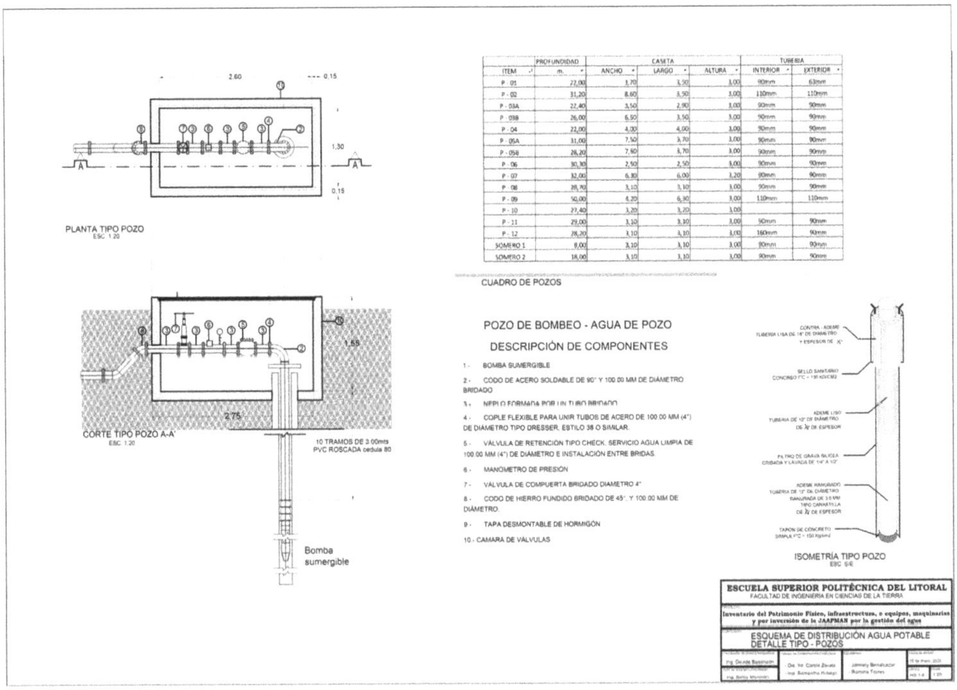

Figure 2: Typical well plan with a detailed description of its components to estimate its economic value.

years of the good's life, to decrease over the course of the asset. For the truck it was determined that the initial purchase value in 2015 was USD26,990, with a useful life of 10 years, a present value of USD 9323.82 was obtained. For the motorcycle, it was necessary to consider that its initial value was USD3090.00 purchased in 2014, with a 5-year useful life, which had already been fulfilled at the date of depreciation, resulting in a value of USD 618.00.

c) Real estate value: For real estate, the office equipment was analyzed, giving a total value with replacement of USD 11,338.94 and with depreciation of USD 8,326.76.
d) Land valuation: The most of the land where JAAPMAN has its structures is not owned by it. Currently only they have deeds of five plots of 1000 m2 each. Obtaining a total value of USD18,467.52, data obtained with the information of the property tax payment; However, to estimate its commercial value, it was necessary to investigate the real value of the land in the area, resulting in between USD16 and USD25 the value of square meter. The lowest value was taken, USD 16.00. The global valuation of the company's land is presented in Table 3.
e) Bank savings: The JAAPMAN bank account has a savings of USD 11,000.00, according to data provided by the company to the date October 28, 2020.
f) Debts: JAAPMAN has a loan, used to buy construction materials of USD 2,500.00.
g) Equity estimation: To determine equity, it was necessary to classify the values into assets and liabilities. The estimate is presented in Table 4.

Table 1: Estimation of the present value of elements of the company (wells, reservoirs, dams, civil distribution networks) with the estimation of depreciation.

Structure		Year of construction	Estimated age	Useful life (years)	% in age	Condition	Replacement value	Current value
Wells	Well 12	2010	11.00	25.00	44.00	OK	$37,446.58	$26,190.14
	Well 11	2015	6.00	25.00	24.00	OK	$38,623.47	$32,703.65
	Somero 1	2014	7.00	25.00	28.00	OK	$19,264.80	$15,798.87
	Well 10	2014	7.00	25.00	28.00	OK	$23,328.84	$19,131.75
	Well 9	2005	16.00	25.00	64.00	OK	$65,596.88	$33,905.72
	Well 8	1986	35.00	25.00	100.00	OK	$39,461.86	$4,464.71
	Well 7	1986	35.00	25.00	100.00	OK	$51,330.41	$5,807.52
	Well 6	1986	35.00	25.00	100.00	OK	$42,292.74	$4,785.00
	Well 5a	2000	21.00	25.00	84.00	OK	$41,377.39	$12,386.32
	Well 5b	2020	1.00	25.00	4.00	Optimum	$24,781.52	$24,317.61
	Well 4	2006	15.00	25.00	60.00	OK	$37,447.36	$20,825.23
	Well 3a	1988	33.00	25.00	100.00	OK	$41,477.04	$4,692.71
	Well 3b	2020	1.00	25.00	4.00	Optimum	$52,621.34	$51,636.27
	Well 1	1979	42.00	25.00	100.00	OK	$39,988.16	$4,524.26
	Well 2	1985	36.00	25.00	100.00	OK	$24,539.8	$2,776.43
	Somero 2	2017	4.00	85.00	4.71	OK	$32,044.16	$30,579.10
Manglaralto reservoir	Capacity storage	1987	34.00	20.00	100.00	Ok	$67,635.83	$4,319.90
	Capacity storage	2018	3.00	20.00	15.00	Ok	$64,074.91	$57,421.69
	Capacity storage	1997	24.00	20.00	100.00	Ok	$51,497.24	$3,289.13
	Pumping	1987	34.00	85.00	40.00	Ok	$12,527.58	$8,979.83
Montañita reservoir	PVC tanks	2011	10.00	20.00	50.00	Ok	$56,115.13	$34,337.25
	Tank	2019	2.00	40.00	5.00	Optimum	$105,901.02	$103,265.14
Dykes	Land 1	2018	3.00	25.00	12.00	Ok	$2,348.45	$2,348.45
	Land 2	2018	3.00	25.00	12.00	Ok	$2,692.82	$2,692.82
	Concrete	2019	2.00	85.00	2.35	Optimum	$99,752.34	$98,292.96
Distribution networks		1979 to 2020	1.00	20.00	5.00	Ok	$753,706.14	$719,246.70
JAAPMAN headquarter		2015	6.00	85.00	7.06	Very good	$130,954.67	$125,542.97
Total							$1,958,828.48	$1,454,262.14

Table 2: Summary of the valuation of the structures.

Structure	Value	
	With depreciation	With replacement (USD)
Well valuation	294,525.29	611,622.35
Assessment of the Manglaralto reservoir	74,010.55	195,735.56
Valuation of the Montañita reservoir	137,602.40	162,016.15
Dam valuation	103,334.23	104,793.61
Valuation of distribution networks	719,246.70	753,706.14
JAAPMAN headquarters valuation	125,542.97	130,954.67
Total	1,454,262.14	1,958,828.48

Table 3: Land valuation.

Name	Area (m^2)	Property tax valuation (USD)	Commercial valuation (USD)
Pit 9 plot	1,000	4,122.25	16,000.00
Pit 6 plot	1,000	4,081.00	16,000.00
Pit 5 plot	1,000	4,122.25	16,000.00
Pit 4 field	1,000	4,081.02	16,000.00
Pit 3 plot	1,000	2,061.00	16,000.00
Total	5,000	18,467.52	80,000.00

Table 4: Equity estimation of the company.

Item considered	With replacement and property tax land values (USD)	With depreciation (USD)	With replacement and market land values (USD)
Land valuation	18,467.52	18,467.52	80,000.00
Valuation of movable property	11,338.94	8,326.76	11,338.94
Equipment and machinery valuation	30,080.00	9,941.82	30,080.00
Infrastructure assessment	1,958,828.48	1,449,220.87	1,958,828.48
Savings in the bank	11,000.00	11,000.00	11,000.00
Total assets	2,029,714.94	1,496,956.97	2,091,247.42
PASSIVES			
Building materials loans	2,500.00	2,500.00	2,500.00
Patrimony			
Total patrimony	2,027,214.94	1,494,456.97	2,088,747.42

To estimate the operational balance, the following steps were followed:

a) Income: Past due portfolio: The communes attached to the JAAPMAN have an overdue portfolio as of 2020 of USD 320,137.21. In Table 5, the debt is detailed for each commune.
   Collections: To determine the collections for 40 years, data from the JAAPMAN database were used and are represented in Table 6.
b) Expenses: Expense estimate (salaries, services): To determine the total value of salaries paid by JAAPMAN for 40 years, the division of workers who have all the benefits of the law (social Security), called "affiliated workers", and the others who only received the basic salary, called "workers not affiliates". Giving as a result the data in Table 7.

Table 5: Overdue portfolio data.

Past due portfolio	
Communes	Value (USD)
Cadeate	27,524.50
Liberator Bolivar	49,849.22
Manglaralto	49,117.60
Montanita	158,770.04
little River	14,285.60
San Antonio	20,590.25
Total	320,137.21

Table 6: Company collections.

Collections over 40 years (USD)	
Water consumption	3,211,081.11
Installation right	147,765.67
Installation and maintenance	44,519.02
Maintenance	89,652.28
Total	3,493,018.09

Table 7: Estimate of salaries of members and non-members.

Workers' salaries for 40 years (USD)	
Total salary of affiliated workers	434,396.68
Total salary of unaffiliated workers	519,339.48
Total	953,736.16

The cost estimate was made by obtaining data from the year 2020 and a weighting until the year 1981. The data are presented in Table 8.

Profitability and profit estimation: In addition, the gross and net profitability of the company was analyzed, a value of 50% and 46% respectively was obtained. Gross profits amounting to USD 1,923,367.98 (Table 9).

Table 8: Expenses generated over 40 years by the company.

Expenses during the 40 years (USD)	
Salary of affiliated workers	434,396.68
Salary of unaffiliated workers	519,339.48
Office supplies expenses	59,990.03
Water costs for the consumption of workers	6,757.11
Maintenance of wells and access roads	88,521.91
Transportation cost for procedures	1,180.29
Electric energy expenditure	752,725.39
Phone expense	15,511.94
Internet spending	8,234.27
Gasoline expense	3,130.22
Total	1,889,787.32

Table 9: Profitability and utility of the company.

Gross profit (USD)	
Income	3,813,155.30
Expenses	1,889,787.32
Total	1,923,367.98
GROSS profitability	50%
NET profitability	46%

3.3 Global analysis

Additionally, an analysis of strengths, opportunities, weaknesses, and threats–SWOT, of the Company was developed. The result is presented in Table 10.

The SWOT analysis, allowed to develop strategies for the development of JAAPMAN:

1. To sensitize the population about the responsible use of the water resource.
2. Make alliances with nearby water boards, national and international organizations that allow reducing water distribution problems in times of drought.
3. Address initiatives that allow obtaining national and international loans, to improve the productivity of the community enterprise.
4. Initiate automation processes of accounting systems to keep accounts updated and optimize economic and technical performance.
5. Exploit nationally and internationally to be a reference of efficient community management, the service provided is of quality.
6. Continue with the strategic alliances currently in force with Universities on issues of water management, geotourism, social aid, etc.

Table 10: Analysis SWOT.

SWOT analysis for JAAPMAN	
STRENGTHS	Since 1980 JAAPMAN has supplied water for 6 communities in the Manglaralto parish in an independent way.
	Managed by local users, under the principles of joint work and self-management, with the support of users.
	It has generated important assets in its operating infrastructure with the effort of the community.
	In times of covid the service was not affected.
	It continues to cany out works to improve the water supply service to the community.
OPPORTUNITIES	They have agreements with universities and Research Centers that contribute with studies and development of designs for new projects.
	Surrounding tourism sector, allowing economic and social development through the integral supply of water resources JAAPMAN is recognized nationally and internationally for its good management of self-management.
	JAAPMAN is recognized nationally and internationally for its good management of self-management.
WEAKNESSES	Systematize files where you find passive information and documents, to improve monitoring, maintenance, and expenses.
	Little technical knowledge in administration areas.
	Lack of implementation of artificial recharge for supplying water to aquifers during times of drought.
	They do not have a desalination plant to deal with extreme drought situations.
THREATS	There are plans to build a regional drinking water treatment plant that would end the activities of the board and his 40 years of self-management, and the low prices that they have, without a doubt, the new plant will have at higher costs. However, it could also be viewed as an opportunity for the development of farming and not depend only on tourism for sustenance, it could even be combined with the development of farms and tourism, in development agroecological.
	There is the possibility that the Board will be managed by the public water company that belongs to the central government.

4 INTERPRETATION AND DISCUSSION

It should be noted that the oldest structure (Well 1) dates from 1979 while wells 3B and 5B were built in 2020. Regarding the state of the pipes, they are considered to be in good condition, but in need of maintenance. The structure with the highest depreciation was the 300 m^3 Manglaralto reservoir. Its value went from USD 67,635.83 to USD 4,319.90 (year 2020). The least depreciated structure was well 5B built in 2020 whose value increased from USD 24,781.62 with a current value of USD 24,317.41.

The analysis allowed to define the current total value for wells with USD depreciation 294,525.29 and with replacement value of USD 611,622.35, for the Manglaralto reservoirs with depreciation of USD 74,010.55, with replacement of USD 195,735.56, for the Montañita reservoirs with depreciation is USD 137,602.40, with replacement is USD 162,016.15, for the distribution networks of the six communes the value with depreciation is USD719,246.70, with replacement of USD 753,706.14, for the JAAPMAN headquarters with depreciation of

USD125,542.97, with replacement of USD130,954.67 and for the docks a value with depreciation of USD103,334.23, with replacement of USD104,793.61. Giving a total of USD1,958,828.48

The inventory of the goods and estimated valuation, allowed to define a quantitative value to each one of the elements through the field visits, elaboration of detailed plans, making budgets with their respective depreciation calculations, where a considerable difference of approximately USD509,607.61 was obtained. This is due to the state of conservation of some wells, where the lack of preventive maintenance or abandonment was evidenced because the land was not their own. A total equity of USD 2,088,747.42 was obtained with a profit of approximately 50%, this proves that despite the fact that the administrative and operational processes can be reviewed and improved, the joint management of 40 years has managed to supply vital fluid to more than 23,000 people, and in times of pandemic continue with the construction of new works such as 4 more wells, which are located on the two-sleeve road due to self-management.

From a methodological point of view, la preparation of the physical inventory, infrastructure, equipment, and machinery of the JAAPMAN, allowed community enterprise evaluated have a knowledge of the economic market value of all his patrimony built daily for 40 years. Although this work indeed focuses on financially valuing the investment and the total assets, it should be noted that this place is a natural laboratory, where several studies-investigations have been carried out, which is part of the know-how developed in this area. Even if it is not valued, it is an addition to all the investment in the sector.

Good governance in the water sector is associated with "integrated management" based on conservation and sustainable development [26]. JAAPMAN, with its participatory model [19] based on governance and administration by local people, delegates from each community and partners, show resilience to cope with population explosion, climate change, and tourism growth [15].

The SWOT analysis carried out provided a comprehensive analysis of the JAAPMAN and the definition of strategies to enhance the administrative management of the company. In addition, the evaluation of the results obtained allowed to have a real knowledge of the situation to tackle future improvement tasks and activities, the option to obtain credits and, fundamentally, to improve the management of the community company.

The traditional economic valuation (i.e., economical balance sheet) [2]–[4] of the evaluated community company has shown that, in addition to offering an efficient service (that are not provided in a timely manner by the state or a private company), it shows positive balance. Undoubtedly, community management could be considered an alternative form of governance of the natural resources of rural communities, as the case studied shows and other studies [9]–[11].

5 CONCLUSIONS

This research has made it possible to determine the patrimony of JAAPMAN, whose value reaches, to the date, approximately USD 2,088,747.42. This value is the result of work and effort of the community in the last 40 years. The valuation carried out reflects efficient and sustainable community management in drinking water service to the inhabitants. The community enterprise, in the last 40 years, He had an income of USD 3,813,155.30, with expenses of USD 1,889,787.32, which it is indicates a gross profitability of 50%, which shows that the company by self-management has worked efficiently. In general, the assessment obtained indicates that the community company has been managed efficiently. In addition, the economical results allow the option of obtaining complementary financing and international support to improve future tasks as expansion of users receiving water resources, location of greater water reserves, recharge of aquifers, improvement of physical infrastructure-machinery and finally, staff training).

JAAPMAN is a community company that has given much more value to its work and infrastructure with the support of the university and its research. Although these values are not calculated here, it does leave the ethical challenge of recognizing a community's monumental work of supplying water for 40 years to its inhabitants.

ACKNOWLEDGEMENT

We express our gratitude to the Manglaralto Drinking Water Administration Board – JAAPMAN.

REFERENCES

[1] Tarczyński, W., Tarczyńska-Łuniewska, M. & Majewski, S., The value of the company and its fundamental strength. *Procedia Computer Science*, **176**, pp. 2685–2694, 2020. DOI: 10.1016/j.procs.2020.09.331.

[2] Adams, C.A., Conceptualising the contemporary corporate value creation process. *Accounting, Auditing and Accountability Journal*, **30**(4), pp. 906–931, 2017. DOI: 10.1108/AAAJ-04-2016-2529.

[3] Bhullar, P.S. & Bhatnagar, D., Theoretical framework: EV vs stock price – A better measurement of firm value. *International Journal of Business and Management*, **2**(6), pp. 335–343, 2013.

[4] Copeland Thomas, T., Koller, E. & Murrin, J., *Valuation: Measuring and Managing the Value of Companies*. John Wiley and Sons: New York, 1994.

[5] Berzkalne, I. & Zelgalve, E., Intellectual capital and company value. *Procedia Social and Behavioral Sciences*, **110**, pp. 887–896, 2014. DOI: 10.1016/j.sbspro.2013.12.934.

[6] Gan, K. & Saleh, Z., Intellectual capital and corporate performance of technology-intensive companies: Malaysia evidence. *Asian Journal of Business and Accounting*, **1**(1), pp. 113–130, 2008.

[7] Frykman, D. & Tolleryd, J., *The Financial Times Guide to Corporate Valuation*. Prentice Hall: London, 2012.

[8] Binswanger-Mkhize, J.P., de Regt, H.P. & Spector, S., *Local and Community Driven Development: Moving to Scale in Theory and Practice*. World Bank Publications: Washington, DC, 2010.

[9] Feldman, T.R. & Assaf, S., *Social Capital: Conceptual Frameworks and Empirical Evidence. An Annotated Bibliography*, The World Bank Social Capital Initiative: Washington, DC, 1999.

[10] Ostrom, E., *Governing the Commons: The Evolution of Institutions for Collective Action*. Cambridge University Press: Cambridge, 1990.

[11] Palerm Viqueira, T., Martínez Saldaña, J., *Aventuras con el agua. La administración del agua de riego: historia y teoría*. Colegio de Postgraduados: Montecillo, Texcoco, México, 2009.

[12] R. D. E. Asamblea Nacional Constituyente, *Ley Orgánica de recursos hídricos, usos y aprovechamiento del agua*. Quito-Ecuador, pp. 1–43, 2014.

[13] Arroyo Castillo, A.D.P., Análisis de los cambios normativos en el ámbito de la gestión de los recursos hídricos, planteados en la Constitución del 2008 y en la Ley de Aguas del 2014: a la luz de dos visiones de gestión, la pública y la comunitaria. Universidad Andina Simón Bolívar, Sede Ecuador, 2015.

[14] Borbor Roca, A.G. & Vera Mendoza, J.J., Estudio para la Optimización de la Distribución de Agua para Consumo Humano, por Parte de la Junta Regional de Manglaralto, Santa Elena, Ecuador, ESPOL, 2016.

[15] Herrera-Franco, G., Carrión-Mero, P., Aguilar-Aguilar, M., Morante-Carballo, F., Jaya-Montalvo, M. & Morillo-Balsera, M. C., Groundwater resilience assessment in a communal coastal aquifer system. The case of Manglaralto in Santa Elena, Ecuador. *Sustainability*, **12**(19), p. 8290, 2020. DOI: 10.3390/su12198290.

[16] Peel, M.C., Finlayson, B.L. & McMahon, T.A., Updated world map of the Köppen-Geiger climate classification. *Hydrology and Earth System Sciences*, **11**(5), pp. 1633–1644, 2007. DOI: 10.5194/hess-11-1633-2007.

[17] García-Garizábal, I., Rainfall variability and trend analysis in coastal arid Ecuador. *International Journal of Climatology*, **37**(13), pp. 4620–4630, 2017. DOI: 10.1002/joc.5110.

[18] Carrión-Mero, P. et al., Hydrochemical and isotopic characterization of the waters of the Manglaralto River Basin (Ecuador) to contribute to the management of the coastal aquifer. *Water*, **13**(4), p. 537, 2021. DOI: 10.3390/w13040537.

[19] Herrera, G., Carrión, P. & Alvarado, N., Participatory process for local development: Sustainability of water resources in rural communities: Case Manglaralto-Santa Elena, Ecuador. *Handbook of Sustainability Science and Research. World Sustainability Series*, eds L. F. W., Springer: Cham, pp. 663–676, 2018.

[20] Morante Carballo, F., Marcatoma Brito, L., Carrión Mero, P., Aguilar Aguilar, J.M. & Ramírez, T., Urban wastewater treatment through a system of green filters in the Montañita commune, Santa Elena, Ecuador. *WIT Transactions on Ecology and the Environment*, 2019, pp. 233–249. DOI: 10.2495/WS190211.

[21] Carrión-Mero, P. Montalván, F. J., Morante-Carballo, F., Loor-Flores de Valgas, B., Apolo-Masache, C. & Heredia, J., Flow and transport numerical model of a coastal aquifer based on the hydraulic importance of a dyke and its impact on water quality. Manglaralto—Ecuador. *Water*, **13**(4), p. 443, 2021. DOI: https://doi.org/10.3390/w13040443.

[22] Carrión, P., Herrera, G., Briones, J., Sánchez, C. & Limón, J., Practical adaptations of ancestral knowledge for groundwater artificial recharge management of Manglaralto coastal aquifer, Ecuador. *WIT Transactions on Ecology and the Environment*, Sep. 2018, pp. 375–386. DOI: 10.2495/SDP180341.

[23] Herrera-Franco, G. et al., Geosites and georesources to foster geotourism in communities: Case study of the Santa Elena Peninsula Geopark Project in Ecuador. *Sustainability*, **12**(11), p. 4484, 2020. DOI: 10.3390/su12114484.

[24] Franco, G. H., Quinchuelaa, T. G., Macancela, N. A. & Mero, P. C., Participative analysis of socio-ecological dynamics and interactions a case study of the Manglaralto coastal aquifer, Santa Elena, Ecuador. *Malaysian Journal of Sustainable Agriculture*, **1**(1), pp. 19–22, 2017. DOI: 10.26480/mjsa.01.2017.19.22.

[25] Parenti, P. et al., Pasantia como auxiliar de ingeniería: inspección de inmuebles urbanos y elaboración de informes de avalúos. *Jurnal Sains dan Seni ITS*, **6**(1), pp. 51–66, 2017.

[26] Rivière-Honegger, A. Traditional water management in the Mediterranean: Authorized union associations in Languedoc-Roussillon. *Globalized Water*, Springer Netherlands: Dordrecht, pp. 149–159, 2014.

COMMUNICATION METHODS ON WATER CARE DURING THE COVID-19 PANDEMIC AND ITS IMPACT ON THE RESILIENCE OF THE RURAL COMMUNITY OF "LIBERTADOR BOLÍVAR", ECUADOR

GRICELDA HERRERA-FRANCO[1,2], JOSELYN ALVARADO[3,4], PABLO GORDILLO[3,4], LENIN VEINTIMILLA[3,4], BETHY MERCHAN[2,3,4], PAÚL CARRIÓN-MERO[3,4] & EDGAR BERREZUETA[5]
[1]Facultad de Ciencias de la Ingeniería, Universidad Estatal Península de Santa Elena (UPSE), Ecuador
[2]Geo-Recursos Aplicados (GIGA), Escuela Superior Politécnica del Litoral, Ecuador
[3]Facultad de Ingeniería en Ciencias de la Tierra, Escuela Superior Politécnica del Litoral, Ecuador
[4]Centro de Investigaciones y Proyectos Aplicados a las Ciencias de la Tierra, Escuela Superior Politécnica del Litoral, Ecuador
[5]Departamento de Infraestructura Geocientífica y Servicios, Instituto Geológico y Minero de España, Spain

ABSTRACT

The COVID-19 pandemic has had a significant impact on the life of human beings. In rural areas, water supply and wastewater management are a survival challenge. They must be supported by universities sensitive to the current reality in order to achieve the Sustainable Development Goals, in its numeral 6, water and sanitation. In rural areas across the "Santa Elena" province, due to the isolation generated, communication was affected, especially sectors such as "Libertador Bolívar", a dry-semi-arid zone located in the coastal area of Ecuador, which has access to drinking water collected from an aquifer through the service provided by the local entities. However, few know about its uses, benefits, and the maintenance that these systems require. The present study aims to validate communication methods, through the transmission of messages about the use and care of water in times of COVID-19. The methodology to follow is: i) potable and wastewater research, ii) surveys to measure knowledge about drinking water and sanitation, iii) preparation of diffusion material, iv) dissemination of messages through different digital media, and v) monitoring of social networks. It found that the Libertador Bolívar sector has a significant deficiency internet access, considering that only 8 out of 40 students at elementary school can receive virtual classes, and not necessarily through a computer, but through mobile devices; therefore, the most effective method for the information transmission were social media. Then, it proceeded to create an Instagram and Facebook page. In 4 weeks, 589 followers obtained on Facebook and 185 on Instagram, as well as significant interactions with this topic, mainly because a drought occurred (Presence of the climatic phenomenon "La Niña") that aggravated the emergency to all the inhabitants. Finally, in times of pandemic, an excellent method of communication and diffusion are the social networks, due to its easy accessibility, which reflected in the resilience of interaction in the community.
Keywords: drinking water, wastewater, COVID-19, communication, resilience, social media.

1 INTRODUCTION

Communication could be defined as the transmission of verbal and non-verbal messages. It consists of a sender, a receiver and channel of communication. In the process of transmitting messages, the clarity of the message may be interfered or distorted by what is often referred to as barriers [1]. According to [2], communication-education includes both intentional processes of influencing others through a structured teaching process and spontaneous learning processes. Additional factors such as the intentionality of learning and the degree of influence of the learner have to be considered. Communication processes can be classified into two categories namely mass media and group media. The Internet is included in mass media and it is focused on reaching a wide audience while the group media. According to

WIT Transactions on Ecology and the Environment, Vol 250, © 2021 WIT Press
www.witpress.com, ISSN 1743-3541 (on-line)
doi:10.2495/WRM210101

[3], [4], the Internet could be considered the most popular, widespread and ubiquitous (personal) communications technology on the planet.

The use of the new techniques of information in public health is focused on improving the knowledge and transmission of relevant alerts/news [5]. However, public health problems encountered by the community are policy, economic or political related, and no amount of communication would influence change because there is a need for a policy or political decision [1].

COVID-19 is an infectious disease caused by a newly discovered coronavirus (SARS-CoV-2) and is rapidly spreading around the world. According to the World Health Organization [6], COVID-19 has already affected 222 countries and territories around the world. Currently, although the vaccines for this disease are beginning their distribution, the first line of defense to fight against this pandemic is a combination of preventative measures and mitigation strategies [7]. In this way, suggestion of social distancing, environmental surface cleaning, self-isolation, travel restrictions, local and national lockdowns, quarantine, limits on large gatherings, restrictions on opening businesses, and school closures, are as some of the preventive measures that are needed to limit the spread of the disease.

In the case of Ecuador, like other countries, agglomerations are being avoided, schools and colleges are being kept closed and meetings are not allowed because of fears of contagion. In this way, the government has established that classes must be held virtually. However, in the province of Santa Elena, not all inhabitants have access to internet. According to the 2010 census, just 7.2% have access to this service, 13.5% have access to a computer and 72.1% have access to a mobile phone [8]. According to [9], the use of alternative media such as the radio or the peripheral was very popular in rural areas.

In the selected rural area (sector Libertador Bolivar), where the residents live mainly from fishing and tourism in the area, the access and distribution of water requires special care. The area is supplied with drinking water by the community water board JAAPMAN, through the extraction of groundwater [10], [11]. It provides service to 6 communities, including the parish of Libertador Bolivar [12]. The board is also in charge of transmitting information on water supply and management [13], ancestral practices have been relevant mechanisms to handle water shortage problems [14]. However, the university-community interaction has carried out different service projects to the community that range from geology and hydrogeology studies to green filter systems for Urban wastewater treatment [15].

In the case of wastewater, the stabilization ponds system has not been formally handed over to them. However, few are aware of the special care and maintenance that these systems require in sanitation terms. In addition, the area has no official communication system to overcome the existing mobility restrictions. However, the extensive use of mobile phones and applications such as WhatsApp, Instagram and Facebook, could be a suitable means of communication for less playful issues [16].

The Internet is positioning as the platform that accumulates large amounts of knowledge, which contributes to people who sometimes do not have much information or academic knowledge, producing thousands of contents that can be not only accurate, but also surprising and imaginative, and that is why we are talking about "the science of networks or network culture" [17]. Specifically, social networks are a fundamental method of communication in several areas, including education. Moreover, we are aware of its potential, because it not only generates new ways of learning, such as collaborative learning, but also promises infinite information and the opportunity for interconnectivity that could be good or bad, depending on the case [18]. Social networks have demonstrated their usefulness in education, in some studies that have been carried out [19], [20], improving productivity and encouraging people to continue learning.

This background leads us to propose this project whose objective is to validate communication methods, through the transmission of messages about the use and care of water during the COVID-19 pandemic and its impact on the resilience of the Libertador Bolívar commune that allows the generation of awareness about the rational use of water.

2 METHODOLOGY

There were five phases. A general outline of the process followed is shown in Fig. 1.

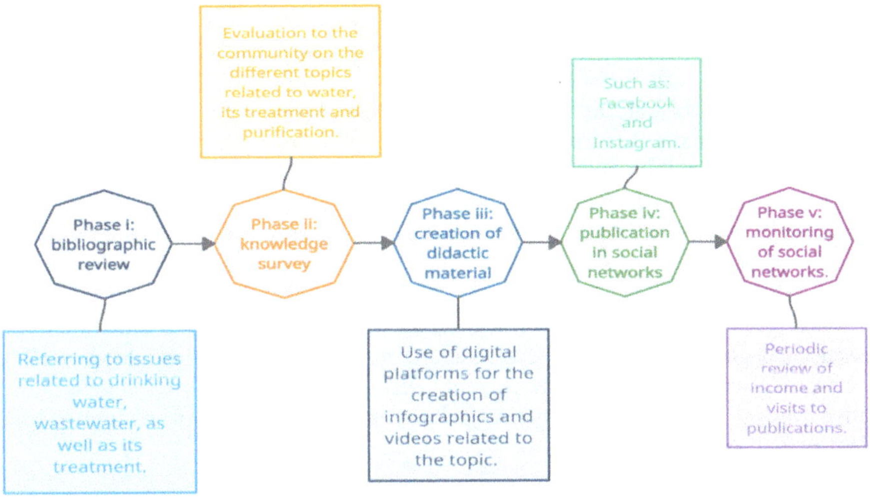

Figure 1: General diagram of the phases followed during the investigation.

Phase i. Bibliographic review. In this phase, an in-depth research was conducted on all topics related to raw water, drinking water, wastewater and its treatment, water use and importance and care of the water resource. In addition, the review of various national and international regulations that regulate the quality of the processes, of different types of wastewater treatment and, finally, information on their impact on the environment and repercussions on health was addressed.

Phase ii. Knowledge survey. Based on the information from the previous phase, two surveys of 10 and 9 questions each were formulated. One survey was for drinking water and the other one was for wastewater and wastewater treatment systems.

For the survey that was about drinking water, there were five levels of knowledge for questions from 1 to 9: level 1: "*No, I'm totally unaware*", level 2: "*Not really, I have little knowledge*", level 3: "*Basic knowledge*", level 4: "*Yes, I have a lot of knowledge*", and level 5: "*Yes, I'm an expert*". In question number 10, was asked "*If they would like to acquire new knowledge about drinking water*", giving options of answers in the selection.

For the wastewater survey, 9 questions were asked about general wastewater issues, with concrete answers, to determine how much knowledge and interest the inhabitants of the commune had about issues related to wastewater and its treatment.

The surveys were conducted online from 27 September to 20 October 2020 using the Google Forms and ArcGIS Survey platforms, which are among the most recommended applications for conducting online surveys [21], [22]. In order to contact the main authorities of the commune, as well as people from neighboring areas, it was assisted by the members of the Administrative Board of Manglaralto Potable Water, JAAPMAN; thus, a total of 42 people were surveyed.

Phase iii. Creation of didactic material. This phase was oriented to the design of infographics and videos. Specifically, the topics researched in the first phase and the results of the second phase were organized to create messages that are easy for the public understanding. The characteristic of the messages is that they should be short, simple, and not consume the audience's computer resources (megabytes). Once this stage is passed, the messages would be ready for publication, diffusion, and monitoring.

Phase iv. Publication and diffusion in social networks. This phase focused on the transmission of all the didactic material. At first it was transmitted through WhatsApp messages to a former official of the municipality who helped with the initial diffusion of the material, thus ensuring that the information reaches the population. In addition, a Facebook and Instagram accounts were created to upload these messages and videos, so that the information is available to all the public interested in issues of care and importance of drinking water, wastewater, and its treatment. Likewise, we sought to add people from the municipality to obtain feedback and improve the content through these platforms.

Phase v. Monitoring of social networks. This last stage focused on monitoring what was published. Specifically, it was analyzed each of the publications on the networks, how many visits there were on both Facebook and Instagram accounts, as well as how many people accessed the page, the views that these generated in each of the publications, as well as the comments that were made about the content. This information was used to determine how effective was the use of social networks as a means of diffusing information in times of pandemic.

3 RESULTS

Phase i. The research was divided into two general topics; the first on drinking water issues and the second on wastewater, to use this information in the following phases, create the didactic material to be published on social networks. The specific results are presented in Table 1.

Table 1: Synthesis of the bibliographic review carried out for drinking water and wastewater.

Drinking water issues	Wastewater issues
National and International Regulations	National and International Regulations
Procedure for obtaining drinking water • Surface water • Groundwater	Purification systems in small communities
Raw water treatment	History of stabilization ponds in the Ecuador
Drinking water distribution system	Wastewater, origin, and definition
Ways to save and take care of drinking water • At home • In the community	Types of stabilization ponds, proper use, and maintenance
Benefits of good water use and care	Optimization of stabilization ponds
Consequences of wasting water	Benefits of good use of stabilization ponds
Water cycle	Consequences of misuse of stabilization ponds

Phase ii. The application of the surveys was carried out from September 27, 2020, to October 20, 2020, obtaining a total of 42 people. The results for drinking water are shown in Table 2, and for wastewater in Table 3.

Table 2: Results of the survey conducted for drinking water.

#	Question	Option	Results	#	Question	Option	Results
1	Do you know where your drinking water comes from?	"No, I'm totally unaware"	8%	6	Are you aware of the consequences of wasting drinking water?	"No, I'm totally unaware"	0%
		"Not really, I have little knowledge"	25%			"Not really, I have little knowledge"	33%
		"Basic knowledge"	46%			"Basic knowledge"	25%
		"Yes, I have a lot of knowledge"	13%			"Yes, I have a lot of knowledge"	17%
		"Yes, I'm an expert"	8%			"Yes, I'm an expert"	25%
2	Do you know the process used to make the water drinkable?	"No, I'm totally unaware"	21%	7	Do you know where drinking water goes after it is used in your home?	"No, I'm totally unaware"	4%
		"Not really, I have little knowledge"	25%			"Not really, I have little knowledge"	21%
		"Basic knowledge"	33%			"Basic knowledge"	17%
		"Yes, I have a lot of knowledge"	17%			"Yes, I have a lot of knowledge"	33%
		"Yes, I'm an expert"	4%			"Yes, I'm an expert"	25%
3	Do you know how to make good use of drinking water?	"No, I'm totally unaware"	4%	8	Do you know about measures to save drinking water consumption?	"No, I'm totally unaware"	4%
		"Not really, I have little knowledge"	13%			"Not really, I have little knowledge"	34%
		"Basic knowledge"	54%			"Basic knowledge"	33%
		"Yes, I have a lot of knowledge"	21%			"Yes, I have a lot of knowledge"	25%
		"Yes, I'm an expert"	8%			"Yes, I'm an expert"	4%
4	Do you understand how important it is to take care of drinking water?	"No, I'm totally unaware"	4%	9	Do you use any drinking water saving method?	"No, I'm totally unaware"	25%
		"Not really, I have little knowledge"	17%			"Not really, I have little knowledge"	13%
		"Basic knowledge"	33%			"Basic knowledge"	33%
		"Yes, I have a lot of knowledge"	25%			"Yes, I have a lot of knowledge"	21%
		"Yes, I'm an expert"	21%			"Yes, I'm an expert"	8%
5	Are you aware of the benefits of taking care of your drinking water?	"No, I'm totally unaware"	8%	10	Would you like to acquire new knowledge about drinking water? (Multiple option)	Water use and care	11
		"Not really, I have little knowledge"	29%			Measures to save drinking water consumption	16
		"Basic knowledge"	17%			Water cycle	6
		"Yes, I have a lot of knowledge"	29%			How does water reach your home?	15
		"Yes, I'm an expert"	17%			How is raw water transformed into drinking water?	19

Table 3: Wastewater survey results.

#	Question	Option	Results
1	What do you understand by wastewater?	Drinking water that has been used for some human daily activity.	17%
		Black water and gray water	12%
		Wastewater	29%
		All the above	24%
		I do not know any of the above	19%
2	Do you know if your home/business has separate wastewater pipes for rainwater and greywater?	Yes	5%
		No	95%
3	Do you believe that all wastewaters should be pre-treated or discharged directly into a body of water?	Yes, it should be treated beforehand	63%
		No, it should be discharged directly	14%
		I am not sure	23%
4	Would you like to know in detail the entire process to which the wastewater is subjected before it is discharged?	Yes	82%
		No	18%
5	Have you ever considered the environmental impact of poor wastewater treatment?	Yes	50%
		No	50%
6	Do you believe that wastewater could be reused?	Yes	27%
		No	14%
		Maybe	59%
7	If so, for which activities do you think they could be used?	Drinking water	0%
		Agricultural activities	10%
		Irrigation	48%
		None of the above	0%
		Not sure	43%
8	Do you know some of the types of wastewater treatment recommended for small communities? Please list them (they could answer several items):	Active sludge	8
		Sequential Discontinuous Reactors	2
		Fixed beds or trickling filters	6
		Stabilization Ponds	14
		I do not know of any	28
9	Did you know that stabilization lagoons are one of the best cost/benefit options for wastewater treatment? (Multiple option)	Yes	18%
		No	82%

(a) (b)

Figure 2: (a) and (b) Some examples of the infographics made.

Phase iii: Some examples of the infographics that were created are presented in Fig. 2(a) and 2(b), in total 23 infographics were created.

Phase iv. After creating and reviewing each didactic material, it was published on each of the Facebook and Instagram pages, as well as shared with community officials via WhatsApp, so that they could share it with their acquaintances and friends in the community. The publication on social networks was diffused on both platforms because they work under the same company, so each of the publications made on the Instagram page were replicated on Facebook. Many factors were considered for the publications to reach a greater number of people, factors such as, for example, that at the time of sharing them, they have hashtags related to the topic to be addressed in the publication so that they reach not only followers, but also other people interested in the topic.

Phase v. We proceeded to monitor each of the networks to observe how much influence, the publications had on people, how they interacted, how many visits each of the publications had, as well as the comments made by the followers. Most of them were people from the Libertad Bolivar community and other areas around.

4 ANALYSIS OF RESULTS

Phase i. After researching the different topics, it was found that there is limited information on stabilization ponds in Ecuador; therefore, it was necessary to resort to information gathered in other Latin American countries. On the opposite, for drinking water the information is vast at the local, regional, and international levels.

Phase ii. A total of 42 people responded to the surveys. The level of knowledge regarding drinking water is within levels 1, 2 and 3, with level 3 being the highest with 34% of people. In addition, it made it possible to determine the scope of messages and what type of information was the most recommendable to transmit. The work also identified that most people (68%) have empirical knowledge of all the topics presented, i.e., they do not have the theoretical foundations. It is important to emphasize that only 32% accepted to have a complete knowledge of all these topics on drinking water.

The results also indicate an interest in the concept of wastewater, the process that wastewater undergoes before being discharged into a water body, the types of wastewater treatment recommended in small communities, the environmental impact of poor treatment, and finally, the ways and means of reusing treated wastewater. The results also indicate a general lack of knowledge of these issues on the part of the population. For example, 63% of those surveyed believe that it is necessary for wastewater to receive treatment prior to final discharge. Along the same lines, 82% of those surveyed would be interested in having a basic knowledge of how wastewater treatment is carried out.

Phase iii. Videos and infographics were created in accordance with the topics determined by the surveys and researched in phase i. For the design, the greatest difficulty was to make the graphic and written content concise and pleasing to the eye for better understanding; this was checked by going through a review phase before publication, then the final designs were approved.

Phase iv. Among the various publications made, some infographics were posted about the types of wastewater treatment in small communities, which did not have a great "acceptance" compared to the other publications, since the quantitative variables that have been used to measure the impact of these, such as number of "likes", comments, posts to conversations and profile visits, were not even like those of previous publications. Perhaps this was since the content is not attractive enough for the user of the platform, therefore, it is advised that, in similar topics, the development of a video could be a better option since it would attract attention, since it is a very interesting topic, but also requires more verbal explanation and not only textual as provided by the infographic. Additionally, it also marks the lack of knowledge of the population about this type of processes, and that they do not generate greater curiosity to know them either.

Phase v. Through network monitoring it was determined that despite the impediment that not all young people have access to the internet and that very few have a smartphone, in one month a total reach of approximately 1908 people was obtained, a very relevant figure considering the size of the population to which the campaign was directed. This represents a total of 185 followers in the Instagram account. Fig. 3 shows the reach (number of people who saw the publications). As well as a total of 580 followers on the Facebook page. At this point the profiles of the followers were checked, and most of them are from neighboring areas of the commune. The results were encouraging, obtaining great social acceptance to the messages disseminated, even receiving suggestions to implement more content, the community was willing to learn more about the subject, thus proving that social networks are a very excellent means of disseminating messages in times of pandemic.

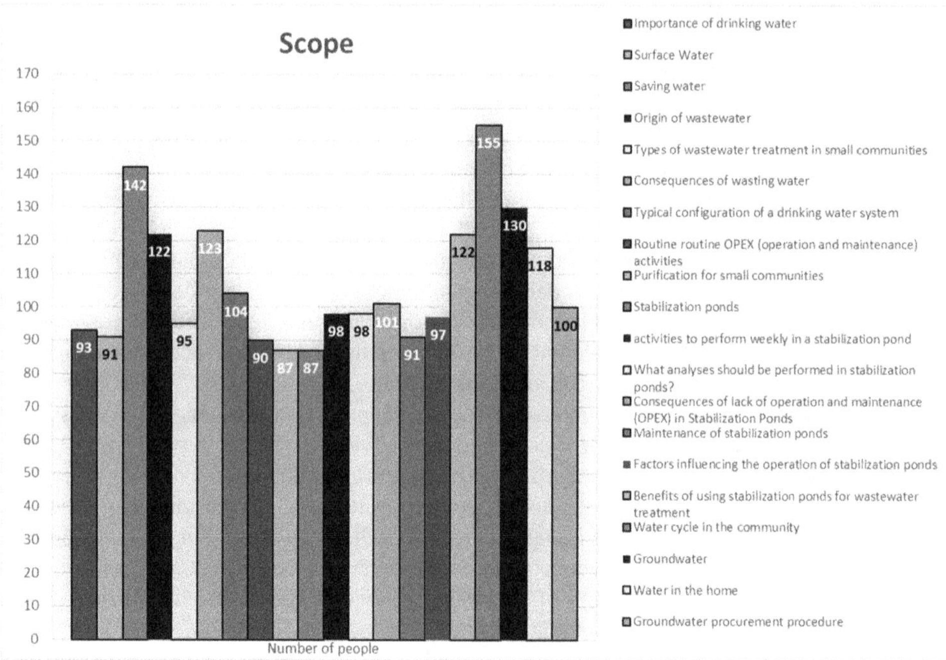

Figure 3: Number of people who watched each post on Instagram, from September 9 to October 6, 2020.

If you wish to review the content published on the pages below are shared each of the direct links to the networks created:

- https://www.facebook.com/cuidadocalidadagua
- https://www.instagram.com/cuidado_calidad_agua/

This initiative is within the framework of a project 'Unit of Linkage with Society' that works on participatory models [10], [20], to promote resilience in the face of different problems around water [11], while recognizing possibilities for geotourism for the sector [23].

5 CONCLUSIONS

From the point of view of the results obtained from the surveys on drinking water, wastewater and stabilization ponds, it is shown that the knowledge of the population is limited, but at the same time there is a great interest in knowing it.

- Due to the pandemic, communication with the inhabitants of the commune was affected and it was not possible to reach a more representative sample of the population, since only 42 people responded to the initial survey. This revealed the problem of getting messages across in times of pandemic.
- Despite all the difficulties in the study period, where there were restrictions and limitations to mobility, this community was resilient: they took the time to pay attention to the research carried out, answer the surveys, showed their concern for the vital liquid, indirectly, people became aware of the importance of water, and the best result was the interaction of people in the community reflected in social networks, which reached very significant values that were not really expected.
- The use of social networks as a channel for disseminating information is a vital tool in times of pandemic. Both Instagram and Facebook, due to their lower consumption of megabytes, were key to receiving information through their mobile phones, to which 70% of the population has access. There was a clear interest among the residents of the commune and surroundings in relation to the published topics.

However, challenges remain, such as further developing interest with new publications, to achieve a community supported with accurate and sustainable information. This project became an initial step that will be projected in various development initiatives to strengthen resilience in rural communities.

REFERENCES

[1] Munodawafa, D., Communication: Concepts, practice and challenges. *Health Education Research*, **23**(3), pp. 369–370, 2008.

[2] del Corral, A., La intersección edu-comunicativa. *Comunicar*, **22**, pp. 13–20, 2004.

[3] ITU-D, *Telecom World Geneva*, 2021. http://www.itu.int/ITU-D/ict/material/Telecom 09_flyer.pdf. 2009. Accessed on: 12 Mar. 2021.

[4] Gagnon, D.J., Mobile learning environments. *Educause Quarterly*, **33**(2), p. 10, 2010.

[5] Gupta, D., Bhatt, S., Gupta, M., & Tosun, A.S., Future smart connected communities to fight COVID-19 outbreak. *Internet of Things*, **13**, p. 100342, 2021. DOI: 10.1016/j.iot.2020.100342.

[6] WHO, Weekly epidemiological update – 22 December 2020. https://www.who.int/publications/m/item/weekly-epidemiological-update---22-december-2020. Accessed on: 12 Mar. 2021.

[7] CDC, COVID-19. https://www.cdc.gov/coronavirus/2019-ncov/index.html. Accessed on: 12 Mar. 2021.

[8] INEC, Fascículo provincial Santa Elena. 2010. https://www.ecuadorencifras.gob.ec/wp-content/descargas/Manu-lateral/Resultados-provinciales/santa_elena.pdf. Accessed on: 12 Mar. 2021.

[9] Moreno, Ó.J.C., Reflexiones sobre la educación rural en el marco de la comunicación-educación. *Civilizar: Ciencias Sociales y Humanas*, **8**(1), pp. 89–102, 2008.

[10] Herrera-Franco, G., Alvarado-Macancela, N., Gavín-Quinchuela, T. & Carrión-Mero, P., Participatory socio-ecological system: Manglaralto-Santa Elena, Ecuador. *Geology, Ecology, and Landscapes*, **2**(4), pp. 303–310, 2018. DOI: 10.1080/24749508.2018.1481632.

[11] Herrera-Franco, G., Carrión-Mero, P., Aguilar-Aguilar, M., Morante-Carballo, F., Jaya-Montalvo, M. & Morillo-Balsera, M.C., Groundwater resilience assessment in a communal coastal aquifer system. The case of Manglaralto in Santa Elena, Ecuador. *Sustainability*, **12**(19), p. 8290, 2020. DOI: 10.3390/su12198290.

[12] Rivera Vinces, G., Diseño de dique para embalsamiento de agua dulce en la comuna Libertador Bolívar, Provincia de Santa Elena. Escuela Superior Politécnica del Litoral (ESPOL), 2019.

[13] Valencia Robles, J.I., Análisis hidrogeológico de la cuenca del río Manglaralto para la caracterización de sus sistemas acuíferos. Escuela Superior Politécnica del Litoral (ESPOL), 2017.

[14] Carrión, P., Herrera, G., Briones, J., Sánchez, C. & Limón, J., Practical adaptations of ancestral knowledge for groundwater artificial recharge management of Manglaralto Coastal Aquifer, Ecuador. *WIT Transactions on Ecology and the Environment*, pp. 375–386, 2018. DOI: 10.2495/SDP180341.

[15] Morante Carballo, F., Marcatoma Brito, L., Carrión Mero, P., Aguilar Aguilar, J.M. & Ramírez, T., Urban wastewater treatment through a system of green filters in the Montañita commune, Santa Elena, Ecuador. *WIT Transactions on Ecology and the Environment*, pp. 233–249, 2019. DOI: 10.2495/WS190211.

[16] ManabiNoticias, Preocupación en sector rural ha provocado el anuncio de retornar a clases, 2020. https://manabinoticias.com/preocupacion-en-sector-rural-ha-provocado-el-anuncio-de-retornar-a-clases/. Accessed on: 12 Mar. 2021.

[17] Vivar, J.M.F., Nuevos modelos de comunicación, perfiles y tendencias en las redes sociales. *Comunicar*, **17**(33), pp. 73–81, 2009.

[18] Buxarrais, M.R., Redes sociales y educación. *Education in the Knowledge Society*, **17**(2), pp. 15–20, 2016 (in Spanish).

[19] Bosch, T.E., Preez, A. & Michell, L., Using online social networking for teaching and learning: Facebook use at the University of Cape Town. *South African Journal for Communication Theory and Research*, **35**(2), pp. 185–200, 2009.

[20] Wang, Q., Woo, H.L., Quek, C.L., Yang, Y. & Liu, M., Using the Facebook group as a learning management system: An exploratory study. *British Journal of Educational Technology*, **43**(3), pp. 428–438, 2012.

[21] Villalobos C., *Los 20 mejores programas para crear encuestas online*. https://blog.hubspot.es/service/crear-encuestas-online. (In Spanish.)

[22] Law D., *5 Reasons to Use Survey123 for ArcGIS*. https://www.esri.com/about/newsroom/arcuser/5-reasons-to-use-survey123-for-arcgis/.

[23] Gricelda, H.F., Paúl, C.M. & Niurka, A.M., Participatory process for local development: Sustainability of water resources in rural communities: Case Manglaralto-Santa Elena, Ecuador. *Handbook of Sustainability Science and Research. World Sustainability Series*, ed. L.F.W., Springer: Cham, pp. 663–676, 2018.

[24] Herrera-Franco G. et al., Geosites and georesources to foster geotourism in communities: Case study of the Santa Elena Peninsula Geopark Project in Ecuador. *Sustainability*, **12**(11), p. 4484, 2020. DOI: 10.3390/su12114484.

SECTION 4
HYDRAULIC MODELLING

GRAPH THEORY AND COMMUNITY DETECTION FOR ELEMENTARY DMA DESIGN

STEFANO MAMBRETTI[1,2], ANITA RAIMONDI[1] & FERNANDO F. STROPPA[3]
[1]Politecnico di Milano, Italy
[2]Wessex Institute of Technology, UK
[3]Università di Pisa, Italy

ABSTRACT

In this work, with the objective of designing elementary district metered areas (eDMAs), graph theory methods and community structured algorithms are applied to the case of the water distribution network of Milano. This network is highly looped and very complex, serving about two million people. Initially, the network is subdivided into 27 eDMAs, which are the number of pumping stations in operation, obtaining three different divided networks (one from community structured algorithms and two from graph theory). Unlike other procedures based on traditional techniques to sectorize networks, the maximum size of each eDMA was not previously restricted, thus generating significantly larger districts. Afterwards, a simple algorithm was applied to aggregate the eDMAs into dynamic DMAs to reduce the energy consumption and overall pressures in the network for the hour of lowest water consumption. Results show improvement in both aspects, thus justifying the study. Further studies are recommended in order to propose other eDMAs designs and other dynamic DMAs configurations.

Keywords: water distribution networks, district metered areas, elementary districts, graph theory, community detection.

1 INTRODUCTION

Water distribution networks (WDN) represent one of the main infrastructure components of society, and, therefore, their operation and management (O&M) have direct impact in energy and water savings [1]. District Metered Areas (DMA) are created from the sectorization of a WDN into subnetworks by installing gate valves or flow meters into the boundary pipes [2]. A permanent pressure control system is facilitated by DMAs, thus reducing the pressure and pressure variation in the WDN, thus also reducing the background leakage [3] and energy consumption [2].

As WDS may consist of thousands to tens of thousands of elements [4], most of which are buried, O&M can be complex [5] and often sub-optimized. In the past, the planning stage for the design of sectors included an evaluation of ground contours, available hydraulic data, existing boundaries, and natural features, combined with a mathematical model [3]. Nowadays, however, with the enhancement of software technology and artificial intelligence, sectorization can be optimized while still considering operational and physical constraints and objectives [5]. The definition of DMAs is usually divided in two phases: clustering, aimed to establish the shape, number of clusters and dimensions of subnetworks, and partitioning, aimed to select the pipes in which the flow meters or gate valves should be installed [5]–[7]. Several methodologies have been proposed to cluster feasible DMAs [5], such as multi-agent approach [8], [9], spectral graph algorithms [10], [11], modularity-based algorithms [12], [13], community structure algorithms [13], [14], graph theory [15]–[17] and multilevel graph partitioning [18]. The partitioning phase is usually done by with evolutionary computation methodologies [5], with single or multi-objective functions designed to optimize hydraulic and/or economic conditions [19]–[22]. The water network partition (WNP) can present some disadvantages, such as a potential negative impact on

www.witpress.com, ISSN 1743-3541 (on-line)
doi:10.2495/WRM210111

water quality, a decrement on energy efficiency and a reduction on the overall hydraulic redundancy of the network [5], [6].

Adaptatively reconfiguring the WDNs is the idea of dynamically changing the DMAs with the objective of reducing these negative impacts [13], [23], [24]. The WNP aimed to enable dynamically reconfigurable DMAs should follow three basis steps:

1. Clustering: Definition of elementary DMAs, which can be dynamically aggregated/segregated in order to achieve a specific goal (e.g. aggregating DMAs in large sectors during high demand periods and segregating DMAs in small sectors for leakage detection during low demand periods [23]).
2. Partitioning: Choice of the pipes in which flow meters, pressure relief valve (PRV) or gate valves should be installed.
3. Iterative refining procedure 1 and 2 to find optimized elementary DMAs.
4. O&M: Dynamically change the gate valves and/or PRVs configurations.

In this paper, the cluster stage of WNP will be applied to the WDN of Milano to select elementary DMAs using graph theory and community structure algorithms. Afterwards, the elementary DMAs will be joint in order to find the *best* sectors in terms of energy optimization.

2 CASE STUDY: THE MILANO WATER SUPPLY NETWORK

2.1 The water supply network

The water supply system of Milano is composed of several reservoirs located at ground level, from which the water is pumped directly highly looped network [1]. The total length of the pipes in the WDN is 2228 km and there are 27 stations in operation, each one equipped with 2–4 pumps [1]. A simplified model of the WDN was built from the available data (such as geometry, mean daily water demand, pumping schedule, etc.) with the well-known software EPANET. Only diameters larger than 300 mm were considered for building the model, and a “fairly accurate reconstruction of the network was reached” [1]. The EPANET file was then transformed in an undirected graph, which consisted of 4994 nodes and 5333 edges. Each node is a graph vertex and represents a reservoir, a tank, or a junction, while each edge represents a pipe, a valve, or a pump. It was assigned a weight for each pipe, based on the estimated conductance defined as [16]:

$$weight = \left(\frac{\lambda}{D.2gA^2}L\right)^{-1}. \tag{1}$$

where λ, D, g, A and L are respectively the resistance coefficient, the diameter, the gravity of Earth, the cross-section area, and the length of the pipe.

2.2 Number of DMAs

The number of DMAs in a WDN has direct impact in the cost of creating them, because when the DMAs sizes decrease, more valves and flow meters are required to be installed and maintained. The *DMA Management Guidance Notes* [3] suggests that each DMA should serve between 500–3,000 properties, and where very poor infrastructure is present, decreasing that value to less than 500 could be ideal. The Guidance Notes also refers that, in DMAs serving more than 5,000 properties, it is difficult to discriminate small bursts during low demand periods. However, there is no imposed limit for a DMA size, as the hydraulic, practical, and economic factors should determine their sizes [3]. Moreover, the important topic to be considered is the reason why the districts are built, which will drive to different definitions and design.

Considering that in the past only traditional methods and empirical approaches were used to O&M of the WDNs, the usage of small DMAs may no longer be required considering the new artificial intelligence approaches. To test this hypothesis, the number of elementary DMAs (eDMAs) to be found should be equal to the number of reservoirs, with one reservoir in each DMA. It is important to notice that during the O&M stage, because of aggregation of elementary DMAs, the number of actual DMAs will be reduced, and their sizes will be increased even further. Hydraulic analysis with EPANET will determine the efficiency of the system.

3 ELEMENTARY DISTRICTS CREATION

3.1 Graph theory

3.1.1 Breadth-first search algorithm

The breadth-first search (BFS) algorithm searches or traverse tree or graph data structures [25]. For the search in graphs, it starts at an arbitrary node and explores all the neighbour's nodes (other nodes directly connected to the present node) at the present depth prior to exploring the nodes on further depths. As the WDN of Milano is highly looped, the source nodes could reach far more than 2 other nodes, thus reducing the total depth during the search. For each station, a list of ordered junctions is generated as illustrated in Fig. 1, which shows an example of the algorithm in the WDN.

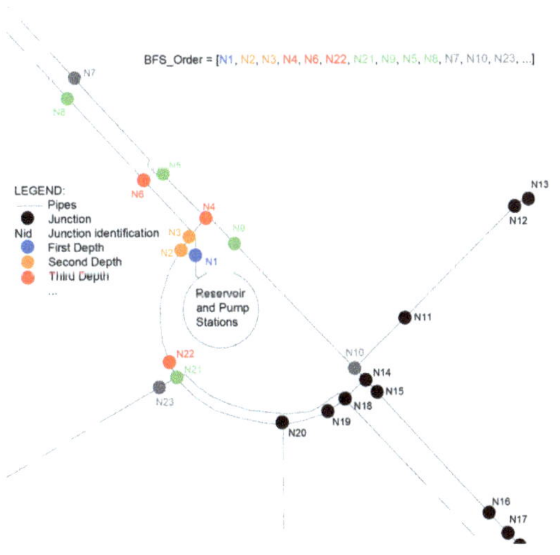

Figure 1: Illustrated BFS applied to a station located in the south-east of the WDN.

3.1.2 Dijkstra's algorithm

The Dijkstra algorithm searches the shortest paths between nodes in a graph [26]. Several variants of the algorithm have been developed along the years. A version that fixes a single node (a junction leaving each station) as the source node and produces a shortest-path tree by finding the shortest paths from the source to all other nodes in the graph will be applied to the WDN. The conductance of the pipes introduced at point 2.1 will be used as the weight of the edges. As in point 3.1.1, a list of ordered junctions is generated for each station, as illustrated in Fig. 2.

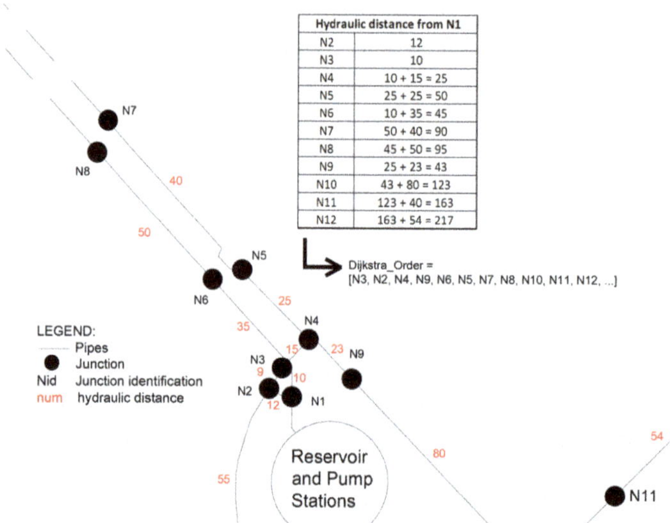

Figure 2: Illustrated Dijkstra applied to a station located in the south-east of the WDN.

3.1.3 Creating elementary districts from Dijkstra and BFS

The proposed algorithm searches for eDMAs with approximate equally demand in each district. Because of the highly complex looped structure of the WDN of Milano, this condition is not always feasible as the reservoirs are usually not spaced regularly along the WDNs. The algorithm is divided in the following steps:

1. Creation of the lists of ordered junctions by BFS and Dijkstra, following Steps 3.1.1 and 3.1.2.
2. Creation of an object structure that stores the currently nodes in each eDMAs, initially containing only the junction leaving of each station, also called control junction.
3. Creation of an object structure that stores the currently total demand of each eDMAs.
4. Selection the eDMAs with the lowest demand.
5. From the list defined in point 1, the next node not yet assigned to an eDMAs is selected if this node is adjacent to the selected eDMA, and:
   a. This node is added to the selected eDMA
   b. This node demand is added to the currently demand of the eDMA.
6. Steps 4 and 5 are repeated until all nodes of the network have been assigned to eDMAS.

3.2 Community structure algorithm

3.2.1 Girvan–Newman algorithm

A community in a network is a structure-property where internal nodes of the community are strongly connected among the group while loosely connected between other communities [27]. The Girvan–Newman algorithm is divisive method used to detect communities in complex systems by progressively removing edges from the original network. The method chooses the edge with the highest betweenness centrality value, which, in the case of a WDN, represents the amount of influence the edge has over the WDN. These edges typically serve as bridges between communities. The algorithm can be summarized as the following:

1. All the values of betweenness centrality of the edges are calculated.
2. The edge with highest betweenness centrality is removed.
3. The values of betweenness centrality are recalculated for edges affected by the previous removal.
4. Steps 2 and 3 are repeated until there are no edges remaining or an established condition is satisfied.

3.2.2 Modified Girvan–Newman algorithm

As in the original Girvan–Newman method does not take into consideration the position of the reservoirs and the total demand of the communities, a modified method is proposed to achieve the goal of attaining 1 reservoir per eDMA. The betweenness centrality of the edges are calculated considering the weight as defined on point 2.1 and the algorithm is summarized as the following:

1. Application of the Steps 1–3 of Girvan–Newman Algorithm
2. Steps 2 and 3 of Girvan–Newman Algorithm is repeated until a new community emerges.
3. When a new community emerges, the following condition is analysed:
   a. If it contains one or more reservoirs: All not removed edges are marked as available.
   b. Else: The edges removed on Steps 1 and 2 are readded and the first edge of Step 2 of Girvan–Newman Algorithm is marked as unavailable.
4. While the number of the communities is lower than the number of reservoirs, repeat Steps 1 3.

The eDMAs are showed in Fig. 3.

4 AGGREGATING ELEMENTARY DISTRICTS

The main idea of aggregating eDMAs is to reduce the energy consumption and to balance the pressure in the WDN. The following equation proposes a simple objective function that take in consideration both factors and can be used as target function to be minimized. *MinPres* is the minimum acceptable pressure in the WDN, which is equal to 25 m for Milano, *CompPres* is the computed pressure in a specific junction and *AvgCompPres* is the average of the computed pressures of the control junctions, which for Milano are the closest junctions to the reservoirs.

$$Objective = EnergyCons + \sum_{n=1}^{NumStations}(MinPres - AvgPres - MinPres)^2. \quad (2)$$

where Objective, EnergyCons, NumStations, are respectively the objective value to be retrieved, the total hourly energy consumption by the pumps and the number of stations presents in the WDN.

A simple algorithm is proposed to aggregate eDMAs into bigger dynamic DMAs during the O&M stage. The aggregation follows the steps.

1. A random percentual of demand and a random number of districts are calculated.
2. All stations are marked as not visited, regardless of the current state of the pumps.
3. A not visited station is selected and marked as visited.
4. One by one, the nearest stations are aggregated to the selected station until the percentual demand is reached. All these stations are marked as visited.

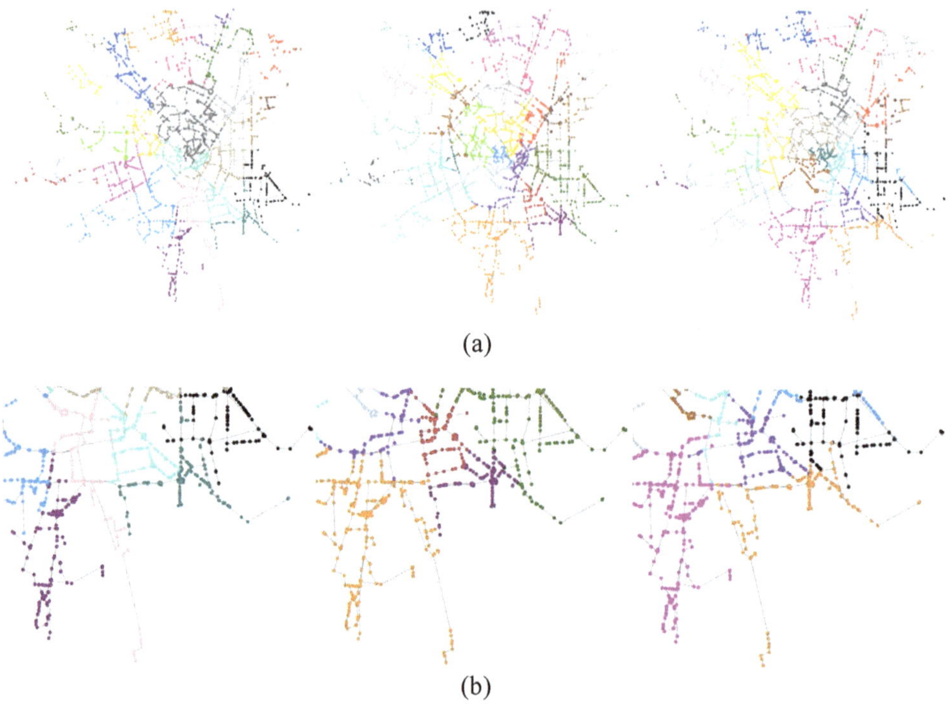

Figure 3: Created eDMAs in the WDN of Milano. Left: Modified Girvan–Newman. Middle: Graph theory using BFS. Right: Graph theory using Dijkstra. (a) Overall WDN; and (b) Southeast of the WDN.

5. Steps 3–4 are repeated until the number of districts is reached.
6. All stations are marked as not visited, and the Step 5 is repeated in order to select all stations in Step 3.
7. Steps 1–6 are repeated for a predetermined number of times and the best DMAs is retrieved based on the proposed objective function.

The proposed algorithm presents two major flaws. Firstly, it can suffer from the *curse of dimensionality* as there are too many possible combinations to be analysed, thus it is much likely that the algorithm will not be able to find the best solution. Secondly, due to the high number of random generations along with the high number of possible combinations, the algorithm is computationally expensive. Further works are recommended to optimize the aggregation of eDMAs.

For achieving the results, firstly an evolutionary computation algorithm developed in previous works [1] was applied to optimize the pumps speeds and status, then the proposed aggregating algorithm was applied for the 3 WDN with different eDMAs to determine what valves should be closed, and finally, the evolutionary computation algorithm was applied once again to optimize the pumps with the established dynamic DMAs.

5 RESULTS

Simulations have been performed for the lowest demand hour of the day (3.00 a.m.). Results show that dividing the WDN of Milano in two or three dynamic DMAs (which is far greater than 3,000 properties) have the possibility to be already highly beneficial regarding energy saving. Fig. 4 illustrates the three dynamic DMAs generated and Fig. 5 shows the pressure in each control junction (only junctions with more than 10% of pressure difference are shown).

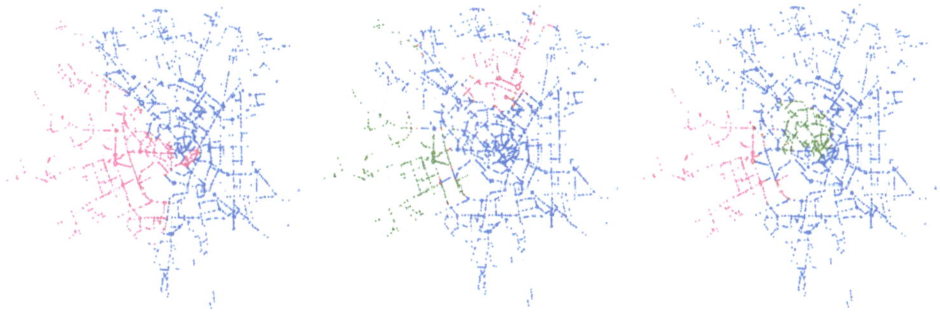

Figure 4: Created dynamic DMAs in the WDN of Milano. Left: Modified Girvan–Newman. Middle: Graph theory using BFS. Right: Graph theory using Dijkstra.

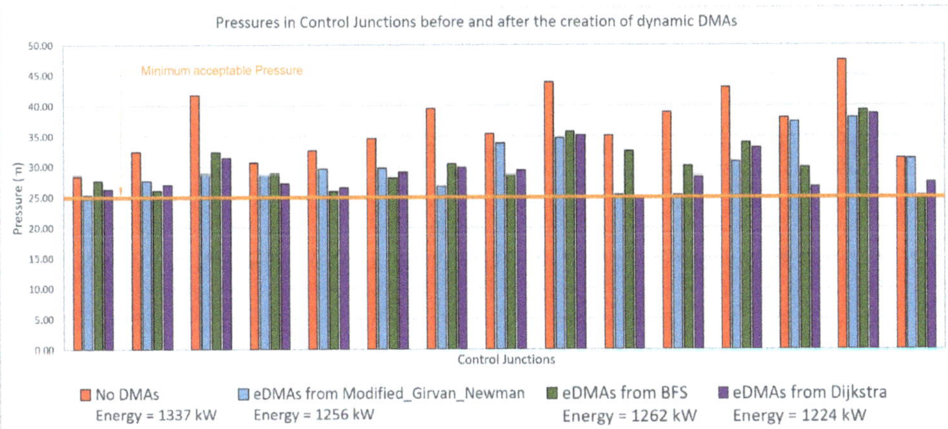

Figure 5: Pressures in control junctions and total energy.

Although the results show the benefits of dynamic DMAs, further works are recommended to determine a better algorithm that aggregates eDMAs, along with a detailed study including the performance during all hours of the day. Lastly, with the introduction of the DMAs, the previous points used for controlling the pressure could become obsolete and could not represent the new state of the WDN. A work to verify that hypothesis and to create an algorithm to find the best new points of pressure control is also recommended.

6 CONCLUSIONS

In the paper, the use of several algorithms has been applied to the creation of eDMAs in the WDN of the city of Milano. Afterwards, thousands of different dynamic DMAs combinations have been tried, led by the simple algorithm previously described, aiming to find optimal configurations in which energy consumption and overall pressures are reduced. Results show that improvements are possible even when the dynamic DMAs are dramatic bigger than traditional DMAs. Although the proposed algorithm is extremely simple, it was able to find configurations where pressure in almost all control junctions were reduced in all the three eDMAs scenarios.

With the simplifications made to generate the model of the WDN in EPANET, field tests are required to study the actual effect on the WDN. Further research is suggested to improve the actual algorithm the generate dynamic DMAs. Further research is also suggested to find other methods to generate eDMAs for complexed looped WDNs.

REFERENCES

[1] Wrachien, D., Mambretti, S. & Orsi, E., Optimization of pump operations in a complex water supply network: New genetic algorithm frameworks. *International Journal of Sustainable Development and Planning*, **12**, pp. 79–88, 2017. DOI: 10.2495/SDP-V12-N1-79-88.

[2] Di Nardo, A., Di Natale, M., Giudicianni, C., Santonastaso, G.F., Tzatchkov, V. & Varela, J.M.R., Economic and energy criteria for district meter areas design of water distribution networks. *Water*, **9**, pp. 1–13, 2017. DOI: 10.3390/w9070463.

[3] Morrison, J., Tooms, S. & Rogers, D., *DMA Management Guidance Notes*. IWA Publishing: London, 2007.

[4] Lifshitz, R. & Ostfeld, A., Clustering for analysis of water distribution systems. *Journal of Water Resources Planning and Management*, **144**, 2018. DOI: 10.1061/(ASCE)WR.1943-5452.0000917.

[5] Khoa Bui, X.S., Marlim, M. & Kang, D., Water network partitioning into district metered areas: A state-of-the-art review. *Water*, **12**(4), 2020. DOI: 10.3390/w12041002.

[6] Giudicianni, C., Herrera, M., di Nardo, A. & Adeyeye, K., Automatic multiscale approach for water networks partitioning into dynamic district metered areas. *Water Resources Management*, **34**, pp. 835–848, 2020. DOI: 10.1007/s11269-019-02471-w.

[7] Di Nardo, A., Di Natale, M., Giudicianni, C., Musmarra, D., Santonastaso, G.F. & Simone, A., Water distribution system clustering and partitioning based on social network algorithms. *Procedia Engineering*, **119**, pp. 196–205, 2015. DOI: 10.1016/j.proeng.2015.08.876.

[8] Herrera, M., Izquierdo, J., Pérez-García, R. & Ayala-Cabrera, D. Water supply clusters by multi-agent based approach. *Water Distribution Systems Analysis*, pp. 861–869, 2010. DOI: 10.1061/41203(425)79.

[9] Hajebi, S., Barrett, S., Clarke, A. & Clarke, S., Multi-agent simulation to support water distribution network partitioning, *Modelling and Simulation 2013 – European Simulation and Modelling Conference*, ESM, 2013.

[10] Di Nardo, A., Di Natale, M., Giudicianni, C., Greco, R. & Santonastaso, G.F., Weighted spectral clustering for water distribution network partitioning. *Applied Network Science*, **2**, 19, 2017. DOI: 10.1007/s41109-017-0033-4.

[11] Lui, J. & Han, R., Spectral clustering and multicriteria decision for design of district metered areas. *Journal of Water Resources Planning and Management*, **144**(5), 2018. DOI: 10.1061/(asce)wr.1943-5452.0000916.

[12] Brentan, B.M., Campbell, E., Meirelles, G.L., Luvizotto, E. & Izquierdo, J., Social network community detection for DMA creation: Criteria analysis through multilevel optimization. *Mathematical Problems in Engineering*, **2017**, 2017. DOI: 10.1155/2017/9053238.

[13] Qingzhou, Z., Wu, Z., Zhao, M., Qi, J., Huang, Y. & Zhao, H., Automatic partitioning of water distribution networks using multiscale community detection and multiobjective optimization. *Journal of Water Resources Planning and Management*, **143**, 2017. DOI: 10.1061/(ASCE)WR.1943-5452.0000819.

[14] Campbell, E., Ayala-Cabrera, D., Izquierdo, J., Pérez-García, R. & Tavera, M., Water supply network sectorization based on social networks community detection algorithms. *Procedia Engineering*, **89**, pp. 1208–1215, 2014. DOI: 10.1016/j.proeng.2014.11.251.

[15] Di Nardo, A., Di Natale, M., Santonastaso, G.F., Tzatchkov, V.G. & Alcocer-Yamanaka, V.H., Water network sectorization based on graph theory and energy performance indices. *Journal of Water Resources Planning and Management*, **140**(5), pp. 620–629, 2014. DOI: 10.1061/(ASCE)WR.1943-5452.0000364.

[16] Alvisi, S. & Franchini, M., A procedure for the design of district metered areas in water distribution systems. *Procedia Engineering*, **70**, pp. 41–50, 2014. DOI: 10.1016/j.proeng.2014.02.006.

[17] Scarpa, F., Lobba, A. & Becciu, G., Elementary DMA design of looped water distribution networks with multiple sources. *Journal of Water Resources Planning and Management*, **142**, 2016. DOI: 10.1061/(ASCE)WR.1943-5452.0000639.

[18] Alvisi, S., A new procedure for optimal design of district metered areas based on the multilevel balancing and refinement algorithm. *Water Resources Management*, **29**, pp. 4397–4409, 2015. DOI: 10.1007/s11269-015-1066-z.

[19] Di Nardo, A., Di Natale, M., Greco, R. & Santonastaso, G.F., Ant algorithm for smart water network partitioning. *Procedia Engineering*, **70**, pp. 525–534, 2014. DOI: 10.1016/j.proeng.2014.02.058.

[20] Di Nardo, A., Di Natale, M., Santonastaso, G.F., Tzatchkov, V.G. & Alcocer-Yamanaka, V.II., Watcr Nctwork Scctorization bascd on gcnctic algorithm and minimum dissipated power paths. *Water Science and Technology Water Supply*, **13**(4), pp. 951–957, 2013. DOI: 10.2166/ws.2013.059.

[21] Shao, Y., Yao, H., Zhang, T., Chu, S. & Liu X., An improved genetic algorithm for optimal layout of flow meters and valves in water network partitioning. *Water*, **11**(5), 2019. DOI: 10.3390/w11051087.

[22] Campbell, E., Izquierdo, J., Montalvo, I., Ilaya-Ayza, A., Pérez-García, R. & Tavera, M., A flexible methodology to sectorize water supply networks based on social network theory concepts and multi-objective optimization. *Journal of Hydroinformatics*, **18**(1), pp. 62–76, 2015. DOI: 10.2166/hydro.2015.146.

[23] Wright, R., Stoianov, I., Parpas, P., Henderson, K. & King, J., Adaptive water distribution networks with dynamically reconfigurable topology. *Journal of Hydroinformatics*, **16**(6), pp. 1280–1301, 2014. DOI: 10.2166/hydro.2014.086.

[24] Farah, E. & Shahrour, I., Leakage detection using smart water system: Combination of water balance and automated minimum night flow. *Water Resources Management*, **31**, pp. 4821–4833, 2017. DOI: 10.1007/s11269-017-1780-9.

[25] Pohl, I.S., Bi-directional and heuristic search in path problems. PhD thesis, Stanford Linear Accelerator Center, Stanford University, 1969.

[26] Dijkstra, E.W., A note on two problems in connection with graphs. *Numerische Mathematik*, **1**, 1959.

[27] Newman, M.E.J. & Girvan, M., Finding and evaluating community structure in networks. *American Physical Society, Physical Review E*, **69**(2), 2004. DOI: 10.1103/PhysRevE.69.026113.

COMPARISON OF URBAN RETROFITTING SCENARIOS FOR SUSTAINABLE STORMWATER CONTROL: A CASE STUDY IN MILAN, ITALY

FRANCO RAIMONDI[1,2], MARIANA LOBO MARCHIONI[1],
ANITA RAIMONDI[1] & GIANFRANCO BECCIU[1]
[1]Politecnico di Milano, Italy
[2]ERSAF, Italy

ABSTRACT

The traditional approach of stormwater management in urban areas frequently affected by floods has proven to be unsustainable for the growing urbanization scenario as well as not adapted to provide climate change resilience. In the spirit of water sensitive cities, a stormwater management change is necessary, moving from pipe-based conveyance system to solutions able to manage at its source by small scale structures sprawled over the entire watershed, in a collection of systems and practices known as SuDs (Sustainable Urban Drainage Systems). These techniques provide several benefits not only related to risk mitigation but also to water quality and ecosystem services enhancement that make cities resilient also in terms of climate change. Currently stormwater and land use regulations mostly define a flow-limit to reduce discharge on the sewer systems. In this way, however, the search for further benefits is not encouraged. Flow-limit restrictions are obtained through storage control systems designed with design-storm methods that neglects the effect of pre-filling reducing the storage capacity before a new rainfall event. This research compares different stormwater management solutions for a new developed area located in Bresso municipality, in the metropolitan area of Milan (Italy), within the Seveso River watershed. The proposed scenarios go from no intervention to completely avoiding discharge on the sewer network with the implementation of stormwater control structures (permeable pavements, green roofs, retention basins, among others). The storage facilities are designed with a traditional design method and then their performance is evaluated with rainfall-runoff and routing model continuous simulations, also used to compare the different scenarios. The simulations confirm that is possible to reach more restrictive goals beyond the flow-limit regulations using stormwater controls structures that also provide multiple benefits and have significant effects in terms of hydraulic risk mitigation at watershed level. Continuous simulation shows that the effect of chained rainfall events can however reduce controls performance and should be considered when designing.

Keywords: urban retrofitting, spatial planning, sustainable drainage systems (SuDs), flood risk mitigation.

1 INTRODUCTION

Multiples impacts on environment and their communities come from urban sprawl. Urbanization affects the water cycle increasing hydraulic risk and contamination of watercourses. Land consumption increases runoff peak flow and volume and the multiple non-point source pollutants worsen water quality. The overloaded drainage system, including artificial and natural watercourse often underground, cause floods with significant human and economic damages.

The metropolitan area of Milan observed an expansion of the urban area driven by economic growth post-war. As the urban area increases, so did the number of flood events on the Seveso River, the main river that crosses Milan entirely by an underground route (Fig. 1). A by-pass channel built in 1982 discharging on the River Ticino reduced the flood events for a few years but lost effectiveness as the cities located upstream continued to develop. A study held on the Seveso River basin estimated that to reduce the flood events would be necessary to implement five detention basins with an estimated cost of 130 million Euros [1].

WIT Transactions on Ecology and the Environment, Vol 250, © 2021 WIT Press
www.witpress.com, ISSN 1743-3541 (on-line)
doi:10.2495/WRM210121

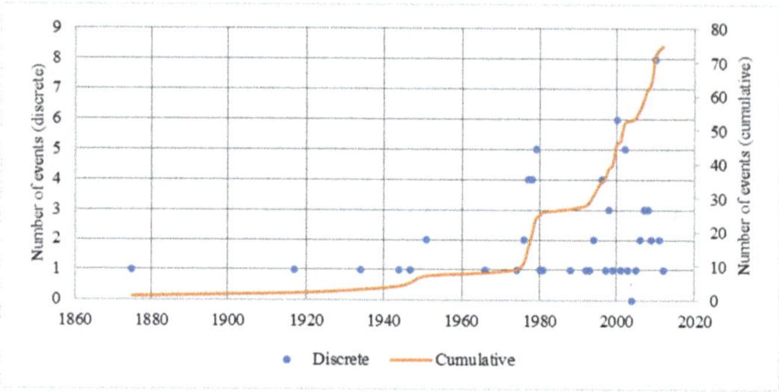

Figure 1: Flood events registered on the Seveso River from 1860.

The traditional approach relies on stormwater discharge to watercourses directly or through combined sewers overflows (CSOs) and large-scale infrastructures to manage flooding. This approach is unsustainable on the current urbanization growth scenario and it is preferable to manage stormwater through the reproduction of natural processes of storage, infiltration and evapotranspiration through small-scale control infrastructures. These small-scaled control structures can efficiently reduce runoff peak flow and volume with several benefits incorporated, such as biodiversity opportunities, heat island mitigation, amenities. In an effort to mitigate flood risk following this philosophy the Lombardy region, where the metropolitan Milan is located, developed in 2017 a policy on stormwater management to new constructions and major retrofits (RR7-2017) [2]. The text includes flow-limits and mandatory storage facilities establishing minimum values and suggesting design methods. The regulation, however, does not include measures to encourage the additional benefits from small-scale control structures, although it mentions that water harvesting for reuse and infiltration should be encouraged.

Design procedure for control storage facilities normally uses a design storm event approach. Lombardy regulation suggests a simple method to obtain preliminary results for small catchments areas. This method considers inflow as the net water depth from Intensity-Duration-Frequency equation (IDF) and outflow as the outlet and soil infiltration rate, neglecting rainfall-runoff transformation. These preliminary results should be confirmed with routing calculations using computational models. An alternative approach is to adopt a probabilistic analytical method that provides long-term simulation accuracy with a simple and low computational effort. This approach has been applied to control storage systems by various authors with good correlation when compared with continuous simulation [3]–[8].

Previous research undertaken by the authors studied the hydrological response on flood area reduction by implementing ecological buffers on a stretch of the Seveso River, Milan [9]. Encouraging source-area controls could provide further flood area reduction. To achieve this goal a retrofitting plan is proposed in an area located in the vicinity of the Seveso river to observe the runoff peak flow and volume reduction for different scenarios with the following systems: permeable pavements (PP), rain gardens (RG), bioretention cells (BC), green roof (GF) and a wetland. The control structures with storage (PP, BC) were designed with a design storm approach and have their performance analyzed with a rainfall-runoff routing model varying the Annual Recurrence Interval (ARI) (2, 10, 50 and 100-years storm) and with registered rainfall-data. The hydrological benefits of the different scenarios were also compared with the no-retrofit scenario.

2 CASE STUDY AREA

The Bresso municipality is the most densely populated (7,597 inhabitants/km^2) in the metropolitan area of Milan. In 2019, the municipal administration of Bresso approved an urban redevelopment project that involves the increase of the existing social and health services. This project represents an opportunity to implement different source control systems (Tables 1 and 2) not only to comply with regional policies but also to provide combined benefits. Furthermore, in the watershed perspective and multidisciplinary approach pursued by the Seveso River Sub-basin Strategic Project [10], it could be the opportunity to take a forward step and realize a neighborhood in which all the stormwater is effectively treated and managed at its source without discharging into the sewer system. In particular, in the redevelopment area of 55,800 m^2, the parking areas de-sealing, the construction of a green roof for a sport hall, the downspouts disconnection and finally the creation of a wetland (1,600 m^2) are assumed.

Table 1: Retrofitting scenario description.

ID	Scenario	Impervious area	
		(m^2)	(%)
A	Current state	23,290	43.4
B	New developments without control systems	25,894	48.3
S1	Permeable pavements (all parking area)	19,965	37.2
S2	Bioretention-cells (10% of impervious area)	25,059	46.7
S3	Rain gardens (10% of impervious area)	25,059	46.7
S4	PP, GF, BC, RG	17,568	32.7
S5	only wetland	25,894	48.3
S6	PP, GF, BC, RG + wetland	17,568	32.7

Table 2: Percentage surface occupied by controls structures respect to the total impervious area (25,894 m^2) for each scenario.

Control structures	Scenario					
	S1	S2	S3	S4	S5	S6
Permeable pavements	22.9	-	-	22.9	-	22.9
Bioretention-cells	-	10	-	3.3	-	3.3
Rain gardens	-	-	10	1.6	-	1.6
Green roofs	-	-	-	8.7	-	8.7
Presence of wetland	No	No	No	No	Yes	Yes

A brief description of the main characteristics of the Suds techniques considered in retrofitting scenarios are summarized below.

Permeable pavement (PP) allows water infiltration through its their surface into the underlying layer infiltrating or discharging through a drainage pipe towards the next system promoting runoff volume, peak flow and load reduction [11].

Green roofs (GF) consist of multiple layers (filter fabric and growing medium layer, water-proof membrane, mechanic protection geotextile, drainage layer) able primarily to retain rainwater and delaying its release [12].

Rain gardens and bioretention cells consist of a vegetated surface layer and a filter media within which the filtering water from roofs, roads and parking areas is treated.

Wetlands are shallow and deep water areas, with the main function to store stormwater for extended periods to facilitate sediments settling and water depuration [13].

3 METHODOLOGY

Three types of comparisons for the scenarios were investigated. The first type of comparison considered each scenario described in Tables 1 and 2 and involved a 24-h duration and one-in-2-year, one-in-10-year, one-in-50-year and one-in-100-year storm. Results are reported in Tables 4 and 5. In particular, the analysis of control structures performance, in urban retrofitting scenarios (S1, S2, S3, S4 and S6), for the one-in-50-year event are shown in Table 6. Scenario 5 is not reported due to the absence of Suds techniques implementation.

The second type of comparison concerned only scenario 1 and scenario 2, which include PP and BC, respectively. A 2-h critical rainfall durations of for PP and of 14 h for BC were considered. These durations were calculated using the storm design approach. The scenarios were then compared for one-in-2-year, one-in-10-year and one-in-50-year event. Results are presented in Tables 7 and 8.

The last comparison was carried out by using two continuous time series (year 2015 and November–December/2014). In this case scenarios 5 and 6 were not considered. Results are shown in Tables 9 and 10 and in Fig. 2 for year 2015, in Tables 11 and 12 and in Fig. 3 for November–December/2014.

3.1 Design-storm approach

The design-storm approach is traditionally used to design control structures that temporarily stores rainfall. A simple design procedure that provides a preliminary estimation consists of finding the critical design-storm, in this case the one that yields the maximum volume considering the water inflow and outflow. The inflow represents the net rainfall depth from the site I-D-F (Intensity-Duration-Frequency) equation and outflow represents the soil infiltration rate and outlets. The maximum water depth, that leads to maximum volume, is obtained through the derivation of the inflow and outflow curves. For infiltration systems, without considering the outlet, the critical storm duration (θ_c) can be obtained with eqn (1), and the water depth (h) with eqn (2). This method considers a rectangular hyetograph and neglects the rainfall-runoff transformation.

The method does not consider the effect of pre-filling [14], where the control system is partially full at the beginning of the rainfall event and in the case of a series of rainfall events. To minimize the risk of pre-filling a limit of 48 h to completely empty the structure, obtained with eqn (3), is recommended.

$$\frac{dh}{d\theta} = \varphi \cdot a \cdot n \cdot R \cdot \frac{\theta^{n-1}}{p} - \frac{f_c}{p} = 0 \rightarrow \theta_w = \left(\frac{f_c}{\varphi \cdot a \cdot n \cdot R}\right)^{\frac{1}{n-1}}, \tag{1}$$

$$h = h(\theta_w) = \frac{\theta_{w,}}{p} \cdot (\varphi \cdot a \cdot \theta^{n-1} \cdot R - f_c), \tag{2}$$

$$t_{emp} = \frac{p \cdot h}{f_c}, \tag{3}$$

where φ stands for the runoff coefficient, a and n are the parameters of an IDF $h = a\theta^n$; R the ratio between the catchment area that discharges on the structure (A_c) and the area available for infiltration (A_i) where $R = A_c/A_i$; θ is the rainfall duration, f_c is the soil infiltration capacity in mm/h, p is the filling material porosity, h is the structure depth for a given area and t_{emp} is the emptiness time. The following IDF parameters were used: 2-years storm – a = 30 mm/h^n;

10-years storm – a = 47 mm/h^n; 50-years storm – a = 63 mm/h^n; 100-years storm – a = 70 mm/h^n. For all return periods n = 0.31.

This procedure was used to design the PP for a one-in-50-year design storm, according to the regional regulation, concomitantly with a structural design using the AASHTO mechanistic-empirical method. The BC was design for a one-in-2-year storm since this structure typically manages small storms, discharging the excess through a spillway. Considering that the Bresso municipality has a sand-gravel soil the infiltration rate estimate was 3.33×10^{-6} m/s (12 mm/h). The method was not applied to GF and RG where typical thickness where adopted, while the wetland maintained an existing design by the municipality.

3.2 Rainfall-runoff model

The preliminary results obtained with the design-storm approach were checked by performing routing calculations using a rectangular storm with 24-h duration and 2, 10, 50 and 100 years-storm. The storm shape and duration selection means to yield maximum volume. A continuous simulation was also undertaken with registered rainfall data from 2015 considered a typical hydrological year, and with November–December from 2014, when a major flood event was registered in the Seveso River. The storage routing was undertaken using the USEPA-SWMM (United States Environmental Protection Agency – Storm Water Management Model). SWMM is a dynamic rainfall-runoff simulation model used for single event or long-term (continuous) simulation of runoff quantity and quality from primarily urban areas. The runoff component of SWMM operates on a collection of subcatchment areas that receive precipitation and generate runoff and pollutant loads. The routing portion of SWMM transports this runoff through a system of pipes, channels, storage/treatment devices, pumps, and regulators. SWMM tracks the quantity and quality of runoff generated within each subcatchment, and the flow rate, flow depth, and quality of water in each pipe and channel during a simulation period comprised of multiple time steps [15]. The processes modelled were rainfall-runoff and flow routing using kinematic wave, Horton infiltration model and the control structures with LID block option. Ponding and discharge from control structures were observed to verify the systems performance. The parameters adopted for the controls structures were based on the preliminary results obtained with the storm-design approach. The criteria used to compare scenarios considered peak flow, lag time and volume.

4 RESULTS AND DISCUSSION

The PP with storm-design approach resulted in a storage layer thickness of $h = 102\ mm$ for a rainfall duration of $\theta_c = 2\ h$ and emptiness time $t_{emp} = 4\ h$. Structural design considering vehicular traffic resulted in a thickness $h = 300\ mm$, being the critical thickness adopted. For a one-in-a-100-years storm $h = 120\ mm$, within the thickness selected for the PP.

The BC storage layer equals $h = 1{,}074\ mm$ for a rainfall duration of $\theta_c = 14\ h$ and emptiness time $t_{emp} = 31\ h$. Notice that a one-in-50-years storm would require a BC with $h = 3{,}148\ mm$ and $t_{emp} = 91\ h$. The storage layers thickness obtained with the design-storm approach were used for the rainfall-runoff simulations, the other parameters were obtained following typical values (Table 3).

Volume and peak flow obtained with the rainfall-runoff model for a one-in-24-h storm duration and constant rainfall intensity (rectangular storm) for the current scenario (A), post-intervention (B) and the retrofitting with stormwater controls are presented in Table 4. The hydrologic performance comparing the initial scenario (A) with the post-retrofitting scenario

Table 3: LID parameter for rainfall-runoff simulations.

Control simulation parameters	PP	GF	RG	BC
Surface				
Berm height [mm]	2.5	75	300	300
Vegetation volume fraction [%]	0.0	0.2	0.15	0.15
Roughness	0.1	0.3	0.3	0.3
Slope [%]	0.5	0.5	0.5	0.5
Pavement layer				
Thickness [mm]	100	-	-	-
Void ratio	0.2	-	-	-
Permeability [mm/h]	1,000	-	-	-
Soil layer				
Thickness [mm]	200	150	1,200	1,200
Porosity	0.3	0.6	0.52	0.52
Field capacity	0.2	0.45	0.25	0.25
Wilting point	0.05	0.05	0.05	0.05
Conductivity [mm/h]	750	10	110	110
Storage layer				
Thickness [mm]	300	-	-	900
Void ratio	0.5	-	-	0.35
Seepage rate [mm/h]	12	-	12	12

Table 4: Volume and peak flow for a 24 h duration storm for different scenarios (ARI = 2, 10, 50 and 100 years).

	Scenario							
	A	B	S1	S2	S3	S4	S5	S6
One-in-2-year storm								
Volume (m^3)	1,778	1,977	1,514	17	17	65	0	0
Peak flow (l/s)	20.64	22.95	17.57	0.2	0.2	5.41	0	0
One-in-10-year storm								
Volume (m^3)	2,926	3,254	2,492	170	687	669	853	0
Peak flow (l/s)	33.93	37.72	28.89	21.34	29.67	17.37	27.32	0
One-in-50-year storm								
Volume (m^3)	3,924	4,363	3,341	1,149	1,738	1,347	1,899	0
Peak flow (l/s)	45.48	50.56	38.72	42.75	42.75	27.99	40.1	0
One-in-100-year storm								
Volume (m^3)	4,361	4,848	3,713	1,620	2,208	1,667	2,366	0
Peak flow (l/s)	50.53	56.18	43.03	48.47	48.47	32.1	45.69	0

(S1-6) is shown on Table 5. The scenario S6, with a combination of controls systems and a wetland, manages to retain all stormwater for all ARI. Local regulation establishes a flow-limit of 10 l/s/ha_{imp} for the Bresso municipality, resulting in a total allowed flow of 23 l/s for the current scenario (SA) and 26 l/s for the post-interventions (SB).

Table 5: Hydrologic performance of the retrofit scenarios for a 24 h duration storm at the assigned return period (ARI = 2, 10, 50 and 100 years).

	Scenario					
	S1	S2	S3	S4	S5	S6
One-in-2-year storm						
Volume reduction [%]	14.8	99.0	99.0	96.3	100.0	100.0
Peak reduction [%]	14.9	99.0	99.0	73.8	100.0	100.0
Lag time [h]	0	-	-	20	-	-
One-in-10-year storm						
Volume reduction [%]	14.8	94.2	76.5	77.1	70.8	100.0
Peak reduction [%]	14.9	37.1	12.6	48.8	19.5	100.0
Lag time [h]	0	21	16	11	-	-
One-in-50-year storm						
Volume reduction [%]	14.9	70.7	55.7	65.7	51.6	100.0
Peak reduction [%]	14.9	6.0	6.0	38.5	11.8	100.0
Lag time [h]	0	15	12	8	10	-
One-in-100-year storm						
Volume reduction [%]	14.9	62.9	49.4	61.8	45.7	100.0
Peak reduction [%]	14.8	4.1	4.1	36.5	9.6	100.0
Lag time [h]	0	14	10	7	9	-

The controls performance parameters for a one-in-50-year storm are presented in Table 6. When the depth of ponded water is above the maximum storage level then surface outflow occurs, outflow reduction performance compares inflow and surface outflow. When the control structures are considered individually, PP (S1) represent the best alternative because show no outflow for the one-in-50-year event. Instead, when the overall runoff volume is considered, scenarios with combined control structures (S2, S3, S4 and S6) are the best solutions. A pre-filling (initial storage) was considered for the storage facilities and for all scenarios. After the rainfall event, control structures were partially full (final storage) and the runoff probability could increase if an event occurs without completely emptying the structure.

Table 6: Stormwater controls performance for retrofitting scenarios for one-in-50-year storm.

	S1	S2	S3	S4	S6
Total inflow (mm)	506	16,771	16,771	23,156	23,156
Surface outflow (mm)	0	4,086	6,363	10,129	10,129
Outflow reduction (%)	100	76	62	56	56
Infiltration loss (mm)	416	3,459	3,459	3,656	3,656
Initial storage (mm)	30	600	600	578	578
Final storage (mm)	115	9,810	7,533	9,821	9,821

The rainfall-runoff and routing simulation was also held with the storm duration obtained with the storm-design approach for PP (S1) and BC (S2) with a rectangular shape storm (Table 7). The S1 exceeds the regional flow-limit (25.89 l/s) for all storms ARI, while for S2 only for one-in-50-year storm. Considering that the regional regulation establishes a one-in-

50-year storm additional control structures should be include for both S1 and S2. Storms with 2-years and 10-years ARI were completely manage by the controls with residual water depth after the end of the rainfall event (Table 8).

Table 7: Volume and peak flow routing simulation results for a storm duration (θ_c) obtained with the storm-design method (ARI = 2, 10 and 50 years).

	Scenario	
	S1	S2
θ_c (hours)	2	14
2 years return period event		
Volume (m^3)	704	0
Peak flow (l/s)	98.7	0
10 years return period event		
Volume (m^3)	1,149	0
Peak flow (l/s)	160.46	0
50 years return period event		
Volume (m^3)	1,544	672
Peak flow (l/s)	221.66	64.93

Table 8: Stormwater controls performance for a storm duration (θ_c) obtained with the storm-design method s (ARI = 2, 10 and 50 years).

ARI	2-years		10-years		50-years	
	S1	S2	S1	S2	S1	S2
Total inflow (mm)	107.52	6,445.28	174.81	10,580.31	234.3	14,187.52
Surface outflow (mm)	0	0	0	0	0	2,427.16
Outflow reduction (%)	100	100	100	100	100	83
Infiltration loss (mm)	17.34	858.06	68.88	1,364.07	78.12	1,564.41
Initial storage (mm)	30	600	30	600	30	600
Final storage (mm)	119.76	6,185.22	136.11	9,812.44	185.76	10,792.13

Table 9 presents volume and peak flow and Table 10 presents the controls performance for the continuous rainfall simulation for the year 2015 (Lambrate rain gauge station). The total rain depth for this year was h_{tot} = 778 mm with N = 54 rainfall events for Inter Event Time Duration equals IETD = 6 h and Initial Abstractions IA = 2 mm resulting on an average depth of μ_h = 13 mm, average duration μ_Θ = 14 h and average previous dry hours μ_{pdh} = 138 h. Scenarios S5 and S6 where the outflow node is connected with a wetland did not produce runoff. The flow-limit is not met for both scenarios as for the PP (S1) scenario (Table 9). For the entire year PP (S1) did not generate surface outflow, but to meet regulations a combination with other controls systems is necessary to manage roof areas. The cumulative runoff volume increases 11% after the interventions planned for the area, while when adopting control structures cumulative volume reduces from the initial scenario 15% for S1 (PP), 99% for S2 (BC), S3 (RG) and S4 (combined PP, BC, RG and GF) (Fig. 2). Controls manage inflow almost entirely with residual water depth on the storage facilities for all scenarios with exception of S1 (PP) (Table 10).

Table 9: Volume and peak flow for continuous rainfall simulation (reference year: 2015).

	Scenarios					
	SA	SB	S1	S2	S3	S4
Volume (m^3)	17,622	19,590	14,998	137	137	132
Peak flow (l/s)	150.07	166.85	127.78	1.65	1.65	3.6

Table 10: Stormwater controls performance for retrofitting scenarios during continuous rainfall simulation (reference year: 2015).

	S1	S2	S3	S4
Total inflow (mm)	2,335	75,866.31	75,866.31	104,794.49
Surface outflow (mm)	0	0	0	1,891.34
Outflow reduction (%)	100%	100%	100%	98%
Infiltration loss (mm)	398	62,112.37	62,076.9	82,488.04
Initial storage (mm)	30	600	600	577.5
Final storage (mm)	30	3,186.21	3,188.98	8,256.65

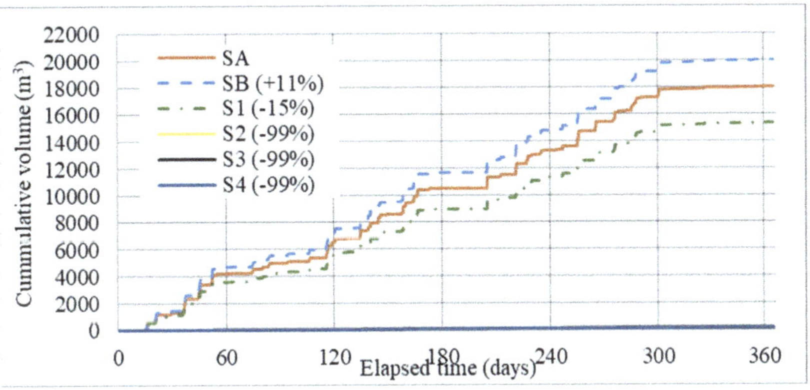

Figure 2: Cumulative volume for the continuous rainfall simulation (reference year: 2015) for initial scenario (SA), post-intervention (SB) and retrofitting (S1, S2, S3, S4). S2, S3 and S4 don't generate outflow volume, so the cumulative volume curves are equal to the horizontal lines with ordinates equal to 0.

Continuous simulation was also undertaken for November–December/2014 (Lambrate rain gauge), typically wet season with a particular series of events in that year that caused intense flooding. Table 11 presents volume and peak flow. The period registered a total rainfall depth of ht_{ot} = 756 mm and maximum rainfall intensity of i_{max} = 90 mm/h. The effect of concatenated rainfall events, where the controls storage capacity where not completely available at the beginning of a new rainfall event, reduce their performance (Table 12). PP (S1) did not generate surface runoff for the previous simulations (Tables 6 and 10) but generate 33.36 mm of surface outflow for the November–December/2014 period (Table 12). The controls structures that rely on smaller infiltration areas like BG and RG presented a poor performance with a series of rainfall events. Fig. 3 shows the cumulated volume and volume reduction comparing with the initial scenario (SA). With respect to previous scenarios (Fig. 2) the poor performance of control systems leads to a smaller volume reduction.

Table 11: Volume and peak flow for continuous rainfall simulation November–December 2014.

	Scenarios					
	SA	SB	S1	S2	S3	S4
Volume (m^3)	18,250	20,011	15,606	6,326	7,799	6,727
Peak flow (l/s)	641	697	551	507	480	402

Table 12: Stormwater controls performance for retrofitting scenarios for November–December 2014.

	S1	S2	S3	S4
Total inflow (mm)	2,267	72,974.53	72,974.53	100,843.5
Surface outflow (mm)	0	19,388.11	24,414.61	39,028.6
Outflow reduction (%)	100	73	66.5	61.3
Infiltration loss (mm)	1,752	47,866.41	42,819.25	50,224.02
Initial storage (mm)	30	600	600	577.5
Final storage (mm)	30	4,732.03	4,668.64	9,411.04

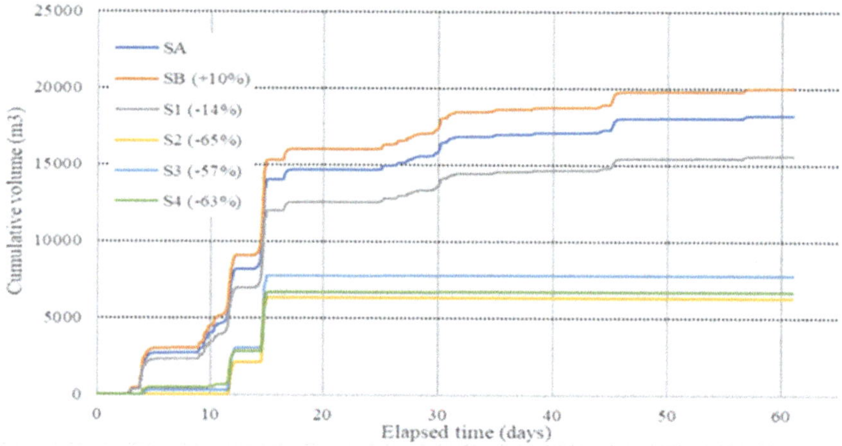

Figure 3: Cumulative volume for the continuous rainfall simulation for November-December/2014 for initial scenario (SA), post-intervention (SB) and retrofitting (S1, S2, S3, S4).

5 CONCLUSION

As a start-up for a research project on stormwater management and river restoration on the Seveso River basin, retrofitting scenarios were proposed for the densely urban area of the Bresso municipality, part of the Milan metropolitan area. The controls were pre-designed with a design-storm approach and then synthetic storms and continuous simulations were used to evaluate their performance and compare the scenarios proposed.

Efficiency of control structures on reducing runoff and meeting regional flow-limit regulations were discussed highlighting the importance of combining various types of control structures. Combination of controls present in each land use (rooftops, parking lots, roads,) gives better performance on volume and peak flow reduction.

Some ponding from the control structures were observed with continuous simulation. Traditional storm-design approach and rainfall-runoff and flow routing simulation with synthetic storm does not represent the effect of two close rainfall events that significantly reduces stormwater storage controls performance. To proceed with continuous simulation or design methods instead of the probabilistic approach allows better representation of the effects of pre-filling.

The following up on this research will further detail the controls performance considering the additional benefits on biodiversity, climate resilience and including stormwater load partitioning performance, a critical issue for the Seveso River that struggles to meet the European Union river quality standards for state members.

REFERENCES

[1] Becciu, G., Ghia, M. & Mambretti, S., A century of works on River Seveso: From unregulated development to basin reclamation. *International Journal of Environmental Impacts*, pp. 461–472, 2018. DOI: 10.2495/EI-V1-N4-461-472.

[2] Regolamento regionale 23 novembre 2017 n.7, Regolamento recante criteri e metodi per il rispetto del principio dell'invarianza idraulica ed idrologica ai sensi dell'articolo 58 bis della legge regionale 11 marzo 2005, n. 12 (Legge per il governo del territorio). https://www.regione.lombardia.it (in Italian).

[3] Becciu, G. & Raimondi, A., Probabilistic analysis of the retention time in stormwater detention facilities. *Procedia Engineering*, **119**, pp. 1299–1307, 2015. DOI: 10.1016/j.proeng.2015.08.951.

[4] Becciu, G. & Raimondi, A., Probabilistic analysis of spills from stormwater detention facilities. WIT Transactions on the Built Environment, vol. 139, WIT Press: Southampton and Boston, 2014. DOI: 10.2495/UW140141.

[5] Becciu, G. & Raimondi, A., Probabilistic modeling of the efficiency of a stormwater detention facility. *International Journal of Sustainable Development and Planning*, **10**(6), pp. 795–805, 2015. DOI: 10.2495/sdp-v10-n6-795-805.

[6] Raimondi, A. & Becciu, G., Probabilistic modeling of rainwater tanks. *Procedia Engineering*, **89**, pp. 1493–1499, 2014. DOI: 10.1016/j.proeng.2014.11.437.

[7] Becciu, G., Raimondi, A. & Dresti, C., Semi-probabilistic design of rainwater tanks: A case study in Northern Italy. *Urban Water Journal*, **15**(3), pp. 192–199, 2018. DOI: 10.1080/1573062X.2016.1148177.

[8] Becciu, G. & Dresti, C., Uncertainties in flow duration curves in anthropized catchments. *WIT Transactions on Ecology and the Environment*, vol. 199, WIT Press: Southampton and Boston, pp. 123–134, 2015. DOI: 10.2495/RAV150111.

[9] Raimondi, F., Dresti, C., Marchioni, M.L., Kian, D., Mambretti, S. & Becciu, G., Integrated strategies for river restoration and land re-naturalization in urban areas: A case study in Milan. *WIT Transactions on the Built Environment*, vol. 194, WIT Press: Southampton and Boston, pp. 23–34, 2020. DOI: 10.2495/FRIAR200031.

[10] Progetto Strategico di Sottobacino del torrente Seveso. www.contrattidifiume.it.

[11] Marchioni, M. & Becciu, G., Experimental results on permeable pavements in urban areas: A synthetic review. *International Journal of Sustainable Development and Planning*, **10**(6), pp. 806–817, 2015. DOI: 10.2495/SDP-V10-N6-806-817.

[12] Raimondi, A. & Becciu, G., Performance of green roofs for rainwater control. *Water Resources Management*, **35**, pp. 99–111, 2021. DOI: 10.1007/s11269-020-02712-3.

[13] Rizzo, A. et al., Constructed wetlands for combined sewer overflow treatment: A state-of-the-art review. *Science of the Total Environment*, 2021. DOI: 10.1016/j.scitotenv.2020.138618.

[14] Raimondi, A. & Becciu, G., On pre-filling probability of flood control detention facilities. *Urban Water Journal*, **12**(4), pp. 344–351, 2015. DOI: 10.1080/1573062X.2014.901398.
[15] Rossman, L.A., *Storm Water Management Model User's Manual*, version 5.0. National Risk Management Research Laboratory, Office of Research and Development, US Environmental Protection Agency: Cincinnati, p. 276, 2010.

MODELING THE BEHAVIOR OF RAIN GARDENS

HARSHO SANYAL[1], YARA YASSER ELBOROLOSY[2] & JOSEPH CATALDO[1]
[1]The Cooper Union for the Advancement of Science and Art, USA
[2]United States Army Corps of Engineers, USA

ABSTRACT
Globally, many cities have an overreliance on gray infrastructure, which does little to prevent flooding, and further contaminates the runoff, exacerbating its environmental effects. Green infrastructure (GI) has emerged as an alternative, which works by utilizing natural hydrology through the incorporation of permeable surfaces, retention, and water filtration. GI practices can include structures such as green roofs, permeable pavements, and rain gardens. Rain gardens are among the highest-priority GI practices in urban areas, due to their soil suitability, space requirements, and high nitrogen removal from stormwater. They are built with layers of gravel, geotextile fabric, engineered soil, and a hardy plant species. Stormwater is captured and retained, before eventually exiting via outlets, sewer piping, or evapotranspiration. This analysis of rain gardens includes an examination of stormwater infiltration and storage of a hypothetical design at a NYC location. Furthermore, "greened acres" (equivalent impervious area for stormwater) are presented for the project, to highlight the effectiveness of stormwater quantity management. Additionally, runoff quality was experimentally measured using a physical model of a rain garden, where nitrogen concentration is measured before and after to determine change in runoff quality. This concept is based on a previous green roof study, where nitrogen removal in soil was determined after wood mulch (which is known to use up excess nitrogen) was added. This previous experiment showed a 22% reduction in nitrates after a week of soil mixed with wood mulch. The modelled rain garden was able to capture 80% of the nitrogen in the inflow for a single storm event, but only 60% for a second storm event. Additionally, previously present nitrogen leached from the soil into the runoff, significantly reducing its effectiveness. Further research of the nitrogen removal by rain gardens is recommended.
Keywords: rain garden, green infrastructure, nitrogen capture, runoff, stormwater, green roof, stormwater management.

1 INTRODUCTION

Beginning at the turn of the century, the New York government began to prioritize improved water and air quality. From 2002 to 2010, $6 billion was invested in water quality alone. In 2010, the New York City Department of Environmental Protection (NYCDEP) released its first Annual Green Infrastructure report, initiating the focused development of green infrastructure in New York City. This would include porous pavements, green and blue roofs, bioswales, and many more [1].

The purpose of this report is to study and develop a green infrastructure project in New York City. This GI practice is a rain garden. Rain gardens can either be connected to sewer systems, or separate, and serve the primary purpose of reducing sewage overflows, and filtering the water which flows into the sewers. The rain garden is analyzed to determine its effectiveness of water retention and filtration, using resources provided by the NYCDEP.

To further study a rain garden's effect on soil retention and filtration, a model rain garden was designed and constructed. Various parameters were recorded and analyzed to experimentally determine its effectiveness at nitrogen removal. This study was carried out in the Fluids Laboratory at The Cooper Union for the Advancement of Science and Art.

WIT Transactions on Ecology and the Environment, Vol 250, © 2021 WIT Press
www.witpress.com, ISSN 1743-3541 (on-line)
doi:10.2495/WRM210131

2 BACKGROUND

2.1 Emergence of green infrastructure

The environmental health of the globe faces many struggles, whether it be with water quality, ecosystem health, or rising temperatures from climate change. Many of these can be attributed to a heavy overreliance by societies on grey infrastructure, the conventional stormwater management system. Gray infrastructure, which uses impervious structures such as concrete, storm drains, and pipes, transfers (often untreated) stormwater from the streets into natural waterways via the sewer system. Such designs do little to ease flood management, and actually reduce runoff water quality as contaminants such as pathogens, heavy metals, and sediment are picked up from impervious surfaces [2]. The Natural Resources Defense Council reports that 10 trillion gallons of this polluted runoff eventually enters natural bodies of water, harming the ecosystem and water source for human consumption.

Green infrastructure (GI) has emerged as a viable alternative. It is designed to imitate natural hydrology, incorporating porous surfaces which can absorb up to 90% of the stormwater runoff that reaches them. In absorbing this runoff, GI reduces the stress on sewer systems and mitigates risk of flooding [2]. Furthermore, the quality of the runoff is filtered through processes such as adsorption, filtration, and plant uptake. GI can also generate positive effects such as air quality improvement (by absorbing pollutants from the air), preservation of ecological habitats (by reducing erosion-causing runoff], and increased resilience against climate change (by combating urban heat island effect and reducing temperatures, while absorbing potential floodwater) [3].

2.2 Criteria for green infrastructure

There are numerous forms of GI which can be incorporated in an urban setting, which are useful for stormwater management. Criteria for these can include space availability and soil suitability. Fig. 1 from NYCDEP's Stormwater Management Program shows numerous green infrastructure practices such as green roofs, sand roofs, and rain gardens prioritized by these parameters.

Furthermore, total nitrogen (TN) removal can be used as a measurement of a GI type's effectiveness. Nitrogen occurs naturally in aquatic ecosystems, but in excess it can be harmful. While green roofs have a TN removal rate of 35%, sand filters are more effective at 40%, and rain gardens are expected to remove 100% of total nitrogen [5].

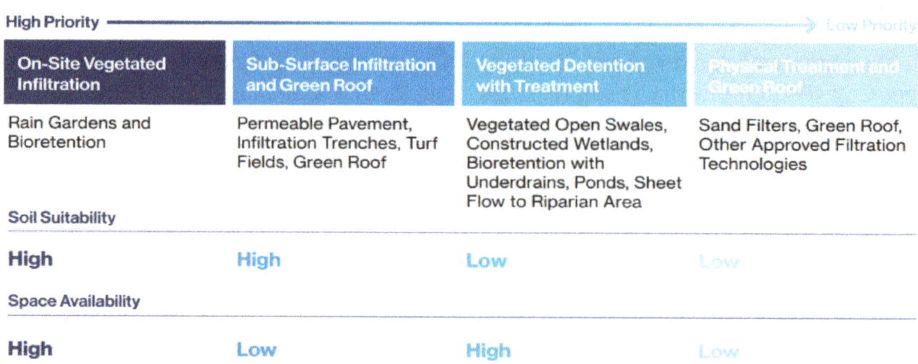

Figure 1: Hierarchy of stormwater practices by preference [4].

Due to their high efficiency and suitability, rain gardens were chosen for the purpose of this project. Bioswales and rain gardens are lengths of vegetation usually found on sidewalks. Functionality-wise, they are nearly identical, with the key difference being that bioswales have 4 feet of equally divided engineered soil and stone subbase underneath (with 9 feet separating the surface from the bedrock), whereas rain gardens have only 3 feet of engineered soil and stone subbase (with 7 feet between the soil surface and bedrock) [6]. The primary purpose of rain gardens and bioswales is to collect rainwater from impervious surface runoff, preventing it from putting excessive stress on sewer systems. Only when a rain garden/bioswale has reached full capacity, will stormwater begin overflowing from the outlet, and enter the sewers via the catch basin [3].

3 PREVIOUS MODELLED GREEN ROOF STUDY

The model rain garden study was preceded by another study at the Cooper Union, which examined a model green roof based on the green roof at New York City's Javits Center. The purpose of that study was to investigate the effects on soil of excess nitrogen or trace heavy metals, the latter of which can be linked to Javits' proximity with smoke from the Lincoln Tunnel. Strategies were then designed for remediation of both contaminants, but due to time and resource constraints, only the nitrate studies were carried out in the laboratory.

3.1 Research findings

The effects of nitrates and trace heavy metals were studied from a variety of sources, and compiled in an unpublished report. As with other salts, nitrates in soil increase osmotic pressure outside of the plant roots, reducing the amount of water they can take in against the concentration gradient. Furthermore, excess nitrogen can be leached out of the soil by runoff water, which can enter aquatic systems and cause rapid algal growth. This process, known as eutrophication, can lead to reduced dissolved oxygen levels and limited penetration of sunlight into the water. Regarding heavy metals, harmful effects include a change in enzyme production, reducing plant biodiversity and richness. Additionally, heavy metals can accumulate in the food chain, affecting animals and humans.

Processes for soil remediation vary. For nitrate remediation, usage of wood mulch to tie up excess nitrogen was seen to be an effective strategy. Heavy metals were primarily managed by bioremediation, planting crops which have rapid growth, far reaching roots, high biomass, and can handle high levels of trace metals. These plants soak metals out of the soil, before being extracted and incinerated in a closed furnace [7].

3.2 Setup

The study was designed to examine the effect of wood mulch on nitrate concentration in soil. It was carried out over the month of December 2020 in the Fluids Laboratory at 41 Cooper Square, New York, NY. The project setup is shown in Fig. 2.

Initially, two 2 feet × 4 feet boxes had holes drilled along one of the shorter ends. This end would be covered by a PVC pipe, to allow a path for runoff. These boxes were coated with felt to cover any cracks, after which they were both filled up to a height of 1 foot with soil. The upper layer of soil for Box A (the experimental box) was mixed with four quarts of wood mulch. Each box then had two moisture probes placed at the points shown in Fig. 2. The first trial date established nitrogen levels in runoff for both boxes, and the second trial date determined the effect of wood mulch on nitrogen capture [7].

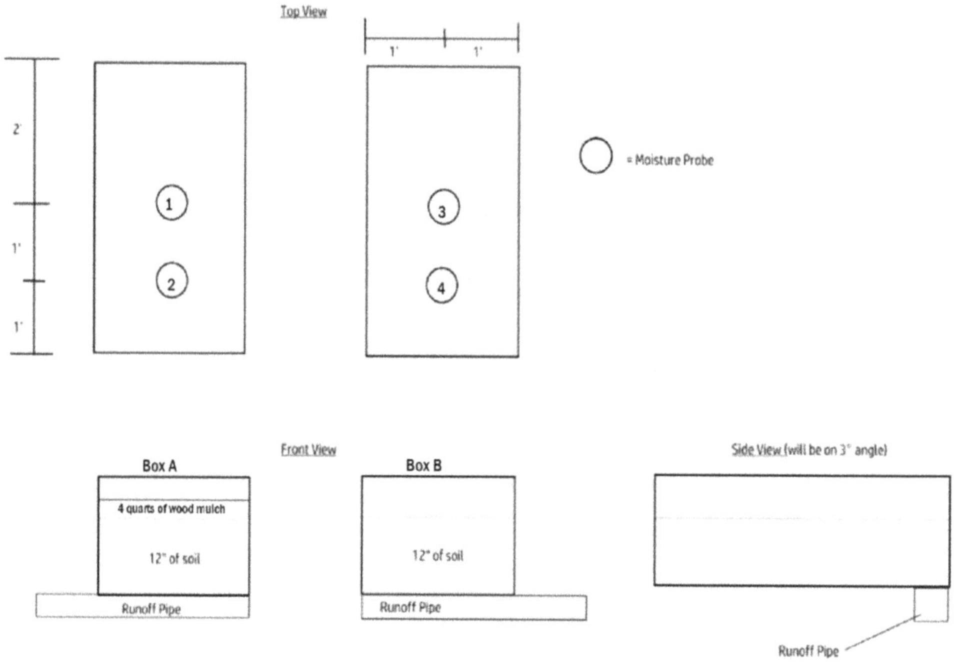

Figure 2: Diagram of experimental setup for model green roof base.

3.3 Analysis

The parameters measured included soil moisture, soil temperature, and nitrate concentration of runoff. The experimental trials were carried out on 1 December 2020 and 11 December 2020. As seen in Table 1, soil moisture increased after running water through the system, while soil temperature remained consistent on the first date and increased slightly on the second date. Table 2 shows how runoff tended in larger quantities from the box with wood

Table 1: TEROS soil probe data for tap water inflow into green roof.

		Trial 1 (1 December 2020)		Trial 1 (11 December 2020)	
	Probe number	Water content (m^3/m^3)	Soil temperature (°C)	Water content (m^3/m^3)	Soil temperature (°C)
Before inflow	1	0.108	20.40	0.109	19.60
	2	0.164	20.40	0.065	19.60
	3	0.049	20.50	0.035	19.50
	4	0.038	20.60	0.079	19.50
After inflow	1	0.134	20.80	0.130	19.80
	2	0.055	20.40	0.069	20.80
	3	0.053	20.50	0.055	19.60
	4	0.064	21.60	0.051	21.20

Table 2: Flow data for tap water inflow into rain garden.

		Trial 1 (1 December 2020)		Trial 1 (11 December 2020)	
		Quantity (mL)	Nitrate concentration (ppm)	Quantity (mL)	Nitrate concentration (ppm)
Wood mulch-mixed soil	Inflow	10,500	N/A	10,500	N/A
	Runoff	3,750	1,333	3,000	1,042
Only soil	Inflow	10,500	N/A	10,500	N/A
	Runoff	1,850	833	1,180	750

mulch mixed in (likely due to the additional mixing of soil, leading to larger voids). Table 2 also illustrates nitrate/nitrite levels decreasing slightly more in the soil from the first to the second date in the wood mulch-mixed box, in comparison to the milder decrease in the control box. This variation may be attributed to a lack of precision associated with nitrate test strips, as the samples were diluted prior to nitrate measurement. The standard deviations of the nitrate measurements ranged from 144 ppm (for both soils on 1 December 2020) to 72 ppm (for both soils on 11 December 2020). Overall, this study provided evidence for the claim that wood chips can reduce the amount of nitrates in soil [7].

4 RAIN GARDEN DESIGN

The purpose of this section is to initialize the design for a rain garden at a real-world location in New York City, choosing a site from the GIS green infrastructure map available publicly [8]. This design was incorporated at 30-02 48th Ave, Long Island City, NY 11101, where there is currently a 13 feet × 5 feet right-of-way bioswale. It is a Type A design, which means it is built on stone columns [9]. Columns are typically used during rain garden/bioswale construction for affordability, ease of use, or when the water table is too high. However, they make sacrifices when it comes to load handling and retention.

The alternative design presented by this report is for a hydraulically connected rain garden (HCRG), which consists of two rain gardens connected by an underground layer of open base. This is a strategy used to increase the size (and therefore retention capacity) of rain gardens, while complying with the imposed 20 feet maximum. The separation between the two rain gardens is 5 feet long, allowing for pedestrians to cross. The planting schedule is taken for a mixed sun/shade site at a west facing location [9]. A diagram of the HCRG is seen in Fig. 3.

Each half of the HCRG will have transverse dimensions of 20 feet × 5 feet, leading to a combined surface area of 200 square feet, corresponding to the local impervious tributary area of 7,450 feet. Furthermore, it will be a Type C practice, and use a stormwater chamber, which will allow for bonus retention and better be able to handle any loads. Storage calculations were done, assuming maximum retention, with NYCDEP's stormwater calculation table for on-site GI Practices [10]. With 1.5 feet of porous engineered soil and 1.5 feet of sub gravel, the HCRG is designed to allow 238.3 cubic feet of infiltration, with an additional 287.5 cubic feet of storage, for a total of 525.8 cubic feet of rainfall captured. This will account for 38% of the corresponding impervious surface for a 1.25 inches rainfall storm.

Additional calculations were done to determine several parameters for the rain garden practice. Flow due to the New York 5-year storm of 5.95 inches/hour was found to be 0.81 cfs of inflow. The effective infiltration depth of water was determined to be

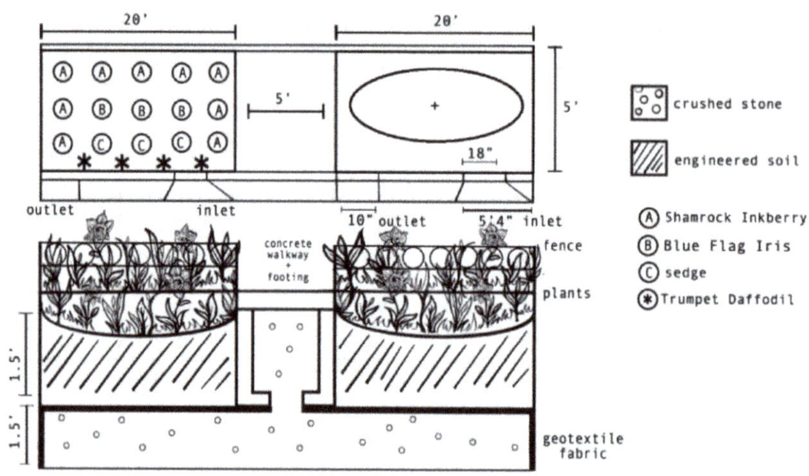

Figure 3: Overhead and cross-sectional views of hydraulically connected rain garden, including planting schedule and grading plan. Low point of grade is 3 inches below edges.

17.25 inches, and designed to drain fully over 57.5 hours. With an inlet with a base of 18 inches and a curb height of 6 inches, the maximum allowable flow was calculated to be 17.24 cfs [11]. Finally, the greened acres were determined. Greened acres signify the area impervious land which, when covered evenly with 1 inch of rainwater, manages an equivalent amount of water to the given GI practice. For this rain garden, the greened acres were found to be 0.145 acres [12].

5 MODELLED RAIN GARDEN SETUP

The theory behind the hypothetical rain garden was combined with the experimental procedure of the green roof model to begin a further study. This study examines various parameters regarding stormwater retention and filtration through the construction and analysis of a model rain garden. The study was carried out over the month of March 2021 in the Fluids Laboratory at 41 Cooper Square, New York, NY. The rain garden model was constructed as per NYCDEP Green Infrastructure standards [9]. The model layers during construction can be seen in Fig. 4.

(a) (b) (c) (d)

Figure 4: Layers of model rain garden. (a) 1.5 cubic feet of gravel; (b) Geotextile fabric; (c) 1.5 cubic feet of soil; and (d) Six sweet basil and two blue flag irises.

During setup, holes were first drilled at the base of one of the shorter ends of the box. The PVC pipe was then affixed over this, to allow for a path for runoff. Beyond this point, the model setup mirrored that of a real rain garden. Rain gardens are typically constructed from the bottom up, with a crushed stone base, a layer of geotextile fabric, a layer of engineered soil equivalent to the volume of the crushed stone, and finally a layer of plants [6]. For the model 1.5 cubic feet of gravel was poured over the base, after which it was covered with a layer of geotextile fabric, which was then coated with 1.5 cubic feet of engineered soil, which was then planted with two blue flag iris plants and six sweet basil plants. This process can be seen in Fig. 4. A clip-on lamp which featured a combination of blue and red light (for maximum photosynthesis) was affixed to the box, and set to run for 12 hours daily to allow for plant growth. This lamp, as well as the runoff pipe, can be seen in Fig. 5.

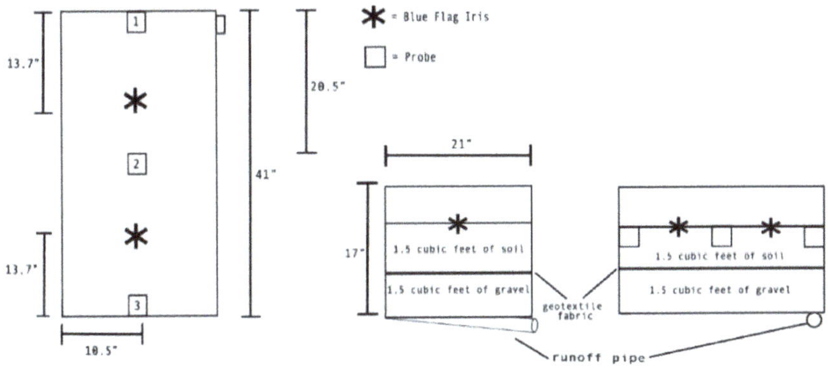

(a)

(b)

Figure 5: Views of model rain garden. (a) Dimensioned diagram showing locations of probes and plants; and (b) Photograph of complete rain garden with visible probe cables, lamp, and runoff pipe.

While the setup and procedure were similar to the previously mentioned green roof model study, several improvements were incorporated to improve the significance of the results. One key difference is the usage of plants in this model. Blue flag iris was chosen as the main plant for stormwater absorption, as is the primary choice of perennial for NYCDEP rain gardens at sites with mixed sun and shade [9]. Sweet basil plants were added, to complement the blue flag iris, but unfortunately did not survive the duration of the experiment due to an irregular watering schedule. TEROS moisture probes were added as seen in Fig. 5.

Furthermore, additional parameters were measured to increase the validity of the study. In addition to soil moisture/temperature, quantity of inflow/runoff, and runoff nitrates, measurements were recorded for pH of inflow/runoff, temperature of inflow/runoff, and electrical conductivity. Electrical conductivity (EC) is important for understanding salt concentrations in soil, which can affect plant growth in the rain garden. EC can also be used to estimate soil nitrate levels in non-saline soils with the formula $N = 140 \times EC$. Nitrogen fertilizer is known to increase salinity, and consequently EC as well [13].

Preliminary runs showed that tap water runoff from the rain garden was not nitrogen free, illustrating that the soil already had a certain amount of nitrogen present (as plants thrive in nitrogen-rich soil). The following strategy was therefore devised to determine the rain garden's capture of nitrogen from contaminated inflow. (1) A baseline nitrate concentration was established from the outflow after running tap water through the rain garden. (2) After this was known, a solution with known nitrogen levels was run through the rain garden, and this outflow also had its nitrogen levels determined. (3) The difference between the soil baseline nitrogen + the known nitrogen levels and the runoff from the nitrogen-contaminated water through the rain garden was the amount of nitrogen which the rain garden capture.

6 RESULTS

For all trials, 9,000 mL of water was poured over the rain garden, corresponding to 1 inch/hour of rainfall, which is the 1 hour 2-year storm in New York City [14]. Trial 1 used tap water, Trial 2 used water with 1,000 ppm of nitrogen for a singular storm event, and Trial 3 used 1,000 ppm nitrogen inflow after an initial 1 inch/hour tap water storm 2 hours previously. Of these inflows, 46%, 54%, and 59% of the flow emerged for Trial 1, 2, and 3, respectively. The remaining inflow was either absorbed by the plants and exited the system via evapotranspiration, or filled voids in the soil. Evapotranspiration is assumed to be roughly negligible for this experiment [15]. It is seen that with a second storm event, the retention capacity of the rain garden may decrease. Eqn (1) illustrates an example water balance equation for the model. Of the measured parameters, soil moisture behaved as expected. From Tables 3, 5 and 8, all trials had three probes with lower moisture at the beginning, much higher immediately after inflow, and a slight drop following runoff drainage. Trials 1 and 3 had warmer inflow than Trial 2 and this showed an effect in the soil temperature changes; the temperatures gradually rose for Trials 1 and 3 (as seen in Tables 3 and 7), while they dropped during Trial 2 (from Table 5). In Trials 1 and 2, the runoff temperature was 18°C, despite inflow being warmer for Trial 1 and cooler for Trial 2, as seen in Tables 4 and 6, respectively. In all cases, the pH decreased from inflow to outflow, although adding the nitrogen fertilizer decreased the initial pH by 0.5. This shows that the rain garden is slightly acidic, as suits the plant growth. However, the pH only dropped by 0.5 for Trial 3, unlike the first two trials where it dropped by 1, showing that multiple storm events may reduce acidity. Furthermore, despite Trial 3 undergoing a second storm event, the soil water content did not appear to differ drastically from Trial 2. Tables 3, 5, and 8 also display the electrical conductivity, which followed no particular pattern in Trial 1 to hover around the same values at all three data points. For Trials 2 and 3, however, there was a marked increase in EC as expected, as EC corresponds with salinity, which nitrates increase. Regarding nitrogen runoff, 500 ppm, 800 ppm, and 1,000 ppm emerged for Trials 1, 2, and 3, respectively. This shows that the soil for this rain garden had a baseline of 500 ppm, and was able to capture 70% of the nitrogen from singular storm event and only 50% from the second storm. This nitrogen was captured by the soil, and is either consumed by the plants, or transformed into nitrous gas via denitrification (as occurs in standard bioretention systems) [16]. A sample nitrogen balance equation was done in eqn (2), accounting for the nitrogen already present in the rain garden prior to contamination. This highlights the fact that the base nitrogen from the rain garden which contaminated the runoff cannot be ignored; 1,000 ppm flow entered the rain garden, and 800 ppm emerged after one storm, while 1,000 ppm emerged after two storms. This would bring the nitrogen reduction percentage to 20% for one storm and 0% for two storms. This model does not approach 100% nitrogen capture as cited by NYCDEP.

Example equation for water balance of model rain garden. Values from Trial 2 are used.

$$P = R + ET + \Delta S \tag{1}$$

$$9{,}000\ mL = 4{,}850\ mL + 0\ mL + 4{,}150\ mL$$

Equation for nitrogen balance of model rain garden. Values from Trial 2 are used.

$$N_{inflow} + N_{soil\ baseline} = N_{runoff} + N_{soil\ storage} \tag{2}$$

$$9\ g\ N + 21.25\ g\ N = 3.88\ g\ N + 26.37\ g\ N$$

Table 3: TEROS soil probe data for tap water inflow into rain garden.

Trial 1 (30 March 2021)				
	Probe number	Water content (m³/m³)	Soil temperature (°C)	Bulk electrical conductivity (mS/cm)
Before inflow	1	0.171	17.70	0.438
	2	0.244	16.30	0.311
	3	0.152	17.10	0.585
Immediately after inflow	1	0.302	20.60	0.438
	2	0.367	18.50	0.216
	3	0.293	18.10	0.989
After runoff drainage	1	0.271	20.90	0.407
	2	0.338	19.80	0.233
	3	0.256	18.90	0.958

Table 4: Flow data for tap water inflow into rain garden.

Trial 1 (30 March 2021)				
	Quantity (mL)	pH	Temperature (°C)	Nitrate concentration (ppm)
Inflow	9.000	6.5	22	5
Runoff	4,150	7.5	18	500

Table 5: TEROS soil probe data for nitrogen-contaminated water inflow into rain garden.

Trial 2 (7 April 2021)				
	Probe number	Water content (m^3/m^3)	Soil temperature (°C)	Bulk electrical conductivity (mS/cm)
Before inflow	1	0.277	18.50	0.108
	2	0.336	17.90	0.113
	3	0.260	18.40	0.160
Immediately after inflow	1	0.318	18.1	0.164
	2	0.378	17.8	0.205
	3	0.292	17.2	0.178
After runoff drainage	1	0.296	16.30	0.134
	2	0.354	16.20	0.142
	3	0.275	17.10	0.157

Table 6: Flow data for nitrogen-contaminated water inflow into rain garden.

Trial 2 (7 April 2021)				
	Quantity (mL)	pH	Temperature (°C)	Nitrate concentration (ppm)
Inflow	9,000	6	17	1,000
Runoff	4,850	7	18	800

Table 7: TEROS soil probe data for nitrogen-contaminated water inflow into rain garden, following an initial storm event.

Trial 3 (26 April 2021)				
	Probe number	Water content (m^3/m^3)	Soil temperature (°C)	Bulk electrical conductivity (mS/cm)
Before inflow	1	0.280	19.80	0.129
	2	0.322	19.40	0.238
	3	0.259	19.40	0.167
Immediately after inflow	1	0.317	20.00	0.186
	2	0.359	19.60	0.274
	3	0.306	19.50	0.214
After runoff drainage	1	0.294	21.00	0.141
	2	0.336	20.90	0.238
	3	0.273	20.70	0.167

Table 8: Flow data for nitrogen-contaminated water inflow into rain garden, following an initial storm event.

Trial 3 (26 April 2021)				
	Quantity (mL)	pH	Temperature (°C)	Nitrate concentration (ppm)
Inflow	9,000	6	22	1,000
Runoff	5,250	6.5	20	1,000

7 CONCLUSION

The initial purpose of this report was to design a theoretical rain garden, as an improved alternative for an existing green infrastructure practice in Long Island City, New York. This alternative design was for a Type C hydraulically connected rain garden, and was found to be able to capture 38% of all of the rainwater which the corresponding impervious area would receive during a 5-year storm. Its corresponding green acreage was calculated to be 0.145 acres of impervious surface.

Following this design, an experiment was designed to research the effectiveness of rain gardens through the construction and study of a model rain garden practice. Drawing from a previous study which examined nitrate treatment in a green roof model, this rain garden study was structured to determine and analyse a rain garden's capability for water retention and filtration. However, there was a certain amount of nitrogen already present in the soil used, most likely added to promote plant growth. During the trials, it was seen that nitrogen leached from the soil into the runoff. While the rain garden was able to reduce nitrogen from contaminated inflow to an extent, it was unable to come close to the 100% reduction cited by the NYCDEP. Further testing of this claim is recommended, to determine the true filtration efficiency of rain garden practices.

REFERENCES

[1] NYC Department of Environmental Protection (NYCDEP), NYC Green Infrastructure Plan, 2010. https://www1.nyc.gov/assets/dep/downloads/pdf/water/stormwater/green-infrastructure/nyc-green-infrastructure-plan-2010.pdf. Accessed on: 1 Sep. 2020.

[2] United States Environmental Protection Agency, Benefits of green infrastructure. https://www.epa.gov/green-infrastructure/benefits-green-infrastructure. Accessed on: 1 Oct. 2020.

[3] Denchak, M., Green infrastructure: How to manage water in a sustainable way. National Defense Research Council, 2019. https://www.nrdc.org/stories/green-infrastructure-how-manage-water-sustainable-way. Accessed on: 1 Feb. 2020.

[4] NYC Stormwater Management Program, NYCDEP, 2020. https://www1.nyc.gov/assets/dep/downloads/pdf/water/stormwater/ms4/nyc-swmp-plan-full.pdf. Accessed on: 1 Sep. 2020.

[5] NYCDEP, New York City Stormwater Design Manual. 2016. https://www1.nyc.gov/assets/dep/downloads/pdf/water/stormwater/ms4/stormwater-manual-final.pdf. Accessed on: 1 Oct. 2020.

[6] NYCDEP, NYC Green Infrastructure: 2014 Annual Report. 2014. https://www1.nyc.gov/assets/dep/downloads/pdf/water/stormwater/green-infrastructure/gi-annual-report-2014.pdf. Accessed on: 1 Jun. 2020.

[7] Sanyal, H., Crop Nitrates Research. Unpublished manuscript, The Cooper Union for the Advancement of Science & Art, 2020.

[8] NYCDEP, DEP Green Infrastructure Program Map. https://www.arcgis.com/home/webmap/viewer.html?webmap=a3763a30d4ae459199dd01d4521d9939&extent=-74.3899,40.497,-73.3757,40.9523. Accessed on: 1 Aug. 2020.

[9] NYCDEP, Standard designs and guidelines for green infrastructure practices, 2020. https://www1.nyc.gov/assets/dep/downloads/pdf/water/stormwater/green-infrastructure/green-infrastructure-standard-designs.pdf. Accessed on: 1 Aug. 2020.

[10] NYCDEP, Stormwater calculation table for on-site GI practices. https://www1.nyc.gov/assets/dep/downloads/pdf/water/stormwater/green-infrastructure/nyc-green-infrastructure-onsite-design-manual-appendix-d-stormwater-calculation-table.xlsx. Accessed on: 1 Aug. 2020.

[11] NYCDEP, NYC Green Infrastructure On-site Design Manual, 2019. https://www1.nyc.gov/assets/dep/downloads/pdf/water/stormwater/green-infrastructure/nyc-green-infrastructure-onsite-design-manual-v1.pdf. Accessed on: 1 Aug. 2020.

[12] NYCDEP, NYC Green Infrastructure 2018 Annual Report, 2018. https://www1.nyc.gov/assets/dep/downloads/pdf/water/stormwater/green-infrastructure/gi-annual-report-2018.pdf. Accessed on: 1 Jul. 2020.

[13] United States Department of Agriculture, Soil electrical conductivity. https://www.nrcs.usda.gov/Internet/FSE_DOCUMENTS/nrcs142p2_053280.pdf. Accessed on: 1 Mar. 2021.

[14] Cornell University, Intensity duration frequency curves for New York state. http://ny-idf-projections.nrcc.cornell.edu/. Accessed on: 21 Apr. 2021.

[15] Alvizuri, J., Cataldo, J., Smalls-Mantey, L. & Montaldo, F., Intensity duration frequency curves for New York state. Cornell University, Green Roof Thermal Buffering and Buildings, Period 151, 2017.

[16] Stander, E.K. & Borst M., Bioretention monitoring: Designing rain gardens to promote nitrate removal. 2008 International Low Impact Development Conference, 2008.

SECTION 5
EROSION AND SEDIMENT TRANSPORT

ESTIMATING STORAGE CAPACITY LOSS DUE TO SEDIMENTATION FOR MURUM RESERVOIR IN SARAWAK, MALAYSIA

JOYCE JANGGU[1], SUSIE NADYA[1] & MUBASHER HUSSAIN[2]
[1]Sarawak Energy Berhad, Malaysia
[2]Stantec UK Limited, UK

ABSTRACT
Sarawak state of Malaysia has three large storage dams conserving fresh water for the mega hydro plants, i.e. Bakun Hydroelectric Plant (HEP) (2,400 MW), Murum HEP (944 MW) and Batang Ai HEP (108 MW). Murum HEP and Bakun HEP are cascaded development where Murum HEP regulates about 20% of Bakun HEP catchment. The catchment of Bakun HEP is experiencing extensive land development through logging activities and palm oil plantation. Due to these activities, the rate of sedimentation should be assessed in all major river tributaries flowing into Murum Reservoir. The Hydro team has been conducting sedimentation sampling in these river tributaries over the last few years and the total suspended sediment load in the river flow is being analysed. The objective of this study is to estimate the storage capacity loss at Murum reservoir by analysing the sediment build-up rate along with time and the expected reservoir life by calculating the sediment trap efficiency for Murum reservoir. The results will also help to highlight the sedimentation issue at Murum reservoir.
Keywords: sedimentation rate, storage capacity loss, reservoir, hydroelectric plant, dam, Bakun HEP, Murum HEP, Sarawak, Malaysia.

1 INTRODUCTION

Sarawak state of Malaysia has three large storage dams conserving fresh water for the mega hydro plants i.e. Bakun Hydroelectric Plant (HEP) (2,400 MW), Murum HEP (944 MW) and Batang Ai HEP (108 MW). Murum and Bakun HEP are cascaded development where Murum HEP regulates about 20% of the Bakun HEP catchment. The catchment of Murum HEP is experiencing some land development through logging activities and palm oil plantation. Due to these activities, the rate of sedimentation should be assessed in all major river tributaries flowing into Murum Reservoir. The Hydro Department team has been conducting sedimentation sampling in the river tributaries of Murum Reservoir since 2015 and the total suspended sediment load in the river flow is being analysed. The objective of this study is to estimate the storage capacity loss at Murum reservoir by analysing the sediment build-up rate along with time and the expected reservoir life by calculating the sediment trap efficiency for Murum reservoir. The results will also help to highlight the sedimentation issue at Murum reservoir.

It is estimated that after the 100 years of operation, the reservoir storage would decrease by 1.6% adopting Churchill curve method and 1.8% decrease in storage loss under Brune Envelop Curve method. The Churchill curve method estimated sediment volume of 150 million cubic metre (MCM) to be deposit in Murum Reservoir in 100 years while Brune Envelop Curve method estimated 165 MCM to be deposit within the same period. The sediment build-up rate for Murum reservoir is very low compared to the average global water storage loss of 1% per year due to sedimentation in the storage facilities [1]. However, it is recommended to implement a catchment management plan to minimise the sedimentation and erosion in the Murum catchment.

WIT Transactions on Ecology and the Environment, Vol 250,
www.witpress.com, ISSN 1743-3541 (on-line)
doi:10.2495/WRM210141

2 BACKGROUND

Dams interrupt the continuity of sediment transport through rivers resulting in reservoir storage loss [2]. Reservoir sedimentation is a critical issue worldwide causing a reduction is storage volumes and reservoir efficiency [3]. Sediment transport studies are essential to estimate the total sediment inflow to the Dam, trap efficiency of the reservoir, trapped sediment load and the life of the reservoir with and without flushing or sluicing option [4]. Moreover, sediment study is essential to explore ways to minimize the entry of sediments into the reservoir and techniques to de-silt the deposited sediments to enhance its life [5].

The loss of storage due to sedimentation exacerbates the problem of providing enough storage for rising population with rising aspiration and standards. Annual reservoir storage loss due to sedimentation in United States is 0.22% while 2.3% in China, which is higher than the global average reservoir storage loss [6], [7]. The 20th century was concerned with the development of reservoir storage; more emphasis will be required in the 21st century on the conservation of storage and hence sediment management has become crucial.

Three strategies can be used to control reservoir sedimentation. First, sediment yield delivered from tributary basins can be reduced by erosion control or upstream sediment traps. This includes watershed management practices [8]. Second, sediments may be removed by dredging [9]. However, high costs and environmental consequences of large-scale spoil disposal are major obstacles to large scale dredging. Third, hydraulic regime can be designed to route inflowing sediment through or around the storage pool to minimize deposition, and to remobilize and flush out previously deposited sediments [10].

Murum Reservoir is the second largest storage reservoir in Malaysia after the Bakun Reservoir with a total storage capacity of 9,247 million cubic metre (MCM) at its full supply level of 540 masl. The reservoir filling was started in September 2013 and it was fully impounded in November 2014. There are two main river tributaries, Plieran River and Danum River contributing to the Murum Reservoir with other small river tributaries. A catchment of 2,750 km^2 feeds into the Murum Reservoir. The location of Murum and Bakun HEP is illustrated in Fig. 1 and the location of Plieran and Danum rivers is shown in Fig. 2.

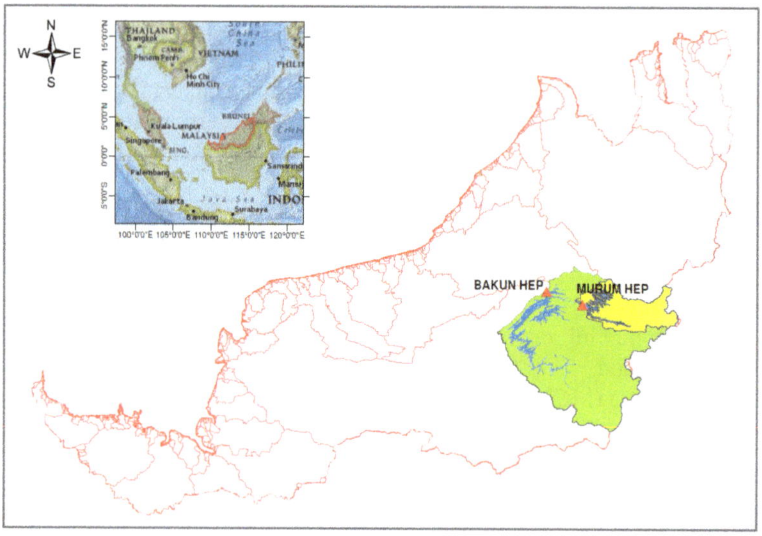

Figure 1: Location of Murum and Bakun HEP on the Malaysian island of Borneo.

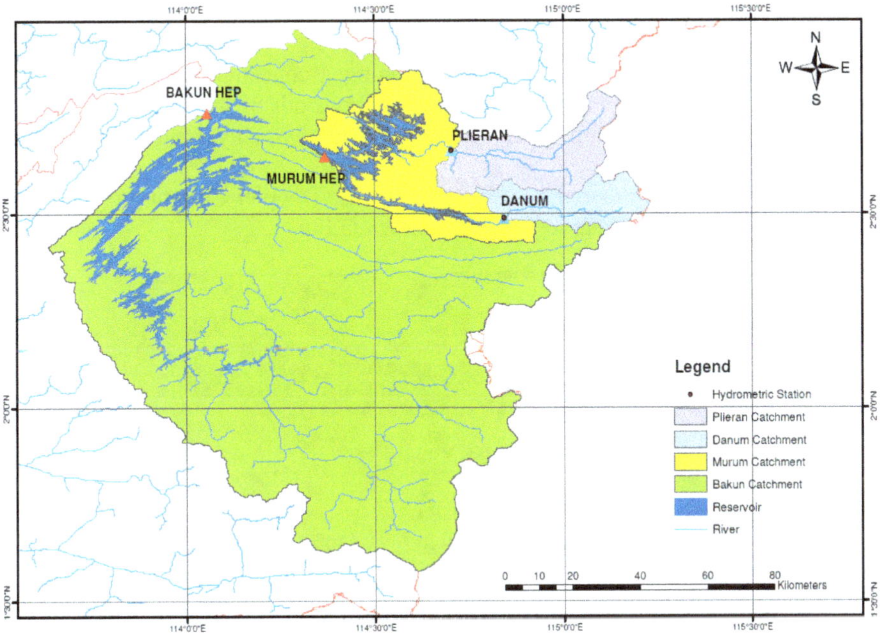

Figure 2: Location of Plieran and Danum hydrometric stations and their catchment boundaries.

3 MATERIALS AND METHODS

The methodology of this study is summarized in Fig. 3.

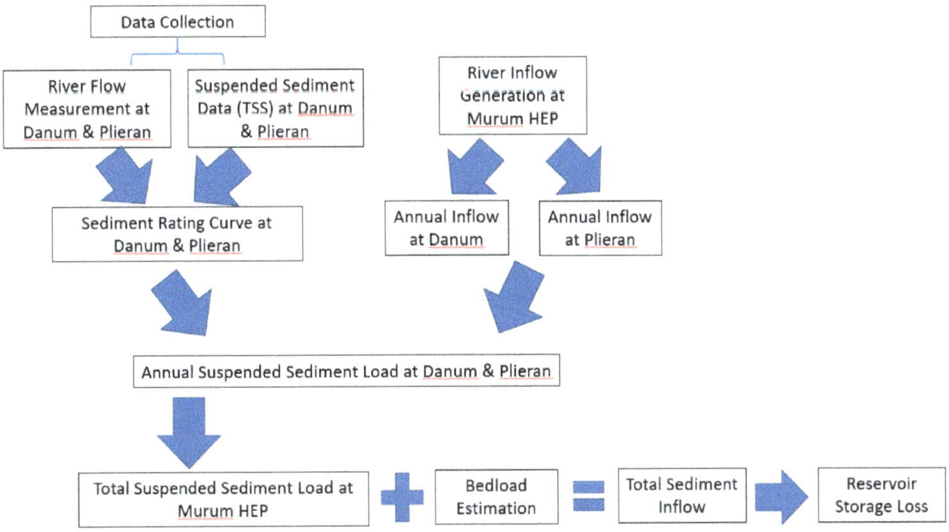

Figure 3: Summary of methods applied in this study.

3.1 Data collection

For this study, the river inflow and suspended sediment concentration data was collected nearby the hydrological station of Plieran and Danum; both of which are upstream river tributaries contributing to Murum Reservoir. The location of the hydrometric stations are as shown in Fig. 2.

The river inflow was measured by using an Acoustic Doppler Current profiler (ADCP); in particular, the Sontek RiverSurveyor M9 as shown in Fig. 4.

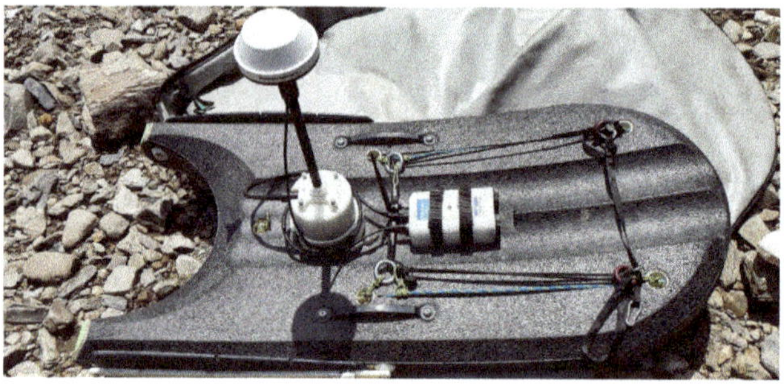

Figure 4: Sontek RiverSurveyor M9 used for inflow measurement.

3.2 River inflows generation

Hydrological model was established for the Bakun catchment to simulate daily river inflows at Bakun HEP. A rainfall-runoff model was developed using the observed rainfall data from the rainfall gauges in the Bakun catchment. The model was calibrated with the recorded river inflows at the Bakun Dam. After the calibration, the model was run for the Murum HEP to generate long term daily inflows at Murum Dam. After the generation of inflows at Murum Dam, the inflows were scaled to Plieran and Danum hydrometric stations as these two stations are the location points for the monitoring of suspended sedimentation data to Murum Reservoir. The inflows at Plieran and Danum were derived from the Murum Dam using the scale factor derived from the catchment area ratio.

3.3 Suspended sediment

Suspended sediment data is being collected at Plieran and Danum hydrometric stations since 2015. 42 water samples were collected from the Plieran and 50 samples from the Danum during the period of 2015 to 2019. River water samples were collected using a 500 ml water bottle at a depth of about 30 cm below the water surface and samples were taken in the middle of the river.

Total suspended solids (TSS) in mg/l and part per million (PPM) were analysed for each water sample to assess the suspended sediment concentration in water. With the TSS results for both stations, the suspended sediment concentration rating curves were developed.

3.4 Suspended sediment rating curves

With the TSS results for both Plieran and Danum hydrometric gauging stations, the suspended sediment concentration rating curves were developed as shown in Figs 5 and 6.

These suspended sediment rating curves were further used to calculate the suspended sediment load at both rivers.

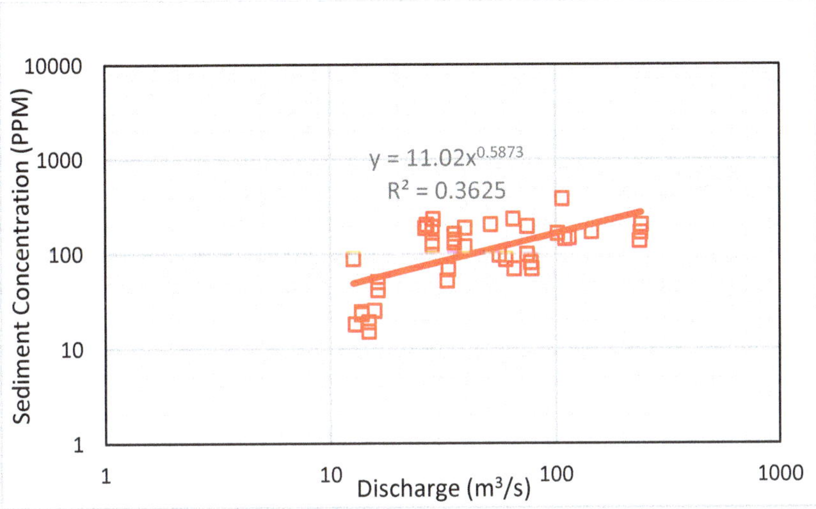

Figure 5: Suspended sediment concentration rating curve for Plieran.

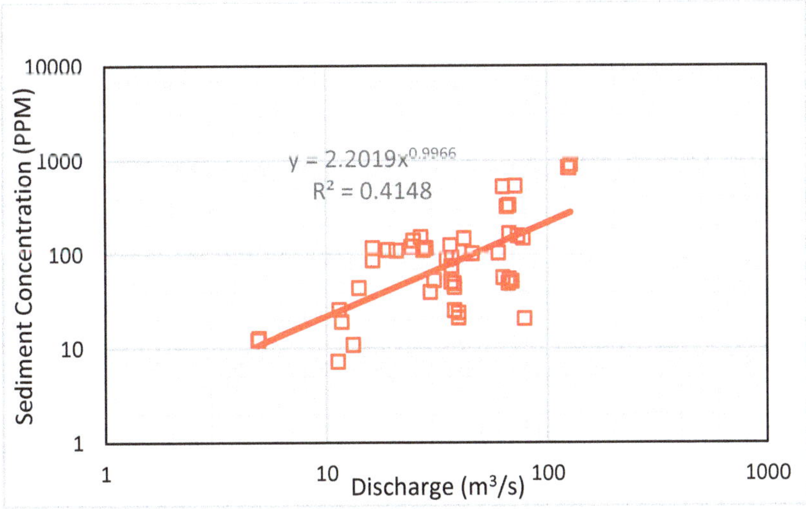

Figure 6: Suspended sediment concentration rating curve for Danum.

3.5 Total suspended sediment load

Annual suspended sediment load at Plieran and Danum were calculated using the sediment rating curves derived for each station and using the long-term daily inflows at both stations. To calculate the suspended sediment load at Murum Dam, the following relationship was used based on the proportionate of each river contributing to the Murum catchment, whereby Q_s represents suspended sediment load:

$$(Q_s)_{Murum} = 0.56\,(Q_s)_{Plieran} + 0.44\,(Q_s)_{Danum}. \quad (1)$$

3.6 Bed load estimation

Currently, the bed load sample at Murum, Plieran and Danum River has not been extracted yet. Therefore, the bed load was estimated using the Modified Einstein procedures for unmeasured sediment load [11].

Modified Einstein procedure is a useful guide for evaluating the unmeasured sediment load. The bed load estimation is shown in Table 1. Five conditions; Condition 1, 2, 3, 4 and 5, are given for defining bed load depending upon suspended sediment concentration and size analysis of stream bed and suspended materials.

According to the Modified Einstein procedure, condition 4 of the Table 1 of bed load correction is very close to the site conditions in which stream bed material is composed of compacted clay, gravel, cobbles or boulders.

Therefore, the adopted percentage of bedload estimation for Murum reservoir according to the Modified Einstein procedure is a conservative value of 15% the total suspended load.

Table 1: Bed load estimation.

Condition	Suspended sediment concentration (mg/L)	Stream bed material	Texture of suspended material	Percentage bed load in terms of suspended load
[1] 1	<1,000	Sand	20% to 50% sand	25 to 150
[1] 2	1,000 to 7,500	Sand	20% to 50% sand	10 to 35
3	>7,500	Sand	20% to 50% sand	5
[2] 4	Any concentration	Compacted clay, gravel, cobbles, or boulders	Up to 25% sand	5 to 15
5	Any concentration	Clay and silt	No sand	<2

Notes:
[1] *Special sampling program for Modified Einstein computations required under these conditions.*
[2] *A bed load sampler such as the Helley-Smith bedload sampler may be used, or computations made by use of two or more of the bedload equations when bed material is gravel or cobble size.*

3.7 Total sediment inflow

Finally, the daily total sediment loads (QS_{total}) to the Murum Dam were computed by adding the bed load, as estimated in Section 3.6, to the suspended sediment load and according to eqn (2):

$$QS_{total} = Q_s + Q_{bedload} = 1.15(Q)_s. \quad (2)$$

3.8 Reservoir storage loss

Trap efficiency of a reservoir primarily depends upon fall velocity of sediment particle and on rate of flow through reservoir. Particle fall velocity may be influenced by the size and shape of particles, viscosity of water, and the chemical composition of water.

Methods for estimating reservoir trap efficiency are empirically based on measured sediment deposits in a large number of reservoirs. The famous methods are M. A. Churchill method [12] and Brune Envelope Curve method [13]. M. A. Churchill developed a relationship between percentage of incoming sediment passing through a reservoir and sedimentation index of reservoir. For this study, the trap efficiency of Murum reservoir is computed using the Churchill curve. Using the trap efficiency results from the Churchill curve, Murum reservoir storage loss is computed using both Churchill curve and Brune Envelope method. Results are compared and presented in Section 4.4.

4 RESULTS AND DISCUSSION

4.1 River inflows generation

According to the method defined in Section 3.2, Figs 7–9 shows the annual inflows at Murum Dam, Plieran and Danum hydrometric stations. The average inflow at Murum Dam is 264 m^3/s, whereas Plieran's average inflow is 75 m^3/s and Danum's average inflow is 50 m^3/s. The average inflows are denoted in Figs 7–9.

4.2 Total suspended sediment load

Using the derived rating curves in Section 3.4, the mean annual suspended sediment load at Plieran was calculated as 0.40 million tonne per year, whereas at Danum it was estimated as 0.35 million tonne per year as shown in Figs 10 and 11.

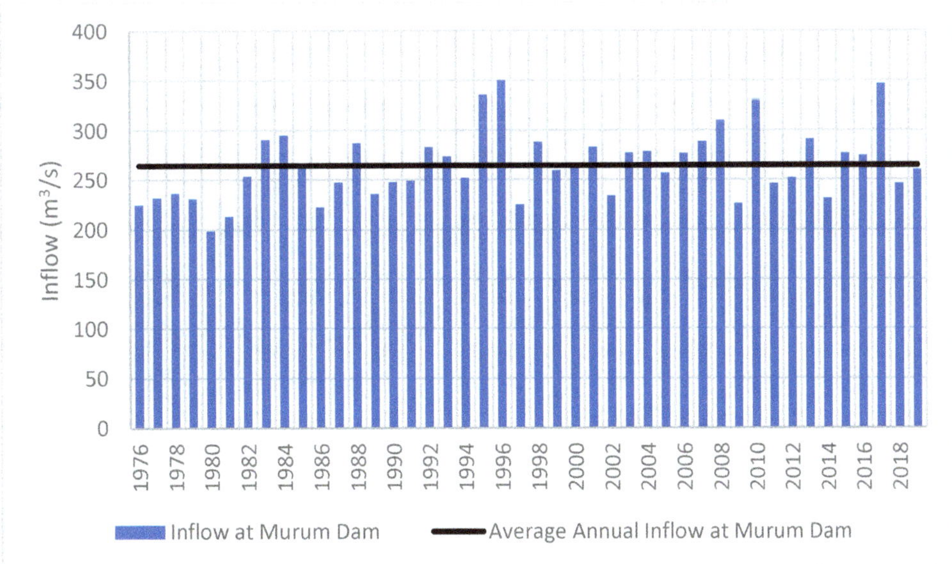

Figure 7: Annual inflows at Murum Dam.

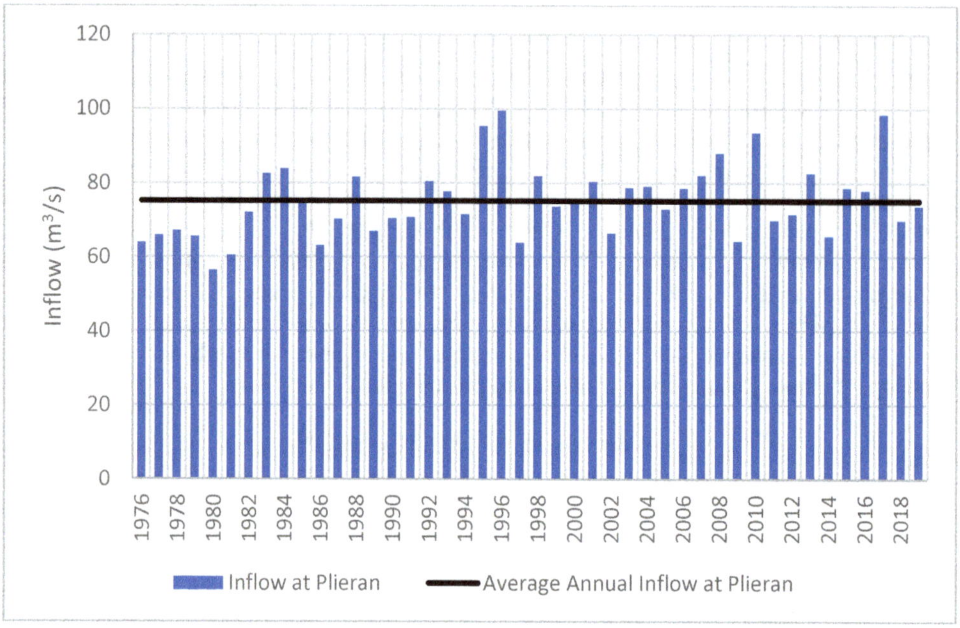

Figure 8: Annual inflows at Plieran.

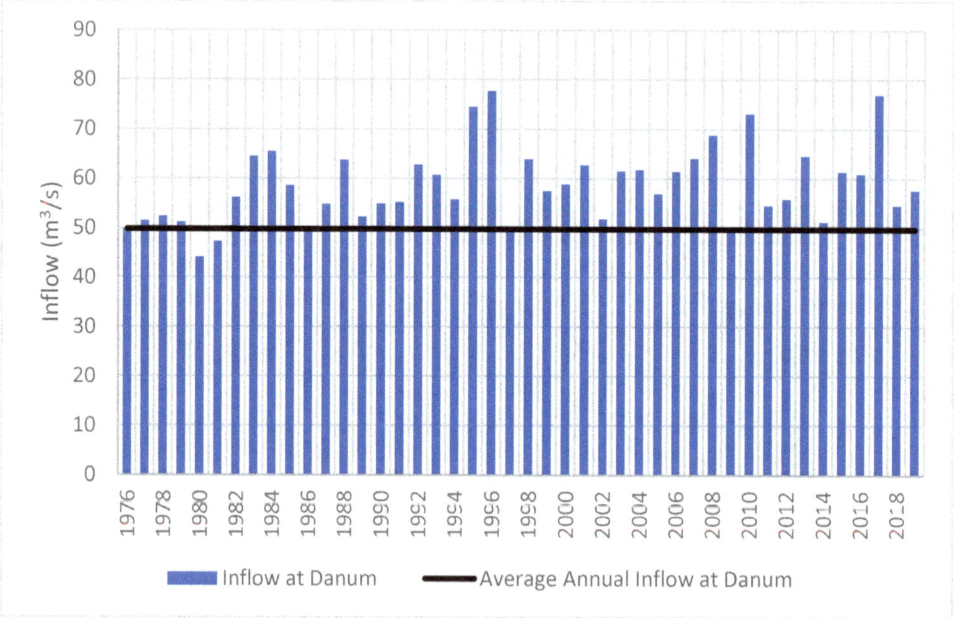

Figure 9: Annual inflows at Danum.

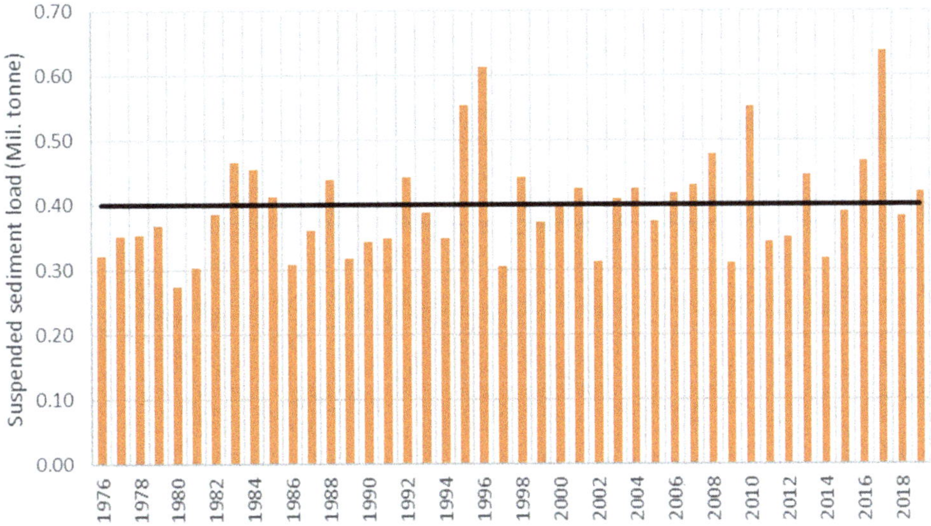

Figure 10: Annual suspended sediment load at Plieran.

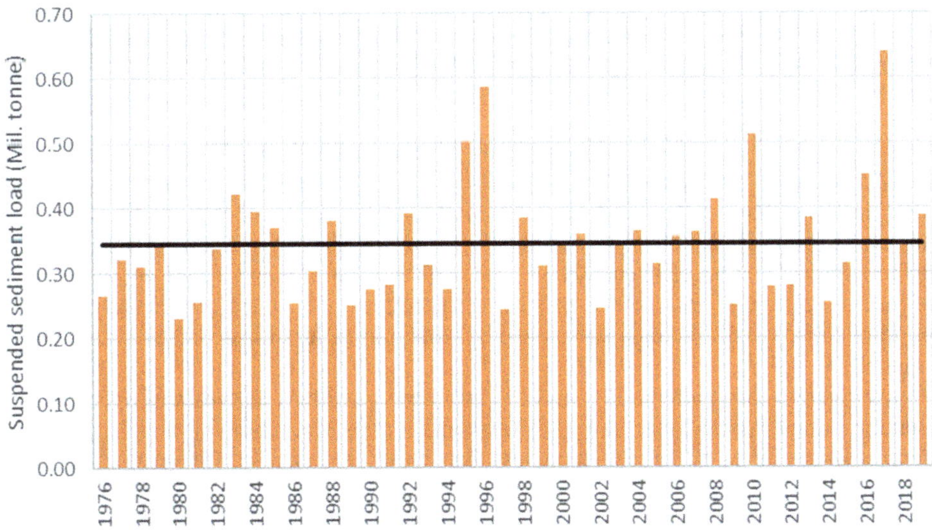

Figure 11: Annual suspended sediment load at Danum.

Using eqn (1) as per Section 3.5, the mean annual suspended sediment load at Murum Dam was calculated as 1.47 million tonne per year as shown in Fig. 12.

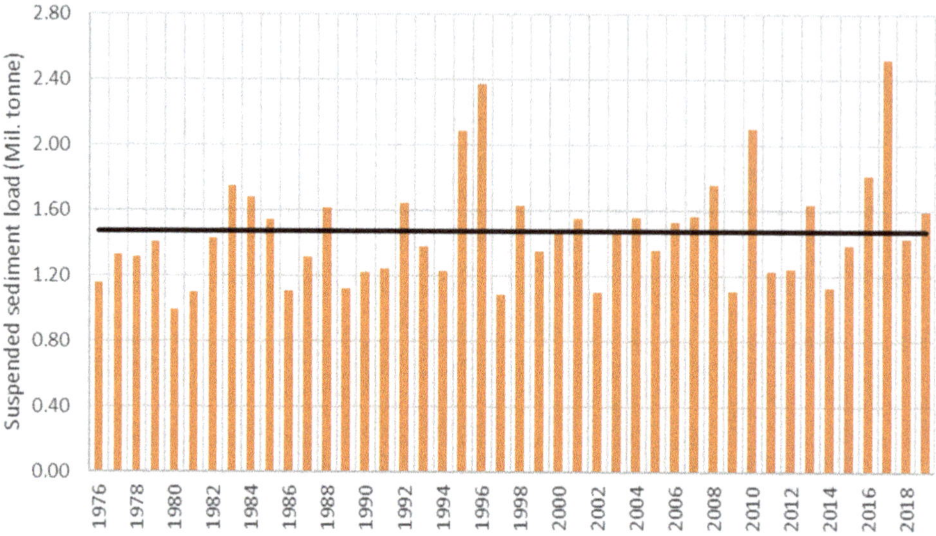

Figure 12: Annual suspended sediment load at Murum Dam.

4.3 Total sediment inflow

The total sediment inflow was calculated according to eqn (2) in Section 3.7. Based on the equation, the total average annual sediment inflow to Murum reservoir comes out to 1.69 million tonnes.

4.4 Reservoir storage loss

The Murum reservoir capacity is 9,247 million cubic metre (MCM), whereas the average annual inflow to Murum Dam is about 8,338 MCM, ratio between the two is 1.11. Consulting Churchill curve [12], trap efficiency comes out equal to 89% while consulting Brune envelop curve method [13], trap efficiency comes out as 98%. The efficiency does not reduce with the increase in age of reservoir as the storage loss does not reduce significantly over the period of 1,000 years. The comparison of reservoir storage loss estimated by both methods is plotted in Fig. 13.

According to the results in Fig. 13, it is estimated that after 100 years of operation, the reservoir storage would decrease by 1.6% when adopting Churchill curve and 1.8% when adopting Brune Envelop Curve method; less than 2% over a period of 100 years of operation and about less than 0.02% annual loss rate, which is about the same as concluded in [14]. On this basis, Murum's reservoir storage loss is low compared to the average global water storage loss of 0.5% to 1% per year due to sedimentation in storage facilities [1]. Compared to a study in Central United States of America, the reservoirs have lost approximately 17% of their original capacity over a period of 52 years, with the highest single reservoir loss of 45% [15]. This equates to an annual loss rate of 0.84%; much higher compared to Murum reservoir's annual loss rate of less 0.02%. Similarly in Brazil, a study found that the storage capacity of the reservoirs in the state of Ceará is experiencing a storage loss of 0.2% per year [16]. Both studies in [15] and [16] concluded that the major cause for reservoir storage loss is due to high sedimentation.

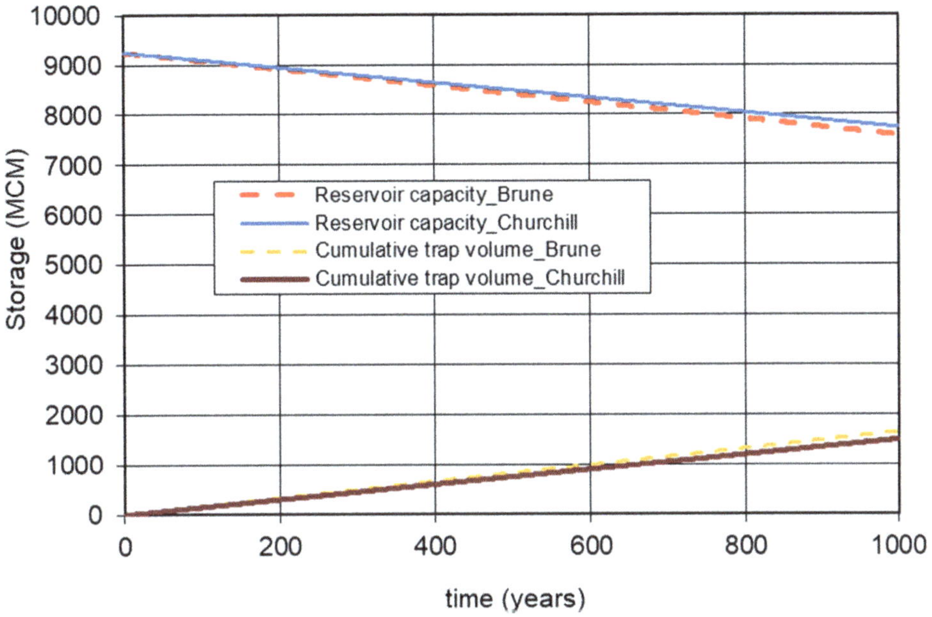

Figure 13: Reservoir storage loss and sediment build-up in Murum reservoir.

The Churchill curve method estimated a sediment volume of 150 MCM to be deposited in Murum reservoir over a period of 100 years while Brune Envelop Curve method estimated 165 MCM to be deposit in Murum Reservoir over the same time period. Using a basis of 1,000 years of operation, the storage loss for Murum Reservoir is expected to be 16% and 18% using the Churchill and Brune Envelop Curve method respectively.

5 CONCLUSION AND RECOMMENDATIONS

It is concluded that the storage loss of Murum Reservoir due to sedimentation is less than 2% over a period of 100 years of operation; about less than 0.02% annual loss rate and similar to what is concluded in [14]. Compared to the average global water storage loss [1], Murum's reservoir storage loss is low.

The estimation in this study is based on a short record of sedimentation data and it is recommended to continue with water sampling to develop matured sediment rating curves at both sites to improve the understanding on the suspended load concentration in the rivers. It is also recommended to conduct bed load sampling by using a sampler such as the Helley-Smith bedload sampler to improve the understanding on the bed load contribution to the sedimentation in the reservoir.

This study used the empirical method to estimate the sedimentation in Murum Reservoir and it is recommended to develop a physical based sedimentation model to understand the catchment behaviour on the sedimentation and erosion in more detail.

It is also recommended to implement a catchment management plan to minimise the sedimentation and erosion in Murum catchment, as well as understanding land-use factor for land-use mapping to identify critical areas prone to high sedimentation rate. In the long-term, it is recommended to manage reservoir sediments by means of bypassing sediment around the reservoir, sluicing or drawdown flushing [2].

REFERENCES

[1] Schellenberg, G., Donelly, C.R., Holder, C. & Ahsan, R., Dealing with sediment: Effects on dams and hydropower generation. *Hydro Review*, **25**(1), 2017.

[2] Kondolf, G.M. et al., Sustainable sediment management in reservoirs and regulated rivers: Experiences from five continents. *Earth's Future*, **2**(5), pp. 256–280, 2014.

[3] Reisenbüchler, M., Bui, M.D. & Rutschmann, P., Reservoir sediment management using artificial neural networks: A case study of the lower section of the Alpine Saalach river. *Water*, **13**(6), 2021.

[4] Gunawan, T.A., Daud, A., Haki, H. & Sarino, The estimation of total sediments load in river tributary for sustainable resources management. *IOP Conference Series: Earth and Environmental Science*, IOP Publishing, 2019.

[5] Dargahi, B., Reservoir sedimentation. *Encyclopedia of Lakes and Reservoirs*, eds L. Bengtsson, R.W. Herschy & R.W. Fairbridge, Springer Netherlands: Dordrecht, pp. 628–649, 2012.

[6] Tan, G., Review and improvement of conventional models for reservoir sediment trapping efficiency. *Heliyon*, **5**, 2019.

[7] Jiang, N. & Fu, L., Problems of reservoir sedimentation in China. *Chinese Geographical Science*, **8**(2), pp. 117–125, 1998.

[8] Xin, Z., Ran, L. & Lu, X.X., Soil erosion control and sediment load reduction in the Loess Plateau: Policy perspectives. *International Journal of Water Resources Development*, **28**(2), pp. 325–341, 2012.

[9] Peterson, S.A. Sediment removal as a lake restoration technique. EPA-600/3-81-013, 1981.

[10] van Rijn, L.C. Sedimentation of sand and mud in reservoirs in rivers. 2013.

[11] *Design of Small Dams*, ed. Interior USDot, Bureau of Reclamation, United States Government Printing Office: Denver, Colorado, 1987.

[12] Churchill, M.A. (ed.), Discussion of analysis and use of reservoir sedimentation data, by L.C. Gottschalk. *Proceedings of Federal Interagency Sedimentation Conference*, Denver, Colorado, Jan. 1948.

[13] Brune, M.G., Trap efficiency of reservoirs. *Transactions of American Geophysical Union*, **34**(3), 1953.

[14] Electrowatt. Murum Hydroelectric Project – Feasibility Study. 1994.

[15] Rahmani, V. et al., Examining storage capacity loss and sedimentation rate of large reservoirs in the Central U.S. Great Plains. *Water*, **10**(2), 2018.

[16] De Araújo, J.C., Güntner, A. & Bronstert, A., Loss of reservoir volume by sediment deposition and its impact on water availability in semiarid Brazil. *Hydrological Sciences Journal*, **51**(1), pp. 157–170, 2010.

SEMIAUTOMATIC ALGORITHM FOR THE INTERPRETATION OF THE BEDFORMS AND STATISTICAL ANALYSIS

LEONARDO SCHIPPA & IRENE CAVALIERI
Department of Engineering, University of Ferrara, Italy

ABSTRACT
This study presents a methodology to simplify the identification of the primary bedforms (medium-large dunes) present on a compound riverbed. The paper refers to a field study conducted on a straight free flow reach of an alluvial river (i.e. River Po in Italy) in January 2021. The surveys were carried out employing a multibeam echosounder which allows a large data collection, which was processed realizing river bed DTMs. The identification of the dunes along transects occurs through the implementation of a semiautomatic algorithm able to analyse the bed elevation profiles and to recognize the troughs and the crests of the primary dunes based on the threshold values of length obtained through to a spectral analysis. Once the dunes are recognized, characteristic parameters of the dunes (wavelength, wave height and steepness) are estimated, by which a statistical analysis is carried out to figure out the most frequently recurring dunes on the riverbed. The semiautomatic algorithm proves to be able to handle a large amount of data, providing a reliable interpretation of the riverbed and it is also consistent with results from previous studies. Starting from the present analysis, it is possible to develop the study of the bedforms kinematic that represents a key issue to quantify the volumes of solid bedload transport, in order to cross the typical limitations derived by the traditional bed load sampling methodologies.
Keywords: dunes bed, river bedform, river morphology, remote sensing riverbed.

1 INTRODUCTION

In recent decades, many Italian riverbeds have suffered significant morphologic changes. These changes, interpretable as a lowering of the riverbed quote, have drawn the attention both in the field of research and spatial planning. The need to apply a, monitoring, and programming activity to manage the evolution trends became evident, thus it is essential to support every planning activity with reliable forecast tools able to assess the evolution trends and river layout at medium to longer horizons, considering the changed hydrological situation due to climate change.

In alluvial river the bed forms often occur on the riverbed, they are complex and dynamic, and represent the most relevant feature governing flow resistance [1]. Complicating matters is the inherent variability of bedform size and shape in natural river systems. Sediment and water discharge conditions vary continuously in natural rivers causing bed morphology to be out of equilibrium with prevailing flow conditions [2]. In fact several field studies suggest that bedform disequilibrium is likely a standard phenomenon in natural rivers, and individual dunes in large sand rivers can have a wide range of geometry patterns corresponding to the same hydrometric condition [2], and it makes inconsistent most of the most accredited relationship among dune dimensions and hydraulic and sedimentary characteristics (e.g. [3], [4]). Therefore, even the estimation of flow resistance during engineering processes is not trivial, since it depends on the bed surface roughness and on the dimension of the bed forms, which in turn depend by flow parameters and sediment properties [5]. River bedforms can be distinguished in principle forms, the dunes, and secondary dunes, ripples and macroripples, which were found on the stoss slope and gentle lee slopes of large dunes [6]. These dunes usually increase in height and decrease in length towards the upper stoss of a large dune but

WIT Transactions on Ecology and the Environment, Vol 250, © 2021 WIT Press
www.witpress.com, ISSN 1743-3541 (on-line)
doi:10.2495/WRM210151

may become less steep and more humpback in profile near the crest of the primary dune if the transition to upper-stage is approached locally in the crestal region crest [7].

The presence of the bed forms, their extension and geometry, influences the kinematic field, the flow resistance, the hydraulic efficiency of the section, and the stability of the fluvial infrastructures. The matter acquires particular importance when calculating the morphological changes they experience under changing flow conditions (e.g. steady and unsteady flow), or the downstream migration of dunes, which is of paramount importance to evaluate the bedload transport component. A popular approach involves calculating transport from the geometry and celerity of migrating bedforms, extracted from time series of bed elevation profiles [8]. It is well understandable that any of the former issue related to bed forms presence involve the knowledge of their geometry. To this purpose a random field approach to analyse spatial structure of bed undulation [9], or spectral analyses approach [10] may be applied.

A novel approach combining topographic, statistic and spectral analysis of large sand river bedforms is presented, and a semiautomatic algorithm aiming to identify bed form characteristic is implemented. To this purpose, a field campaign was launched in different hydrometric conditions along a reach of Po river (Italy) in June 2014, July 2014 and January 2021. The paper at hand particularly describes the last field surveying and shows preliminary results.

2 STUDY AREA, METHODS AND SURVEYS

2.1 Study area

The Po River, with a length of 652 km and a hydrographic basin of about 72,000 km^2, is considered the most important Italian river and one of the principle European rivers. It is situated in norther Italy (Fig. 1(a)) and aided by the fact that its basin covers a quarter of the national territory, its waters represent a hydric resource for the irrigation and an important waterway.

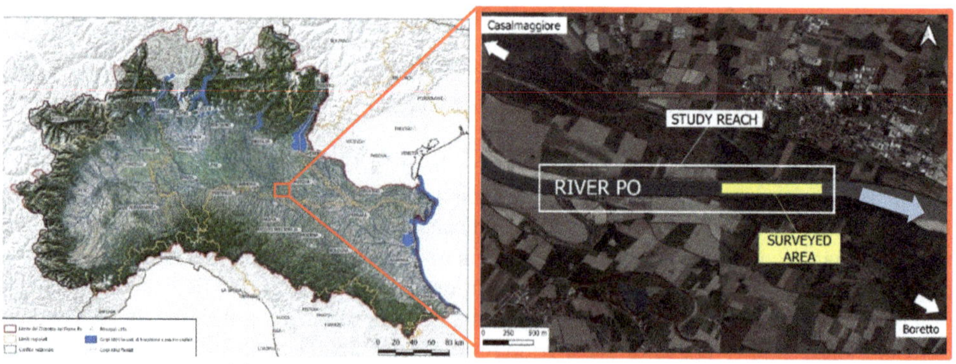

(a) (b)

Figure 1: (a) River Po basin; and (b) Study reach and surveyed area.

The annual hydrological regime is characterized by two low water periods (winter and summer) and by two flood periods (late winter and spring) which are associated, respectively, to rainfall and to the snow melting processes from the mountain district of the basin. The minimum, medium and maximum annual discharges observed at the gauging station Pontelagoscuro are around 600, 1,500 and 5,000 m^3/s, respectively connected to flow peak of 10,000 m^3/s and lower flow at 250 m^3/s.

The study reach (Fig.1(b)) is located 250 km upstream the delta and it is situated between two gauging stations, Casalmaggiore and Boretto. It is a single straight active channel reach having a length of 2.6 km, a width of about 215 m, and a mean bed slope of 0.15 m/km, lower than the ground slope (i.e. 0.26 m/km). The characteristic diameters of the bed sediment are d_{35} = 0.35 mm, d_{50} = 0.39 mm, and d_{90} = 0.76 mm. In particular, the surveyed area (Fig. 1(b)) is 1km long. Water level recording and stage-discharge relation are available for both gauging stations. The discharge duration (1943–2017) in the downstream station is Q_{10} = 3,630 m^3/s, Q_{60} = 1,810 m^3/s, Q_{91} = 1,460 m^3/s, Q_{182} = 933 m^3/s, Q_{355} = 379 m^3/s.

2.2 Methods and surveys

The survey was carried out on 28th January 2021 thanks to a high-speed craft suitable for a high-speed hydrographic survey of the riverbed. It is equipped with a multibeam echosounds system MBES and a GPS system recording the actual position of each swat. According to the hydrometric level observed at the gauging station of Boretto and to the flow rate curve the mean discharge is about Q = 1,327 m^3/s, furthermore the mean water depth on the surveyed area was y = 5.90 m (Table 1).

Table 1: Hydrometric conditions during the field campaign on the 28th January 2021, referring to starting and ending time of surveying.

	Hydrometric conditions (y = 5.90 m)			
	Starting time		Ending time	
	Water stage (m)	Discharge (m^3/s)	Water stage (m)	Discharge (m^3/s)
Boretto	–0.23 m	1,335 m^3/s	–0.27 m	1,320 m^3/s
Casalmaggiore	–1.54 m		–1.56 m	

The surveys started at fluvial harbour of Mezzani, sailing downstream in direction of Boretto to reach the surveying area. Two flanked swaths 50 m wide, and slightly overlapped over a streamwise distance of about 1,000 m were surveyed. The second survey of the same area was repeated in the same way after 40 minutes to gather the dune migration.

3 BEDFORM IDENTIFICATION: DESCRIPTION OF THE ALGORITHM

Once the surveys data were collected, the DTMs were processed. Fig. 2(a) represents the DTM obtained by a grid of 0.2 m surveyed points. Six longitudinal transects spaced few meters each other were defined and riverbed profiles were plotted, starting from the river bed bathymetry. A profile was collected for each line and for each time, so twelve profiles were analysed. The semiautomatic algorithm was applied to each profile and finally a unique sample was achieved. This sample allow the statistical analysis estimating the mean values, the modes, and the standard deviations.

Following the "discrete" approach, the river bottom is considered as a set of discrete two- or three-dimensional bedforms, which can be described by a set of parameters such as length, height, slope [11], thus every dune is theoretically simplified as a triangle identified by two troughs on the bottom and a crest on the top.

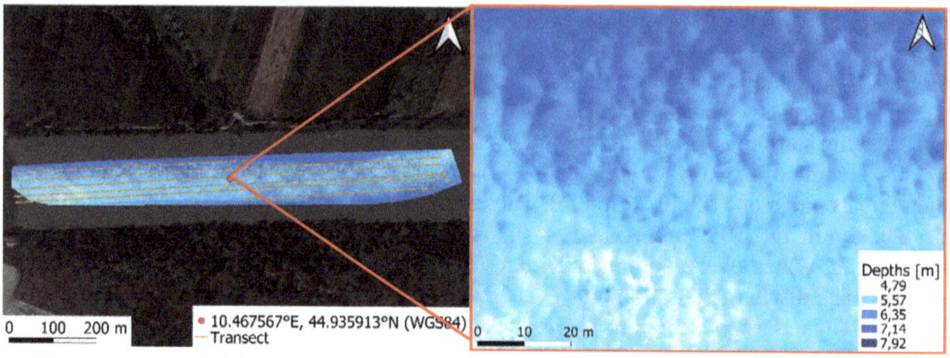

(a) (b)

Figure 2: (a) DTM survey 28 January 2021; and (b) DTM detail.

A preliminary algorithm was implemented, which was able to identify all the local maxima and all the local minima along the profile producing a geometrical interpretation, that did not give an understandable representation of the bed undulation. In fact, secondary (smaller) forms superimposed to primary dunes are present, so losing the data connected to the existence of principal dunes. In fact, the result of the spectral analysis shows a signal characterized by a relevant spectral power around 4–6 m wavelength (Fig. 3).

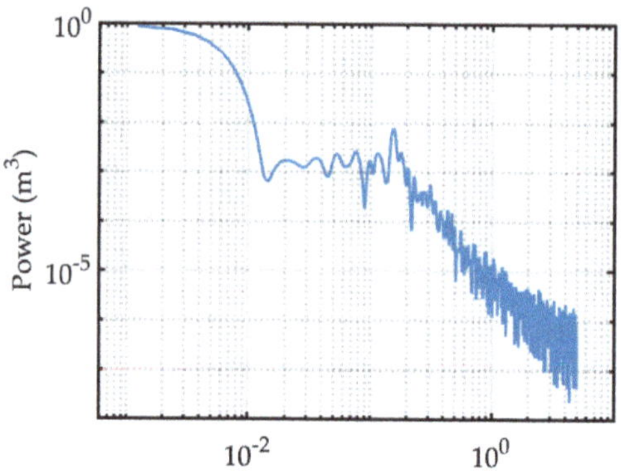

Figure 3: Single-sided spectrum for a representative longitudinal transect.

To overcome the problem, a novel semiautomatic algorithm to identify primary and secondary bed forms and giving a more truthful riverbed interpretation was implemented. Applying filtering parameters, it is possible to isolate those bedforms which are classified as principal forms. Filtering parameters are imposed according to the results of the spectral analysis, whose approach treats the riverbed as a continual moving wave field in which the forms may be described by wavenumber spectra. thus the algorithm was set using these last filtering values.

The semiautomatic algorithm allows to filter the bed elevation profiles which are complex, to recognize the principal bed forms. Based on the results of the spectral analysis, and according to [6], we may distinguish between ripples ($\lambda < 0.6$ m) and small dunes ($0.6 < \lambda < 5$ m) on one hand, and medium to large dunes (5 m $< \lambda$) on the other hand. We consider the latter as the primary features of the river bed, and in our case they result $\lambda_t > 4$–6 m, in relation to a wave amplitude $\Delta_t > 0.15$ m as it results in the filtered bed profiles.

In the first place, the algorithm was applied to each transect along the swath. The proposed algorithm performs these next steps:

1. Bed elevation interpretation through the addition of the geometrical coordinates from the surveys and DTMs. The coordinates are imported to represents the bed elevation. Specifically, the transformation from height profile to bed relative elevation profile occurs.
2. Geometrical identification. Local maxima and local minima are located on the transect profile, thus every crest and every trough are identified (Fig. 4(a)).
3. Definition of the threshold values λ_t and Δ_t. According to the results of the spectral analysis, those values stand for the limit value which marks the clear separation between the existence of the secondary undulations ($\lambda < 4$–6m; $\Delta < 0.15$m) or, on the contrary the presence of a dunes ($\lambda > 4$–6m; $\Delta > 0.15$m). In this case the evaluations were carried out using $\lambda = 4$ m.
4. Setting of a loop able to apply the above-mentioned filter according to threshold values. Scan of each bed form and recognizing only those forms corresponding to filter criterion. Finally, the identification of the filtered dunes through suitable crests and suitable troughs occurs (Fig. 4(b)).
5. In the end, dune (primary bed forms) geometric parameters are computed as following:
   - Dune length. λ: it represents the horizontal distance between two subsequent troughs along the longitudinal transect. Length λ': it represents the horizontal distance between the crests two consecutive dunes. Length λ_{stoss} : it represents the horizontal distance between the initial trough and the crest of a dune. Length λ_{lee} : it represents the horizontal distance between the crest of a dune and its final trough.
   - Dune height. Δ_{stoss}: it represents the vertical distance between the upstream trough and the subsequent crest of a dune. Δ_{lee} : it represents the vertical distance between the crest of a dune and its downstream trough.
   - Dune steepness. Stoss steepness is the ratio between Δ_{stoss} and λ_{stoss}. Lee steepness it is the ratio between Δ_{lee} and λ_{lee}. Representative steepness is the ratio between Δ_{lee} and λ.

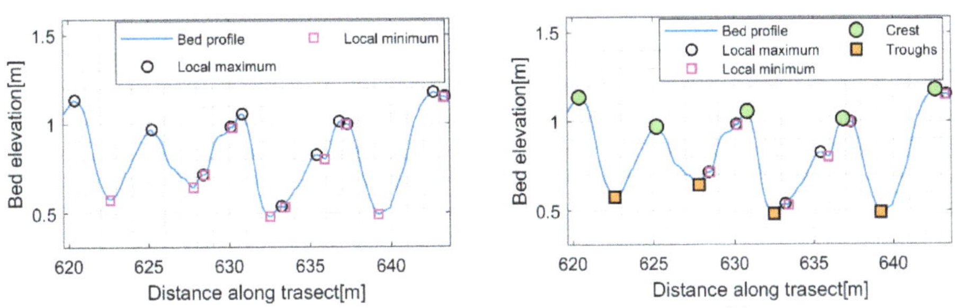

(a) (b)

Figure 4: (a) Geometrical bed profile and local extreme location; (b) Filtered bed profile. Subsequent crest location, and through location, of the primary bed forms (i.e. dunes).

4 ALGORITHM'S APPLICATION, SENSITIVITY ANALYSIS AND DISCUSSION

As a first step, a random subjective comparison between longitudinal transects extrapolated from the MBES and the filtered geometry has been carried out, confirming the consistency of the procedure. Then statistical analysis of the filtered geometry has been carried out assuming as a reference threshold value $\lambda_t = 6$ m as it derives from spectral analysis. Log-normal probability distribution function is depicted in Figs 5 and 6, whereas Fig. 7 plots the whole dataset coming from all the longitudinal transects.

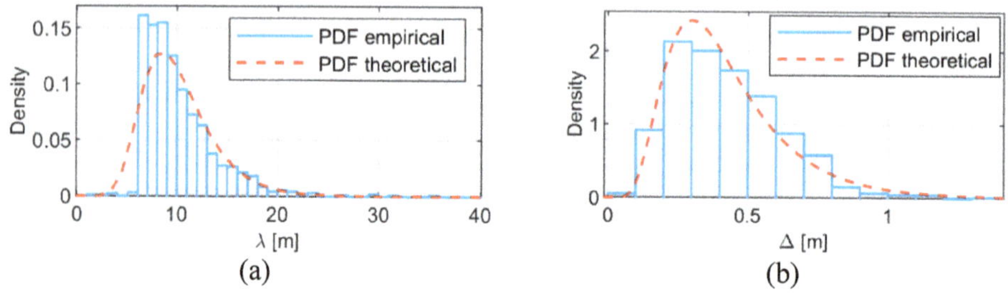

Figure 5: Empirical and theoretical PDF ($\lambda_t = 6$m). (a) Dune length λ; and (b) Dune height Δ_{lee}.

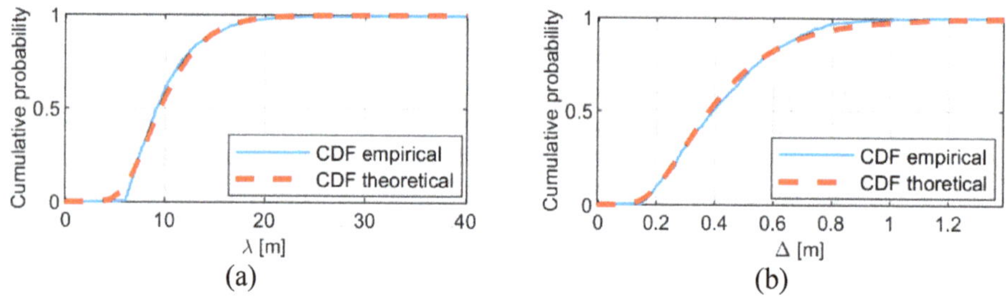

Figure 6: Empirical and theoretical cumulative distribution function CDF ($\lambda_t = 6$m). (a) Dune length λ; and (b) Dune height Δ_{lee}.

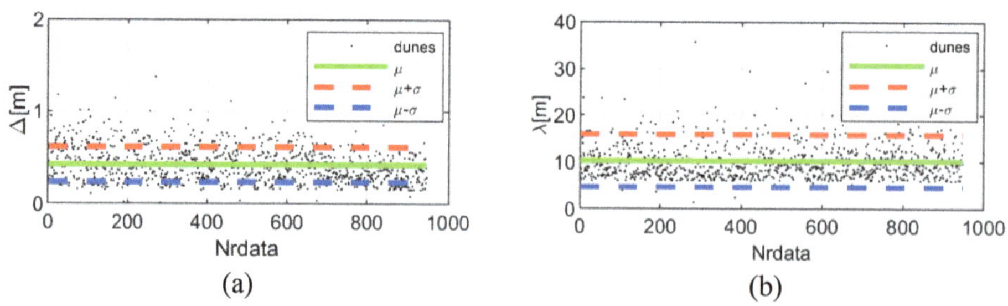

Figure 7: Statistical analysis ($\lambda_t = 6$m). (a) Dune length λ; and (b) Dune height Δ_{lee}.

Sensitivity analysis of the algorithm to the threshold values λ_t and Δ_t, has been carried out, taking advantage of statical analysis, and it is shown in Fig. 8 in terms of dune length λ and dune height Δ_{lee}. Table 2 reports the main statistical parameters of the characteristic dune dimensions based on a unique sample and calculated for the wave length threshold values $\lambda_t = 4$ m, $\lambda_t = 6$ m and $\lambda_t = 8$ m. The frequency peak for λ results equal to 5–6 m, 6–7 m and 9–10 m respectively. In the same way the peaks for λ' are equal to 5–6 m and 9–10 m. The partial dune lengths λ_{stoss} and λ_{lee} result, respectively, equal to 2–4 m and 1–3 m ($\lambda_t = 4$ m), and equal to 2–5 m and 1–3 m ($\lambda_t = 6$ m and $\lambda_t = 8$ m). Similarly, the process is repeated to estimate the wave height, the frequently occurring values can be observed between 0.2 m and 0.5 m in every case. Sensitivity analysis (Fig. 8) shows increasing lengths and heights values when threshold value increase, but the difference referred to their mean values are limited to 21% for lengths λ and 4% for heights Δ_{lee}.

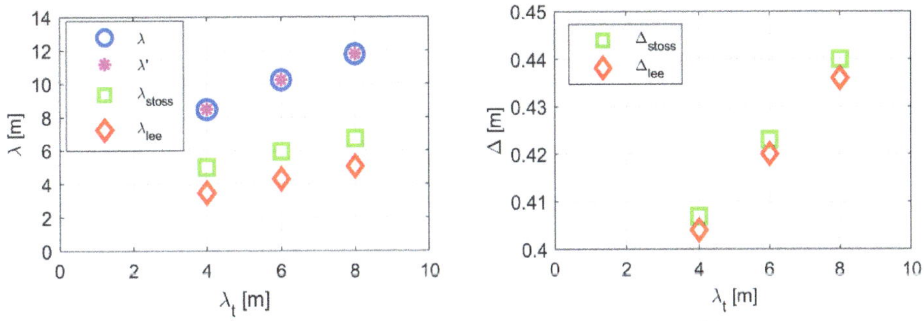

(a) (b)

Figure 8: Sensitivity analysis. (a) Dune length; and (b) Dune height.

Table 2: Statistical parameters of the characteristic dune dimensions corresponding to different threshold value of λ_t. (μ is mean value; σ is standard deviation).

		λ	Λ'	λ_{stoss}	λ_{lee}	Δ_{stoss}	Δ_{lee}	δ_{stoss}	δ_{lee}	Δ_{lee}/λ
$\lambda_t - 4$ m	μ	8.473	8.485	5.008	3.466	0.407	0.404	0.102	0.153	0.054
	σ	5.357	5.795	5.014	2.535	0.181	0.186	0.057	0.086	0.026
$\lambda_t = 6$ m	μ	10.236	10.247	5.935	4.301	0.423	0.420	0.091	0.136	0.045
	σ	5.609	6.524	5.500	3.124	0.190	0.194	0.052	0.086	0.022
$\lambda_t = 8$ m	μ	11.778	11.793	6.718	5.060	0.440	0.436	0.085	0.124	0.040
	σ	5.750	6.981	5.866	3.492	0.194	0.199	0.051	0.084	0.018

The steepness δ_{stoss} and δ_{lee} are focused, respectively, around 0.04–0.09 and 0.07–0.10 ($\lambda_t = 4$ m), 0.05–0.08 and 0.04–0.10 ($\lambda_t = 6$ m), and 0.04–0.08 and 0.03–0.09 ($\lambda_t = 8$ m), total steepness δ result reasonably lower than the above-mentioned values and equal to 0.02–0.07 ($\lambda_t = 4$ m) and 0.02–0.05 ($\lambda_t = 6$–8 m). Average dune steepness reduces increasing the threshold value λ_t, showing a moderate sensitivity to the choice of the threshold wavelength. On the opposite, wave amplitude (stoss or lee side calculated) is almost the same, and they are weakly sensitive to the choice of the threshold dune length value (Table 2). It is worth noting the influence of threshold value λ_t on the dune length statistic. In fact, increasing λ_t from 4 m to 8 m, the average value of λ (calculated trough-to-trough or crest-to-crest) increase by almost 40%, and the standard deviation increase as well. The dune amplitude results almost the same, no matter of it is calculated stoss or lee side, and it moderately increases by 10% if we double λ_t from 4 m to 8 m.

In particular, dune steepness δ_{max} as a function of relative depth $Z = h/d_s$ is investigated (where h is the local water depth, $d_s = 0.430$ mm is the mean sand diameter and $Z \in [1.16 \times 10^4 - 1.61 \times 10^4]$). The variation of the dune steepness δ_{max} related to flow intensity $\eta = \tau/\tau_{cr}$ (where τ is the shear stress acting on the bed, and $\tau_{cr} = 15.0485$ N/m^2 is its critical value calculated by Brownlie criterion [12]) is plotted in Fig. 9 along with data base of large sand rivers [13]. River Po data are consistent with large sand river database, statistical wise, and they confirm dune steepness are not easily featured by the approach originally proposed by Yalin and Karahan [14], mainly based on laboratory flume data. In other words what Yalin and Da Silva argued [13], about dune steepness in a real large sand river, seems to be confirmed by the field data herein presented. In fact, according to $Z >> 1,000$ the maximum dune steepness values could be smaller than its theoretical value 0.06, and the corresponding flow intensity may be larger than 16.

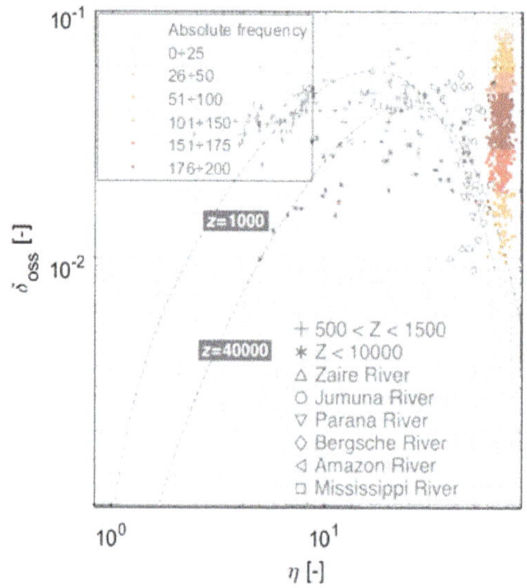

Figure 9: Dune steepness versus relative flow intensity. Po river data (coloured spots) are presented together with large sand river database (black spots) from [13].

Therefore, river Po data are plotted in Fig. 10 where the curve describes δ_{max} derived by eqn (1) [13] and each dot represents a dune based on the calculated steepness δ_{oss} and its related Z.

$$\delta_{max} = 0.00047Z^{1.2}e^{-0.17Z^{0.47}} + 0.04(1 - e^{-0.002Z}). \tag{1}$$

Referring to the examined hydrodynamic conditions, and to the Z range of the river Po field campaign, the maximum dune steepness calculated according to eqn (1) results $\delta_{max} = 0.04$. The average observed dune steepness results $\delta_{oss} = 0.054$; 0.054; 0.040, depending on which λ_t is considered (Table 2), and it is consistent with the theoretical value of $\delta_{max} = 0.04$.

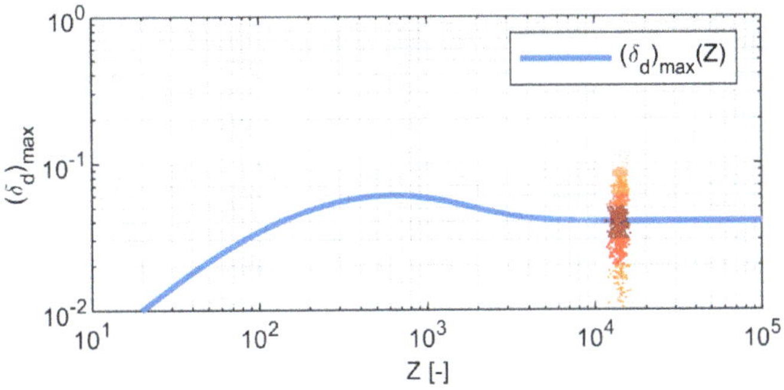

Figure 10: Theoretical maximum dune steepness (eqn (1)) and observed dune steepness.

To conclude this discussion session, it is worth to mention the potentiality of deriving dune kinematics by filtered dune geometry. Quantification of bedform migration by remote sensing has been studied for many years with limited success, both in river and shallow water sea environment, motivating some researchers to conclude that accurate observations require solid markers installed on the sea (river) bed. It introduces operational problems which can be hardly overcome in the natural environment. The main difficulty related to remote sensing seems to be collecting a sufficiently large number of small to medium dunes that will produce the necessary statistical parameters. Fig. 11 shows two MBES subsequent surveys repeated within an hour. The two bed elevation profiles have been overlapped in order to gather the dune migration and therefore being able to appreciate the dune kinematic, which is the standard point to modelling bedload transport, based on dune geometry. As it is evident from Fig. 11 crest and through migration is sharply defined, and the accuracy is favoured by the RTK-GPS surveying.

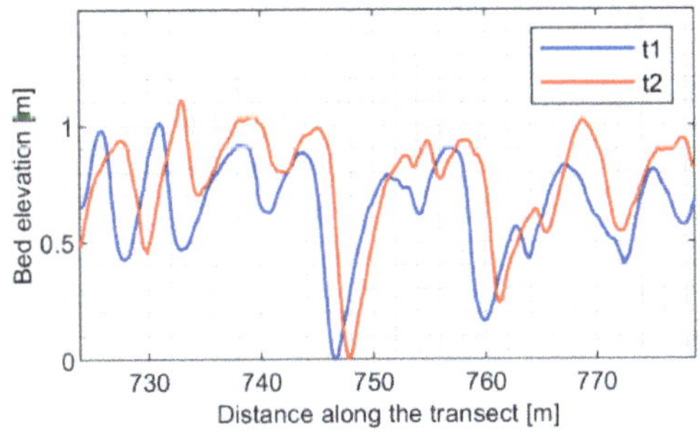

Figure 11: Overlapping of bed profiles obtained by MBES surveys carried out at t1 and t2.

5 CONCLUSION

The field campaign carried out on the middle course of the river Po and the application of the semiautomatic algorithm allowed to analyse the morphology of the riverbed. The algorithm has been tested on a large amount of data obtained with MBES in three survey campaigns, accounting for 39 longitudinal transects extrapolated from DTM covering about 10^5 m^2 of active riverbed. In the reported field campaign, dune lengths range of 4–9 m, and dune heights consist of 0.2–0.5 m and the total steepness ranges between 0.03 and 0.06 for the investigated hydrometric conditions (Q = 1,327 m^3/s and y = 5.90 m). Spectral analysis shows the prevalent wavenumber of the bed undulations, and it provides wavelength threshold value to be used for the filtered geometry. The algorithm developed plausible and truthful results in terms of filtered medium-large dune bed form, and it provides a correct interpretation of the riverbed, as it was confirmed by a random subjective comparison between filtered and original DTM river bed geometry along several random longitudinal transects. Sensitivity analysis demonstrates that dune steepness reduces increasing the threshold value λ_t, showing a moderate sensitivity to the choice of the threshold wavelength. On the opposite, wave amplitude (stoss or lee side calculated) is almost the same, and they are weakly sensitive to the choice of the threshold dune length value.

Further works are necessary to extend the field campaigns at higher flow conditions and during different rising period of selected flood to test any sensitivity to the hydrological condition.

ACKNOWLEDGEMENTS

The work has been realized thanks to the collaboration between Department of Engineering of University of Ferrara, AIPo and ARPAE. A grateful thanks to Alberto Agnetti, Luca Crose, and Paolo Piovani for the collaboration on field campaign organization, performing and data set analysing. Part of the work was developed during MSc thesis of Irene Cavalieri. The study was funded by AIPO contract 137/2019.

REFERENCES

[1] Schippa, L., Cilli, S., Ciavola, P. & Billi, P., Dune contribution to flow resistance in alluvial rivers. *Water*, **11**, 2094, 2019.

[2] Bradley, R.W. & Venditti, J.G., Reevaluating dune scaling relations. *Earth-Science Review Journal*, **165**, pp. 356–376, 2017.

[3] Karim, F., Bed configuration and hydraulic resistance in alluvial channel flows. *Journal of Hydraulic Engineering*. **121**(1), pp. 15–25, 1995

[4] van Rijn, L.C., Sediment transport III: Bedforms and alluvial roughness. *Journal of Hydraulic Engineering*, **110**, pp. 1733–1754, 1984.

[5] van Rijn, L., Equivalent roughness of alluvial bed. Journal of Hydraulic Engineering, pp. 1215–1218, 1982. https://doi.org/10.1061/(ASCE)0733-9429(1983)109:9(1250).

[6] Ashley, G.M., Classification of large-scale subaqueous bedforms: A new look at an old problem. *J. Sedim. Petrol.*, **60**, pp. 160–172, 1990.

[7] Carling, P.A., Golz, E., Orr, H.G. & Radecki-Pawlikà, A., The morphodynamics of fluvial sand dunes in the River Rhine near Mainz, Germany. I. Sedimentology and morphology. *Sedimentology*, **47**, pp. 227–252, 2000.

[8] Leary, K.C.P. & Buscombe, D., Estimating sand bed load in rivers by tracking dunes: A comparison of methods based on bed elevation time series. *Earth Surface Dynamics*, **8**, pp. 161–172, 2020.

[9] Aberle, J., Nikora, V., Henning, M., Ettmer, B. & Hentschel, B., Statistical characterization of bed roughness due to bed forms: A field study in the Elbe River at Aken Germany. *Water Resources Research*, **46**(3), 2010.

[10] Lefebvre, A., Ernstsen, V. & Winter, C., Bedform characterization through 2D spectral analysis. *Journal of Coastal Research*, pp. 781–785, 2011.

[11] Nikora, V.I., Methods for quantitative description of channel bed-forms. *Erosional and Channel Processes in Various Climatic Conditions*, Moscow University, Moscow, pp. 327–328, 1987. (In Russian.)

[12] Brownlie, W.R., Compilation of alluvial channel data: Laboratory and field, 1981.

[13] Yalin, M.S. & Da Silva, A.M.F., *Fluvial Processes*, 2nd ed., Taylor & Francis: London, pp. 51–57, 2017.
[14] Yalin, M.S. & Karahan, E., Steepness of sedimentary dunes. *Journal of the Hydraulics Division, ASCE*, **105**(4), pp. 381–392,1979.

COASTAL EROSION HAZARD AND VULNERABILITY: CASE STUDY OF PORTICELLO, SOUTH CALABRIA, ITALY

GIUSEPPINA CHIARA BARILLÀ[1], GIUSEPPE BARBARO[1], GIANDOMENICO FOTI[1], PIERLUIGI MANCUSO[2], VINCENZO FIAMMA[1], AGNIESZKA MALESINSKA[3], PIERFABRIZIO PUNTORIERI[1] & MARIA MANDALARI[1]
[1]DICEAM Department, Mediterranea University of Reggio Calabria, Italy
[2]Public Works Department of Calabria Region, Italy
[3]Faculty of Building Services, Hydro and Environmental Engineering, Warsaw University of Technology, Poland

ABSTRACT

Coastal erosion is due to natural phenomena and anthropogenic interventions that alter the coastal dynamics and the sediment balance. Therefore, to manage the coastal areas and to correctly design the necessary defence interventions, it is necessary to analyze all these factors. Consequently, in recent years the methodologies for the coastal erosion hazard and vulnerability assessment have increasing importance. However, there is currently a lack of a general methodology that considers all the factors that influence the coastal dynamics and, therefore, the hazard and the vulnerability. This paper, through a case study, describes a new methodology based on indices that considers all the factors that influence the hazard and the vulnerability, thus resulting in general validity. This methodology defines two indices of hazard and vulnerability. The hazard index considers all the natural processes that contribute to the phenomenon of coastal erosion, such as tides, currents, average sea level rise due to climate change, wave climate and both coastal and river sediment transport. The vulnerability index, on the other hand, includes all those characteristics of the coast that can make it more or less vulnerable to coastal erosion, such as its geological and morphological characteristics (both of the emerged and submerged part), its evolutionary trend, the presence of dune systems, vegetation and coastal and/or port defence works, the subsidence rates and the degree of waterproofing of the coast. The case study concerns Porticello, a town near the Strait of Messina characterized by a high coastal erosion risk. Porticello represents an interesting case study due to the peculiarities of the Strait of Messina and the presence of numerous coastal structures. The study area extends for about 1.5 km, inside there are numerous coastal structures, two river mouths and numerous houses built close to the shoreline.

Keywords: coastal erosion, coastal erosion risk assessment, hazard, vulnerability, index-based methodology, coastal structures.

1 INTRODUCTION

Coastal areas are particularly important from a socio-economic point of view. Indeed, about 20% of the world population lives within 25 km of the coast [1] and the number of coastal cities has quintupled in the last 70 years [2]. This anthropogenic pressure has often altered the coastal dynamics and the sedimentary regime, triggering erosive processes [3]. Indeed, most of the world's coastal areas are currently subject to coastal erosion [4].

Along with anthropogenic pressure, various natural factors can also alter coastal dynamics by triggering erosive processes [5], [6]. The main natural factors are the wave climate [7]–[10] and the sedimentary balance between longshore and river transport [11]–[17]. Regarding the anthropogenic pressure, it can concern both rivers and coasts. In the river area it is mainly the construction of dams and weirs, the sediment withdrawal, and the subsidence [18]–[23]. In the coastal areas it is mainly the construction of port and coastal structures [24]–[26].

Furthermore, due to anthropogenic pressure there is an increase in the vulnerability of coastal areas to the action of floods and sea storms [27]–[29], especially in case of concurrent events [30]–[32]. The frequency and the intensity of these events can be increased by climate

WIT Transactions on Ecology and the Environment, Vol 250,
www.witpress.com, ISSN 1743-3541 (on-line)
doi:10.2495/WRM210161

change, which also have consequences on sea level rise and, therefore, on erosive processes [33], [34].

Therefore, to effectively manage coastal areas it is necessary to monitor the above factors and the shoreline changes caused by the alteration of the sedimentary regime [35]–[38].

In recent years, the methodologies for the coastal erosion hazard and vulnerability assessment have increasing importance [39]–[43]. These methodologies allow to effectively manage the risk areas by allocating the funds intended for coastal protection with a certain order of priority [44], [45].

The Italian and international panorama is characterized not only by different evaluation methodologies, variables between the different coastal regions, but above all by a great variety and dispersion of factors taken into consideration by these different methodologies [46]–[49]. Therefore, there is currently a lack of a general methodology that considers all the factors that influence the coastal dynamics and, therefore, the hazard and the vulnerability.

Among the existing methods, the index-based methods are the simplest and most widely used [50]. They are based on a quantitative or semi-quantitative evaluation of different variables or indicators which are combined with each other, through an appropriate set of weights, obtaining as a result an overall index, the coastal vulnerability index (CVI) [51], [52]. In addition, these methods can be supported by GIS applications, therefore it is possible to produce maps that allow to describe the spatial distribution of the CVI index and to overlap these results with additional spatial information [53].

This paper, through a case study, describes a new methodology based on indices that considers all the factors that influence the hazard and the vulnerability, thus resulting in general validity. The case study concerns Porticello, a town near the Strait of Messina which represents an interesting case study due to the peculiarities of the Strait of Messina and the presence of numerous coastal structures. The study area extends for about 1.5 km, inside there are numerous coastal structures, two river mouths and numerous houses built close to the shoreline.

2 SITE DESCRIPTION

The town of Porticello is a fraction of the city of Villa San Giovanni, on the Tyrrhenian side of Calabria. In particular, the study area is located in the terminal part of the Strait of Messina, a portion of the sea that connects the Ionian Sea with the Tyrrhenian one and which, separating the two cities of Messina and Reggio Calabria with their respective urban areas, it separates Sicily from Calabria and therefore from peninsular Italy and the continent. Due to its morphological characteristics, the Strait can be represented as an inverted funnel, reaching, at its narrowest point, a width of about 3 km: at this point there is an underwater saddle, with a minimum depth of about 80 m, which divides the whole area into two sub-basins with different characteristics; to the north of the saddle (Tyrrhenian Sea), in fact, the depths gradually increase, unlike the southern part (Ionian Sea) where the seabed becomes deeper very quickly. Furthermore, the Strait of Messina is the meeting point between two seas with different characteristics in terms of salinity and temperature. All this means that the Strait is home to very strong currents, so much so that in ancient times the legends told of the presence of two monsters (Scylla and Cariddi) that ravaged the navigation between Calabria and Sicily (Fig. 1).

The coast in question is located between Punta Pezzo and Torre Cavallo (Fig. 2(a)), more precisely from the Zagarella river to the Commenda Divale river. The stretch considered has a length of about 1,500 m and is highlighted in red in the Fig. 2(b). The study area is characterized by a high risk of coastal erosion, as numerous houses are built close to the state-

Figure 1: The area of the Strait of Messina with its currents.

Figure 2: The study area. Location of Porticello on the left. Detail of the examined area on the right; the stretch of coast studied highlighted in red.

Figure 3: The coastal defence works completed in 2015.

owned area and in some places, the beach is almost non-existent, so much so that in 2015 coastal defence works were completed (Fig. 3). In addition, in the study area, there are several mouths of valleys and the Santa Trada river.

3 METHODOLOGY

The new methodology proposed therefore arises from the consideration that in the literature there is no risk assessment method that considers all the main factors that influence coastal hazard and vulnerability, thus determining a risk index. For its simplicity of application and for the easy visualization of the results obtained on GIS, a methodology based on indices was chosen, applicable on a regional scale, and which is valid for any Italian region, or which can consider all the factors that influence coastal dynamics.

New indices of coastal hazard and vulnerability were therefore defined, functional to the determination of a standardized coastal risk assessment procedure. For this purpose, the Calabria region, which represents a complex case study for the application of the elaborated procedure, has been divided into three macro-areas according to the different exposure of the coasts to wave motion: the Tyrrhenian area, the area of the Strait of Messina and the Ionian area, characterized not only by different climatic characteristics, above all by different geographical characteristics, in particular in terms of fetch.

All the indices of hazard and vulnerability, appropriately determined, have been classified on a scale from 1 to 5. In particular, based on the level of hazard/vulnerability, with 1 corresponding to a "very low" grade, with 5 corresponding to a "very high" grade. The hazard/vulnerability classes have been effectively calibrated according to the indices considered. With the exception of the factors defined in a range of values already between 0 and 1 (such as, for example, those relating to the wave climate, coastal and river sediment transport, etc.), a score will be assigned for all other factors between 0 and 1, depending on the hazard/vulnerability class resulting from the assessment; in particular, the vulnerability factors that may not come into play in a specific location (such as, for example, the presence of dune systems, port works, the phenomenon of subsidence, etc.) will be assigned a value of 0, without compromising the structure of the overall index.

The overall hazard/vulnerability index can therefore be evaluated using the following equation:

$$I = \sqrt{\frac{v_1^2+v_2^2+v_3^2+\cdots+v_n^2}{n}}, \quad (1)$$

where $v_1, v_2, v_3, \ldots, v_n$ represent all the indices of hazard/vulnerability described below and evaluated as reported in Table 1.

Table 1: Hazard/vulnerability classes and assigned score.

Number of classes	1	2	3	4	5
Classification	Very low	Low	Moderate	High	Very high
Score awarded	0.2	0.4	0.6	0.8	1

3.1 Indices of hazard

The hazard indices are representative of the factors that strongly affect the coastal erosion process, such as tides, currents, average sea level rise due to climate change, wave offshore climate and both coastal and fluvial sediment transport.

As regards the index linked to the variation of the mean sea level (SLR), the classification proposed by Pendleton et al. [54] who conducted a study on the coastal vulnerability linked to the rise in the mean sea level, was adopted.

The index that considers the tide range was defined considering the classification proposed by Davies [55].

The index connected to marine currents was classified according to the average current values of the Italian seas and the Strait of Messina; in fact, in the area of the Strait of Messina the currents, due to the conformation of the Calabrian and Sicilian coasts, take on much higher values than those of the Mediterranean Sea.

The wave climate index was determined by considering a weighted average, with equal weights, of four sub-indices, which take values between 0 and 1, thus defined:

1. Index associated to the flow of energy: it takes into account the energy contribution of the sea states incident on the coast.
2. Index related to the maximum energy flux direction: it takes into account the sectors from which the strongest waves come, as a coast subject to orthogonal sea states is more stressed than inclined sea states.
3. Index linked to the change in significant height with the return period, determined using the Boccotti methodology [56].
4. Index connected to the variation of the run-up with the return period, determined using the model of Stockdon et al. [57].

The index related to coastal sediment transport varies between 0 and 1, as it has been defined as the ratio between the annual average longshore solid material flow and its maximum value calculated in the reference macro-area. Longshore sediment transport was evaluated based on the model of Tomasicchio et al. [58].

The index linked to river sediment transport was defined in a similar way to that linked to coastal sediment transport, therefore it also varies between 0 and 1. It was decided to use the Gavrilovic model [59]: it is a multiparametric empirical model and it is particularly useful for quantifying the solid transport caused by the soil erosion by water (WSE). WSE is very important in territories such as Calabria, characterized by basins with high slopes and irregular hydrological regimes, with alternation between dry periods and sudden flood events.

3.2 Indices of vulnerability

The vulnerability indices allow to describe the degree of susceptibility of the territory to the phenomenon of coastal erosion: for this reason it is essential to evaluate the evolutionary trend of the coast (in the recent, short, medium, long and very long term), the geological and morphological characteristics of the coast (both of the emerged part that of the submerged part), the presence of dune systems, vegetation and coastal and/or port defence works, the subsidence rates of the ground and the degree of waterproofing of the coast.

As regards the index connected to the type of coast, the determination of the vulnerability classes was carried out using the classification proposed by Silva et al. [60], as it is complete and in line with the types of coasts present in Calabria.

The index relating to the morphology of the emerged coast was determined as the average of two sub-indices:

1. Index linked to the slope of the beach, whose classification was based on the slope characteristics of all the coasts of Calabria.
2. Index related to the average width of the emerged beach, whose classification was made according to the maximum value of the average width of the Calabrian beaches.

The index linked to the characteristics of the submerged part of the coast was determined as the average of the following two sub-indices:

1. Index associated to the slope of the seabed, classified in a similar way to that related to the slope of the beach.
2. Index connected to the sedimentology of the seabed, classified according to the various types of sand that can make up the seabed.

The index that identifies the evolutionary trend of the shoreline is very important because it is indicative of the state of health of a coast. This index is based on the estimate of the average shoreline variation rate, v expressed in m/year, according to which it has been classified into 5 classes: one for the advancement ($v > 0.5$), one for the stability ($-0.5 \leq v \leq 0.5$), one for erosion ($-1 \leq v < -0.5$), one for intense erosion ($-2 \leq v < -1$), and one for severe erosion ($v < -2$). This evaluation of the evolutionary trend of the shoreline was carried out for different time intervals: the two most recent available shorelines; a short-term evolutionary trend, with reference to the shorelines available in the last 5 years; a medium-term evolutionary trend, with reference to the shorelines available in the last 20 years; a long-term evolutionary trend, with reference to the shorelines available in the last 30 years; a very long-term evolutionary trend, with reference to the shorelines available in the past 70 years. The index associated with the evolutionary trend of the coast was therefore defined as the weighted average of the various trends referred to the different time intervals considered, with increasing weight as the time interval increases.

The index associated to the subsidence of the land was classified according to the range of variability of the subsidence rates, resulting from studies conducted for the Calabria region.

The index related to the presence of dune systems was evaluated according to the morphological characteristics of the dunes (height and width, considered orthogonally to the coastline) and according to the percentage of the shoreline characterized by the presence of these systems, fundamental for the natural nourishment of the beaches, subject to the action of the storms.

The vegetation index considers both the vegetation present on the beach (in terms of its width behind the beach and the percentage of the coastline characterized by the presence of vegetation), useful for dissipating the energy of the wave motion and reducing erosion in the case of extreme events, and that present in the seabed, which is represented in the Mediterranean basin largely by Posidonia oceanica (Boolean: presence/absence).

The index linked to the presence of port works considers the interruption that such a structure determines in the natural dynamics of the sediments, generating areas of advancement of the coast and areas of retreat.

The index related to the presence of coastal defence works is evaluated according to the ratio between the total length of the protected coast and the length of the stretch of coast considered, thus obtaining the percentage of coverage of the coast by this structure. It is also possible to consider, in defining the index, the degree of conservation of the structures, as a damaged barrier will certainly not perform the same degree of protection as a newly built barrier. Consequently, it is possible to multiply the value of the index by a corrective factor, significant of the state of conservation of the structures, before proceeding to determine the corresponding vulnerability class.

Finally, similarly to how the previous index was defined, it was possible to define an index linked to the portion of the anthropized or cemented coastline, expressed as a percentage, where the coastline is not identified by the shoreline, but by the presence of a structure or infrastructure, often protected by grazing structures.

4 RESULTS AND DISCUSSION

The results of applying the methodology described above in the Porticello area are shown briefly in the following tables (Tables 2–6) which report all the hazard and vulnerability factors relating to the location in question, the evaluation of which appears highlighted. Note how the factors relating to dune systems, ground subsidence and the presence of port works

Table 2: Assessment of the hazard variables.

Hazard variables	Hazard classes				
	Very low 1	Low 2	Moderate 3	High 4	Very high 5
SLR (mm/year)	< 1.8	1.8–2.5	2.5–3.0	3.0–3.4	> 3.4
Mean tidal excursion (m)	< 1	1–2	2–4	4–6	> 6
Sea currents (knots)	< 0.5	0.5–1.5	1.5–3	3–5	> 5
Wave climate	< 0.2	0.2–0.4	0.4–0.6	0.6–0.8	> 0.8
Coastal sediment transport	< 0.2	0.2–0.4	0.4–0.6	0.6–0.8	> 0.8
River sediment transport	> 0.8	0.8–0.6	0.6–0.4	0.4–0.2	< 0.2

Table 3: Evaluation of the wave climate index.

Wave climate index	Hazard classes				
	Very low	Low	Moderate	High	Very high
Total energy flux	< 0.2	0.2–0.4	0.4–0.6	0.6–0.8	> 0.8
Maximum energy flux direction	< 0.2	0.2–0.4	0.4–0.6	0.6–0.8	> 0.8
Variation of significant height with return period	< 0.2	0.2–0.4	0.4–0.6	0.6–0.8	> 0.8
Variation of run-up with return period	< 0.2	0.2–0.4	0.4–0.6	0.6–0.8	> 0.8

Table 4: Assessment of the vulnerability variables.

Vulnerability variables	Vulnerability classes				
	Very low 1	Low 2	Moderate 3	High 4	Very high 5
Type of coast	Rocky, clifted coast	Medium cliffs, indented coast	Low cliffs, alluvional plains	Cobble beaches, estuary, lagoon	Sand beach, dunes system
Emerged coast morphology	0.2	0.2–0.4	0.4–0.6	0.6–0.8	> 0.8
Submerged coast morphology	0.2	0.2–0.4	0.4–0.6	0.6–0.8	> 0.8
Evolutionary trend of shoreline (m/year)	> 0.5	0.5 to –0.5	–0.5 to –1.0	–1.0 to –2.0	> –2.0
Vegetation	Presence of Posidonia Oceanica (0.5)				
Coastal defence works	> 0.8	0.8–0.6	0.6–0.4	0.4–0.2	< 0.2
Degree of anthropization of the coast	<0.2	0.2–0.4	0.4–0.6	0.6–0.8	> 0.8

Table 5: Evaluation of the coastal morphology indices.

Coastal morphology indices		Vulnerability classes				
		Very low	Low	Moderate	High	Very high
Emerged part	Beach slope (%)	> 12	12–9	9–6	6–3	< 3
	Average width of the beach (m)	> 80	80–60	60–40	40–20	< 20
Submerged part	Seabed slope (%)	< 2	2–6	6–10	10–20	> 20
	Seabed sedimentology	Very coarse sand	Coarse sand	Medium sand	Fine sand	Very fine sand

Table 6: Evaluation of the evolutionary trend of the shoreline indices.

Evolutionary trend of the shoreline indices	Vulnerability classes				
	Very low	Low	Moderate	High	Very high
Most recent (m/year)	> 0.5	0.5 to –0.5	–0.5 to –1.0	–1.0 to –2.0	> –2.0
Short-term (m/year)	> 0.5	0.5 to –0.5	–0.5 to –1.0	–1.0 to –2.0	> –2.0
Middle-term (m/year)	> 0.5	0.5 to –0.5	–0.5 to –1.0	–1.0 to –2.0	> –2.0
Long-term (m/year)	> 0.5	0.5 to –0.5	–0.5 to –1.0	–1.0 to –2.0	> –2.0
Very long-term (m/year)	> 0.5	0.5 to –0.5	–0.5 to –1.0	–1.0 to –2.0	> –2.0

are not present in the tables, as they do not characterize the study area. These variables have been assigned a zero value in the application of the eqn (1).

As regards for the tide variations, in the Messina Strait area, the excursion is modest (0.2–0.3 m), as obtained from the analysis of the historical series of the tide gauge located in the cities of Reggio Calabria and Messina. It has been estimated that in the studied area the currents have a value of about 3–4 knots, while in the Tyrrhenian and Ionian they vary between 0.4 and 0.7 knots. The wave climate was analysed starting from the data obtained by applying the ABRC-MACRO software, developed by HR Wallingford Ltd. All the vulnerability characteristics analysed were obtained from various cartography data, which consists of historical shorelines taken from the Calabrian Geoportal, orthophotos taken from the National Geoportal and satellite imagery from 2015 to today provided by Google Earth.

Finally, the overall hazard index is equal to 0.69, while the vulnerability index is equal to 0.50.

5 CONCLUSIONS

In territories such as Calabria (Italy), characterized by approximately 750 km of coastline subjected to considerable anthropogenic pressures which have caused extensive erosion, a methodology for assessing the coastal erosion risk is strongly required.

The paper describes the definition of a new index-based methodology, which unlike those present in the literature, allows to consider all the factors that come into play, with a standardized and generally valid procedure. The methodology developed, in fact, allows to define two indices of hazard and vulnerability. The hazard index considers all the natural processes that contribute to the phenomenon of coastal erosion, such as tides, currents, average sea level rise due to climate change, wave climate and both coastal and river sediment transport. The vulnerability index, on the other hand, includes all those characteristics of the coast that can make it more or less vulnerable to coastal erosion, such as its geological and morphological characteristics (both of the emerged and submerged part), its evolutionary trend, the presence of dune systems, vegetation and coastal and/or port defence works, the subsidence rates and the degree of waterproofing of the coast. The paper also describes the results of the application of the new methodology in Porticello, located in the terminal part of the Strait of Messina (Calabria, Italy).

These results can be combined with the definition of a third index that considers the value of structures, infrastructures, population, cultural heritage, etc. present in the risk area and thus obtain a coastal risk index. From a different combination of the described factors, it will also be possible to evaluate the coastal flooding risk, thus developing a multi-hazard assessment methodology. The results obtained from this methodology can also be plotted on a free access GIS platform, thus offering a useful tool to administrations and stakeholders. For this reason, the methodology described in this paper is of particular relevance and interest in the planning and management of coastal areas.

ACKNOWLEDGEMENT

This work was supported by the Public Works Department of Calabria Region.

REFERENCES

[1] Rangel-Buitrago, N., Williams, A. & Anfuso, G., Hard protection structures as a principal coastal erosion management strategy along the Caribbean coast of Colombia: A chronicle of pitfalls. *Ocean and Coastal Management*, **156**, pp. 58–75, 2018.

[2] Stronkhorst, J., Levering, A., Hendriksen, G., Rangel-Buitrago, N. & Appelquist, L.R., Regional coastal erosion assessment based on global open access data: A case study for Colombia. *Journal of Coastal Conservation*, pp. 1–12, 2018.

[3] Komar, P.D., Coastal erosion-underlying factors and human impacts. *Shore and Beach*, **68**(1), pp. 3–16, 2000.

[4] Barragán, J.M. & de Andrés, M., Analysis and trends of the world's coastal cities and agglomerations. *Ocean and Coastal Management*, **114**, pp. 11–20, 2015.

[5] Bernatchez, P., Fraser, C., Lefaivre, D. & Dugas, S., Integrating anthropogenic factors, geomorphological indicators and local knowledge in the analysis of coastal flooding and erosion hazards. *Ocean and Coastal Management*, **54**(8), pp. 621–632, 2011.

[6] Williams, A.T., Rangel-Buitrago, N., Pranzini, E. & Anfuso, G., The management of coastal erosion. *Ocean and Coastal Management*, **156**, pp. 4–20, 2017.

[7] Günaydın, K. & Kabdaşlı, M.S., Characteristics of coastal erosion geometry under regular and irregular waves. *Ocean Engineering*, **30**(13), pp. 1579–1593, 2003.

[8] Barbaro, G., Foti, G. & Malara, G., Set-up due to random waves. *Transactions of the Royal Institution of Naval Architects Part A: International Journal of Maritime Engineering*, **155**(PART A3), 2013.

[9] Bacino, G.L., Dragani, W.C. & Codignotto, J.O., Changes in wave climate and its impact on the coastal erosion in Samborombón Bay, Río de la Plata estuary, Argentina. *Estuarine, Coastal and Shelf Science*, **219**, pp. 71–80, 2019.

[10] Wang, J., You, Z.J. & Liang, B., Laboratory investigation of coastal beach erosion processes under storm waves of slowly varying height. *Marine Geology*, **430**, article number 106321, 2020.

[11] Walling, D.E., Human impact on land–ocean sediment transfer by the world's rivers. *Geomorphology*, **79**(3–4), pp. 192–216, 2006.

[12] Barbaro, G., Foti, G., Mandaglio, G., Mandaglio, M. & Sicilia, C.L., Estimate of sediment transport capacity in the basin of the Fiumara Annunziata (RC). *Rendiconti Online Società Geologica Italiana*, **21**(1), pp. 696–697, 2012.

[13] Barbaro, G., Foti, G., Sicilia, C.L. & Malara, G., A formula for the calculation of the longshore sediment transport including spectral effects. *Journal of Coastal Research*, **30**, pp. 961–966, 2014.

[14] Barbaro, G., Bombino, G., Foti, G., Borrello, M.M. & Puntorieri, P., Shoreline evolution near river mouth: Case study of Petrace River (Calabria, Italy). *Regional Studies in Marine Science*, **29**, 100619, 2019.

[15] Foti, G., Barbaro, G., Bombino, G., Fiamma, V., Puntorieri, P., Minniti, F. & Pezzimenti, C., Shoreline changes near river mouth: Case study of Sant'Agata River (Reggio Calabria, Italy). *European Journal of Remote Sensing*, **52**(4), pp. 102–112, 2019.

[16] Marchesiello, P., Nguyen, N.M., Gratiot, N., Loisel, H., Anthony, E.J., Dinh, C.S., Nguyen, T., Almar, R. & Kestenare, E., Erosion of the coastal Mekong delta: Assessing natural against man induced processes. *Continental Shelf Research*, **181**, pp. 72–89, 2019.

[17] Valderrama-Landeros, L. & Flores-de-Santiago, F., Assessing coastal erosion and accretion trends along two contrasting subtropical rivers based on remote sensing data. *Ocean and Coastal Management*, **169**, pp. 58–67, 2019.

[18] Andredaki, M., Georgoulas, A., Hrissanthou, V. & Kotsovinos, N., Assessment of reservoir sedimentation effect on coastal erosion in the case of Nestos River, Greece. *International Journal of Sediment Research*, **29**(1), pp. 34–48, 2014.

[19] Zema, D.A., Bombino, G., Boix-Fayos, C., Tamburino, V., Zimbone, S.M. & Fortugno, D., Evaluation and modeling of scouring and sedimentation around check dams in a Mediterranean torrent in Calabria, Italy. *Journal of Soil and Water Conservation*, **69**(4), pp. 316–329, 2014.

[20] Fortugno, D., Boix-Fayos, C., Bombino, G., Denisi, P., Quinonero Rubio, J.M., Tamburino, V. & Zema, D.A., Adjustments in channel morphology due to land-use changes and check dam installation in mountain torrents of Calabria (southern Italy). *Earth Surface Processes and Landforms*, **42**(14), pp. 2469–2483, 2017.

[21] Versaci, R., Minniti, F., Foti, G., Canale, C. & Barillà, G.C., River anthropization, case studies in Reggio Calabria (Italy). *WIT Transactions on Ecology and the Environment*, vol. 217, WIT Press: Southampton and Boston, pp. 903–912, 2018.

[22] Foti, G., Barbaro, G., Manti, A., Foti, P., La Torre, A., Geria, P.F., Puntorieri P. & Tramontana, N., A methodology to evaluate the effects of river sediment withdrawal: The case study of the Amendolea River in southern Italy. *Aquatic Ecosystem Health and Management*, **23**(4), pp. 465–473, 2020.

[23] Nguyen, Q.H. & Takewaka, S., Land subsidence and its effects on coastal erosion in the Nam Dinh Coast (Vietnam). *Continental Shelf Research*, **207**, 104227, 2020.

[24] Sarma, K.G.S., Siltation and coastal erosion at shoreline harbours. *Procedia Engineering*, **116**, pp. 12–19, 2015.

[25] Prumm, M. & Iglesias, G., Impacts of port development on estuarine morphodynamics: Ribadeo (Spain). *Ocean and Coastal Management*, **130**, pp. 58–72, 2016.

[26] Miduri, M., Foti, G. & Puntorieri, P., Impact generated by Marina of Badolato (Italy) on adjacent coast. *Proceeding of the 13th International Congress on Coastal and Marine Sciences, Engineering, Management and Conservation MEDCOAST*, Mellieha, Malta, **2**, pp. 935–945, 2017.

[27] Fiori, E., Comellas, A., Molini, L., Rebora, N., Siccardi, F., Gochis, D.J., Tanelli, S. & Parodi, A., Analysis and hindcast simulations of an extreme rainfall event in the Mediterranean area: The Genoa 2011 case. *Atmospheric Research*, **138**, pp. 13–29, 2014.

[28] Boudet, L., Sabatier, F. & Radakovitch, O., Modelling of sediment transport pattern in the mouth of the Rhone delta: Role of storm and flood events. *Estuarine, Coastal and Shelf Science*, **198**, pp. 568–582, 2017.

[29] Barbaro, G., Foti, G., Nucera, A., Barillà, G.C., Canale, C., Puntorieri, P. & Minniti, F., Risk mapping of coastal flooding areas. Case studies: Scilla and Monasterace (Italy). *International Journal of Safety and Security Engineering*, **10**(1), pp. 59–67, 2020.

[30] Barbaro, G., Petrucci, O., Canale, C., Foti, G., Mancuso, P. & Puntorieri, P., Contemporaneity of floods and storms. A case study of Metropolitan Area of Reggio Calabria in Southern Italy. *Prooceedings of New Metropolitan Perspectives (NMP)* (Reggio Calabria, Italy), Smart Innovation, Systems and Technologies, **101**, pp. 614–620, 2019.

[31] Canale, C., Barbaro, G., Petrucci, O., Fiamma, V., Foti, G., Barillà, G.C., Puntorieri, P., Minniti, F. & Bruzzaniti, L., Analysis of floods and storms: Concurrent conditions. *Italian Journal of Engineering, Geology and Environment*, **1**, pp. 23–29, 2020.

[32] Canale, C., Barbaro, G., Foti, G., Petrucci, O., Besio, G. & Barillà G.C., Bruzzano river mouth damage due to meteorological events. *International Journal of River Basin Management*, 2021.

[33] Yang, Z., Wang, T., Voisin, N. & Copping, A., Estuarine response to river flow and sea-level rise under future climate change and human development. *Estuarine, Coastal and Shelf Science*, **156**, pp. 19–30, 2015.

[34] Toimil, A., Camus, P., Losada, I.J., Le Cozannet, G., Nicholls, R.J., Idier, D. & Maspataud, A., Climate change-driven coastal erosion modelling in temperate sandy beaches: Methods and uncertainty treatment. *Earth-Science Reviews*, **202**, 103110, 2020.

[35] Maiti, S. & Bhattacharya, A.K., Shoreline change analysis and its application to prediction: A remote sensing and statistics based approach. *Marine Geology*, **257**(1–4), pp. 11–23, 2009.

[36] van Rijn, L.C., Coastal erosion and control. *Ocean and Coastal Management*, **54**(12), pp. 867–887, 2011.

[37] Pikelj, K., Ružić, I., Ilić, S., James, M.R. & Kordić, B., Implementing an efficient beach erosion monitoring system for coastal management in Croatia. *Ocean and Coastal Management*, **156**, pp. 223–238, 2018.

[38] Barbaro, G., Fiamma, V., Barrile, V., Foti, G. & Ielo, G., Analysis of the shoreline changes of Reggio Calabria (Italy). *International Journal of Civil Engineering and Technology*, **8**(10), pp. 1777–1791, 2017.

[39] Jaranovic, B., Trindade, J., Ribeiro, J. & Silva, A., Using a coastal storm hazard index to assess storm impacts in Lisbon. *International Journal of Safety and Security Engineering*, 7(2), pp. 221–233, 2017.

[40] Ramieri, E., Hartley, A., Barbanti, A., Santos, F.D., Gomes, A., Hilden, M., Laihonen, P., Marinova, N. & Santini, M., Methods for assessing coastal vulnerability to climate change. ETC CCA Technical Paper, **1**, pp. 1–93, 2011.
[41] Torresan, S., Critto, A., Rizzi, J. & Marcomini, A., Assessment of coastal vulnerability to climate change hazards at the regional scale: The case study of the North Adriatic Sea. *Natural Hazards and Earth System Sciences*, **12**(7), pp. 2347–2368, 2012.
[42] Narra, P., Coelho, C., Sancho, F. & Palalane, J., CERA: An open-source tool for coastal erosion risk assessment. *Ocean and Coastal Management*, **142**, pp. 1–14, 2017.
[43] Satta, A., Puddu, M., Venturini, S. & Giupponi, C., Assessment of coastal risks to climate change related impacts at regional scale: The case of Mediterranean region. *International Journal of Disaster Risk Reduction*, **24**, pp. 284–296, 2017.
[44] Foti, G. & Sicilia, C.L., Analysis, evaluation and innovative methodologies to prevent coastal erosion. *WIT Transactions on Ecology and the Environment*, vol. 169, WIT Press: Southampton and Boston, pp. 219–230, 2013.
[45] Barbaro, G., Master plan of solutions to mitigate the risk of coastal erosion in Calabria (Italy): A case study. *Ocean and Coastal Management*, **132**, pp. 24–35, 2016.
[46] Mavromatidi, A., Briche, E. & Claeys, C., Mapping and analyzing socio-environmental vulnerability to coastal hazards induced by climate change: An application to coastal Mediterranean cities in France. *Cities*, **72**, pp. 189–200, 2018.
[47] Mucerino, L., Albarella, M., Carpi, L., Besio, G., Benedetti, A., Corradi, N., Firpo, M. & Ferrari, M., Coastal exposure assessment on Bonassola Bay. *Ocean and Coastal Management*, **167**, pp. 20–31, 2019.
[48] McLaughlin, S. & Cooper, J.A.G., A multi-scale coastal vulnerability index: A tool for coastal managers? *Environmental Hazards*, **9**(3), pp. 233–248, 2010.
[49] Viavattene, C., Jiménez, J.A., Ferreira, O., Priest, S., Owen, D. & McCall, R., Finding coastal hotspots of risk at the regional scale: The Coastal Risk Assessment Framework. *Coastal Engineering*, 2017.
[50] Satta, A., Snoussi, M., Puddu, M., Flayou, L. & Hoult, R., An index-based method to assess risks of climate-related hazards in coastal zones: The case of Tetouan. *Estuarine, Coastal and Shelf Science*, **175**, pp. 93–105, 2016.
[51] Pantusa, D., D'Alessandro, F., Riefolo, L., Principato, F. & Tomasicchio, G.R., Application of a coastal vulnerability index: A case study along the Apulian Coastline, Italy. *Water*, **10**(9), 2018.
[52] Kantamaneni, K., Phillips, M., Thomas, T. & Jenkins, R., Assessing coastal vulnerability: Development of a combined physical and economic index. *Ocean and Coastal Management*, **158**, pp. 164–175, 2018.
[53] Anfuso, G. & Del Pozo, J.Á.M., Assessment of coastal vulnerability through the use of GIS tools in South Sicily (Italy). *Environmental Management*, **43**(3), pp. 533–545, 2009.
[54] Pendleton, E.A., Thieler, E.R. & Williams, S.J., Coastal vulnerability assessment of Cape Hatteras National Seashore (CAHA) to sea-level rise. US Geological Survey Open-File Report, 1064, 2004.
[55] Davies, J.L., A morphogenic approach to world shorelines. *Zeitschrift für Geomorphologie*, pp. 127–142, 1964.
[56] Boccotti, P., *Wave Mechanics for Ocean Engineering*, Elsevier, 2000.
[57] Stockdon, H.F., Holman, R.A., Howd, P.A. & Sallenger A.H.J., Empirical parameterization of setup, swash, and runup. *Coastal Engineering*, **53**(7), pp. 573–588, 2006.

[58] Tomasicchio, G.R., D'Alessandro, F., Barbaro, G. & Malara, G., General longshore transport model. *Coastal Engineering*, **71**, pp. 28–36, 2013.
[59] Gavrilovic, S., *Méthode de la classification des bassins torrentiels et équations nouvelles pour le calcul des hautes eaux et du débit solide*. Vadoprivreda, Belgrado, Serbia, 1959. (In French.)
[60] Silva, S.F., Martinho, M., Capitão, R., Reis, T., Fortes, C.J., & Ferreira, J.C., An index-based method for coastal-flood risk assessment in low-lying areas (Costa de Caparica, Portugal). *Ocean and Coastal Management*, **144**, pp. 90–104, 2017.

SECTION 6
FLOOD RISK
MANAGEMENT

TOWARDS AN ABSORPTION STRATEGY FOR THE MANAGEMENT OF FLOOD RISK: STUDY OF THE STRATEGY OF RESISTANCE AND THE ARTIFICIALIZATION OF SOILS IN THE FRANCO–BELGIAN CROSS-BORDER TERRITORY

EMELINE COSZACH[1], VINCENT BECUE[1], JÉRÉMY CENCI[1], ISABELLE DE SMET[1], SMAIL KHAINAAR[2], PATRIZIA LAUDATI[3] & CÉDRIC RIVIÈRE[1]
[1]Faculté d'Architecture et d'Urbanisme de l'Université de Mons, France
[2]Laboratoire DeViSu de l'Université Polytechnique Hauts-de-France, France
[3]Laboratoire SIC.Lab de l'Université Côte d'Azur, France

ABSTRACT

Historically, civilizations have settled near streams, rivers and seas. These territories offer a strategic position for the development of trade and transport, an ideal relief for the development of cities and fertile soils, favorable to crops. Idyllic picture if we forget their strong vulnerability in the face of rising waters. Indeed, by this choice of location, humans are also subject to the risk of flooding. Over the course of experienced disasters, human societies have developed protective measures with the aim of reducing the caused damages, both human and material. In the boom of the industrial era, these measures primarily resulted from civil engineering in a strategy of resistance against water. Dams, pumping stations, outlets, pipes, etc. then emerged, exponentially, on all continents. Even though these measures have proved their worth on several occasions and over many years, they present, and in Europe as well, several significant disadvantages. Physical and visual barriers preventing direct contact with water, they have also played a major role in the artificialization of our soils and the loss of biodiversity through the drying out of certain portions of the territory. These infrastructures, legacy of our past, are no longer adequate with the new risks and challenges of the cities of the 21st century. It is therefore necessary, in order to make our cities resilient, to adapt our management of the flooding risk currently under the symbol of water control towards the management of absorption and the "living with water". To manage this, a historical study of the development of these infrastructures but also of their impact on the territory and the landscape will be carried out with a particular focus on the Franco–Belgian cross-border territory formed by the districts of Mons and Valenciennes.
Keywords: risk, inondation, artificialization, resistance, adaptation, nature, biodiversity, Valenciennes, Mons.

1 INTRODUCTION

Water is an essential component for human life, for the development of ecosystems and societies. It is therefore quite natural that since immemorial time, civilizations have settled near streams, rivers and seas. Source of supply and recreation, these territories also offer a strategic position for the development of trade and transport, an ideal terrain for the development of cities as well as rich soils, favorable to crops. An idyllic picture if we forget their high vulnerability to rising water levels.

Indeed, as Terrin [1] explains, the relationships maintained by human societies with regard to water have long exhibited a certain puzzling ambivalence. Sometimes tinged with horror, fear and sadness when men are abandoned idle, sometimes with no more family, culture, livestock or even property. Sometimes tinged with joy and prospect when the land is made fertile and prosperous, suitable to the development of agriculture. First maintained by mythological stories, this dual attitude of societies towards aquatic environments continues to this day. Man is divided between on the one hand the unforeseeable and difficult to

WIT Transactions on Ecology and the Environment, Vol 250,
www.witpress.com, ISSN 1743-3541 (on-line)
doi:10.2495/WRM210171

measure danger that they represent and on the other hand, the unshakeable attraction to water landscapes and their proximity [1].

The concept of flood risk management is relatively recent. But very quickly, on all continents, human communities have continued to develop, throughout disasters, adaptive capacities allowing them to live as close as possible to water, despite the risk that this represents [1]. Rising groundwater, soil liquefaction, subsidence, flooding of rivers, storm surges, torrential rains, … flooding can be extremely variable in nature. Although it was used at one time as a military strategy to push back enemy troops [2], it is generally the result of natural phenomena. While traditional cultures have rather turned to adaptive strategies, Western communities have focused on the culture of zero risk [1]. In the boom of the industrial era, these measures will result mainly from civil engineering in a strategy of water resistance. Dams, pumping stations, outlets, pipes, … will then emerge, exponentially.

Although these measures have proven their worth over and over again for many years, they nevertheless have certain major disadvantages. Physical and visual barriers preventing close contact with water, they have also played a major role in the artificialization of our soils and the loss of biodiversity through the drying up of certain portions of the territory. An insidious utopia to have put an end to natural hazards and to be able to disregard the risk without any consequence according to Bauduceau [3], the natural disasters that have occurred in recent years have sadly reminded us of the growing vulnerabilities of this technology as time goes by.

Thus, these infrastructures, legacy of our past, are no longer up to the new risks and challenges of the cities of the 21st century. It is therefore necessary, in order to make our cities resilient, to adapt our management of the risk of flooding currently under the acronym of water mastery towards management of water absorption and "living with water" with a particular focus on the planning of our territories.

But what is this "resist" flood management strategy and what are these limits that we are talking about? What are the consequences for our territories and human societies, but above all, the possible ways of adaptation for an inclusive risk management? What about the specific territory formed by the districts of Mons and Valenciennes?

Indeed, with the objective of shedding light on the present and guiding this new risk management strategy, a historical study of the development of these infrastructures but also of their impact on the territory and landscape will be carried out with a particular focus on the Franco–Belgian cross-border territory formed by the districts of Mons and Valenciennes. To answer these questions, we will first try to determine what this flood risk management strategy based on water resistance consists of and to highlight its various limits whether they are structural, economic or social. Secondly, we will analyze the consequences that these may have had on the planification of our territories, the composition of our landscapes and the decline of biodiversity before we suggest ways of reflection and actions to develop a reconciliation strategy in terms of flood risk management. In a third and final step, we will carry out an analysis of the cross-border territory formed by the districts of Mons and Valenciennes regarding its evolution following the demergement and containment policies put in place in the two countries and regarding the place left to water.

2 FROM A STRATEGY OF WATER MASTERY

Societies and human activities, by their establishment in the territory, create vulnerability and inherent risk. Artificialization, urban expansion, development of industries and canals, river and rail transport, etc. all create, according to Peil [4], a new topography, a new soil and new climatic conditions which, together, disrupt the cycle of water and soil permeability; in other words, the place of water within this territory. These changes are responsible for the upsurge in flooding episodes and their intensity.

This urban expansion as well as the rise of the population always living as close as possible to the water only increases the stakes associated with flooding. The need to react in order to ensure the safety of populations, infrastructure and property thereby reducing their vulnerability was then felt. It was in this context that the first water resistance structures were created.

2.1 "Resist" strategy

The resistance strategy instinctively developed by human societies for many years is (based)founded on the base idea that "it is desirable and possible to prevent water from rivers and seas from entering exploitable and valuable territories" [3]. The objective is to remove from the whims of nature, the territories threatened by flooding and thus valorize them.

Often referred to as "fight against floodings" or "hazard control", this historical flood risk management is built essentially on the control by technique of natural events. Sometimes still well anchored in the urban policies of our contemporary society, it is established on the absolute confidence in the technology spread by an engineering culture [3].

Until then made in embankments, it is from the industrial period with the development of calculation methods and innovative materials that the engineering works will flourish [5]. All concrete-clad or almost, these boil down to three different types of infrastructure sometimes combined with each other.

The first consists in protecting oneself as well as possible from aquatic flows by setting up barriers such as dikes and dams. The second includes retention areas which are generally built upstream of inhabited areas such as basins, artificial lakes and canals. And finally, in order to allow the fastest and most effective possible evacuation of water, the third category includes evacuation works by pumping, drainage and pipes [1].

The city of Tokyo, an example in this field, has chosen to develop this strategy through the development of water barriers throughout the entire perimeter of the city and even well beyond. Continuously raised, they can, in the event of an imminent flood, be lowered and thus retain the flow of water outside the city. Other barriers, this time sliding, are also present within the city itself. By being deployed on the streets, they create a zone of refuge for all the inhabitants. Finally, the city is also equipped with a colossal drainage system made up of cylindrical reservoirs and monumental underground galleries [6].

The Netherlands, because of its position below sea level, also had to quickly put in place a flood risk management strategy. Although they have since turned away from it, water control had also been favored through the establishment of the Delta National Plan and the many dams that make it up.

2.2 The limits of the "resist" strategy

We cannot argue that this strategy of resistance and water mastery has already proven itself many times over and has made it possible to protect, during numerous climatic episodes, populations and properties located in vulnerable territory. However, it tends to reach its limits.

Indeed, the natural disasters that have occurred in recent years (French floods in 2019, European floods in 2016, British and French floods in 2015, East-European floods in 2014, etc.) have sadly reminded us that the risk of flooding is very real and even increasingly present and disastrous, making entire populations and territories more vulnerable. Of all these events, it was Hurricane Katrina in 2005 in the United States that created the most stir and shook the conscience of the whole world [1]. Bordered by defensive walls and levees for nearly 550 km to protect it from marine and river invasions, New Orleans thinks itself out of reach and out of danger. Its urbanization has, for a non-negligible part, developed in the major bed of the river but also in areas of marshes drained by dammed canals. Thinking of being safe within his fortress, the scenario of a failure of the protection system was never imagined. When, finally, breaches form and allow water to spread over nearly 80% of the territory to heights ranging from two to four meters, the population finds itself trapped. The human and material damages are colossal; transport, energy and telecommunications networks are in jeopardy; the economy is collapsing and the unemployment rate is skyrocketing. Five years after the incident, New Orleans has yet to regain its former glory [3].

This tragic episode, among many others, made people aware of the fragility of this strategy of risk management by water mastery. These infrastructures are certainly robust and imposing but reveal several weaknesses.

The first weakness is the creation of an over-danger situation in the event of a breach of protections by generating a feeling of forgetting the risk or a feeling of overconfidence in technology. Indeed, the danger no longer being visible, a tenacious desire for urban densification without protection ensues. As water and the inherent risk are buried and repelled ever more intensively, the notions of hazard, risk and vulnerability become minimized or even completely erased from political consciousness and local populations. The latter, no longer aware of the danger involved, do not develop a home adapted to the risk of flooding, nor do they know what behavior to adopt in the event of a climatic episode.

The second weakness results in “the frequent difficulty of the territories to manage, maintain, monitor and perpetuate their protection structures” [3]. Indeed, the cost of setting up and then managing and maintaining these infrastructures is colossal. The dike heritage is consequently largely in a degraded state, no longer achieving its initial objective of protection. “In France, one of the most dammed up countries in Europe, seventy percent of the budget dedicated to flood risk management is nevertheless still devoted to the construction, strengthening and raising of these dikes” [3]. Vicious circle of this embankment campaign since it only increases the maximum flow rate of the water and therefore the pressure exerted on the dikes which are more and more weakened and degraded [7].

At the same time, the phenomenon of global warming will also highlight the limits of this flood risk management strategy. Indeed, as explained to us by Gautier [8], due to the increase in the average temperatures of the atmosphere on the one hand which causes the melting of the ice, and the increase in the average temperatures of the oceans on the other hand causing the thermal expansion of the waters, the level of marine waters will be raised. Meteorological phenomena are already on the increase on the different continents. These events will be more and more frequent, fierce and extreme given the changes incremented by the global warming that we are currently experiencing. The infrastructures imagined decades ago on the basis of now obsolete data are therefore no longer commensurate with the risk that our territories are currently experiencing.

We have the technical capacity to build and tirelessly strengthen our dikes in order to make them reach ever greater heights. As Stépahne Hallegate [6] points out: “We can build walls six, eight or even ten meters long, but do we really want to?” This would inexorably tarnish the relationship we have developed with water for ages, so characteristic of our civilization. Moreover, the perception at the foot of these immense walls as a man living in the city, would take a whole new dimension and could give way to a grim feeling of crushing and imprisonment.

Thus, these infrastructures, a legacy of our past, are no longer up to the new risks and challenges of the cities of the 21st century. This management strategy is a dead end, we cannot just erect barriers. Adapting our flood risk management strategy with the objective of making our cities resilient is therefore necessary and must be accompanied by a rigorous analysis and an integrated conception of this risk at the scale of the territory coupled with raising awareness among the population exposed to this risk.

3 TOWARDS AN ABSORPTION STRATEGY

3.1 The consequences of the "resist" strategy

Strongly anchored in the territories, the landscape and management policies, particularly in Europe, this historic flood risk management strategy is far from harmless [3]. By taking place on the territory, these civil engineering infrastructures have in fact had serious consequences and modifications on our soil, our landscape and our biodiversity.

First erased from urban and sometimes even peri-urban landscapes in the 18th century for security, health and economic reasons, this cutoff between water and human societies was accentuated in the 19th and 20th centuries with the development of infrastructure for the water mastery [9]. Dams, dikes, banks, raising of quays, retention basins or even burial under pipeline constitute both physical and visual barriers in the landscape. They prevent any close contact with water for both people and biodiversity. Rather an astonishing and contradictory strategy for a society which has always sought to develop as close as possible to this essential component of human life. Over nearly 200 years, more than 45,000 dams have been erected across the world. These also have accentuated the phenomenon of coastal land recession since the sediments being blocked, they cannot reach the coast anymore [6].

The development of these engineering structures also greatly contributes to the artificialization of our soils. A double problem which resides on the one hand in the works themselves. Almost entirely in concrete as specified above, they will, by their simple implementation, contribute to the artificialization of our soils. On the other hand, the primary objective of this water mastery strategy being to preserve the territories threatened by flooding from natural hazards in order to valorize them, the land gained by their installation will very quickly be urbanized. This urban sprawl, emblematic problem of the second half of the 20th century in Europe, leads to a strong waterproofing of the soil [1]. Sometimes even reducing to void the infiltration of water into the different layers that make up the soil, the phenomenon of runoff appears or is accentuated in both urban and rural areas. These territories are therefore found in some cases and in another way, still and always vulnerable to flooding.

Biodiversity will also be strongly impacted by these flood risk management practices. The drying up of wetlands in the idea of being able to urbanize them as well as the development of drainage systems in the hygienic approach of the 19th century will contribute to the gradual disappearance of these environments and of the particular ecosystem that they constitute [10]. As the infrastructures developed do not present any support for the previously present biodiversity, it will gradually desert this concrete environment. Elements of discontinuity in rivers, they will also have an impact on hydro-systems. Isolating rivers and streams from their alluvial plain, the hydrological rhythm is increased. As the areas of natural expansion of water in the event of flooding are no longer accessible, the risk of flooding is therefore also accentuated [11], [12].

3.2 Leads for an absorption strategy

What if the solution was simply residing in the erasure of these "barrier" measures? What if it consisted, on the contrary, in accepting water as an integral component of the territory rather than containing it and pushing it away with the help of these all-concrete-clad structures?

Over the past 20 years, an alternative has been rising in the flood risk management policy: "Faced with the risks, the territories must adapt and become less vulnerable, more robust, more resilient, that is to say better able to cope with and recover from a catastrophic event" [3]. As Hubert [13] points out, this management encourages us to understand the city as a system and to anticipate on the one hand its possible disturbances and failures during a crisis episode but also on the other hand, to consider its ability to resist, to adapt but also and above all to recover one's functions after the crisis. Risk management, therefore, shifts from a strategy of resistance, established essentially on blind confidence in the various civil engineering works, thus encouraging urban development in total disregard of the existing risk, to an adaptation strategy, where the risk of the flood will be taken into account and integrated into any planning project. The heart of the management strategy therefore becomes territory planning with the design of devices ready to overcome any possible failure of the protective structures [3].

At European level, this paradigm shift is reflected in the creation of an "Inondations" directive. This encourages an essential adaptation of the territories by insisting on the need to "reduce the negative consequences of floods on human health, the economy, the environment and the cultural heritage" [3]. Some countries will therefore gradually develop a new strategy, allowing more room for water and by extension for nature, that of "living with water". It consists of the ingenious, thoughtful and conscious reception of water on the territory or certain portions of it, inducing an adapted design of these zones: natural banks gently sloping towards the water, landscaped floods expansion park, flood plains, etc. "This means that there should be no obstacle to the natural flow of water, because any obstacle, by slowing down the flow, leads to the raising of the water line and thus increases the vulnerability of upstream sectors" [4]. Urbanization and architecture will therefore have to be studied so that, even in the event of a flood, water can retain its place without endangering the populations present. "Cities are no longer pebbles and become sponges" [4].

The "Room for the River" project developed by the Netherlands in 2005 is part of this perspective. Colossal work, its objective was twofold: to reduce the risk of flooding by acting on the water levels of the reference flood and to improve consideration of the environment. In total, thirty different projects have been developed on the arms of the Rhine, the Meuse, the Waal and the IJssel with the common desire to restore water to its full extent [14].

By thus returning to the center of concerns and urban design, water is no longer perceived as a threat but rather as an opportunity, despite the risk it represents. This practice is also in line with the increasingly strong social demand in terms of nature. The fact of integrating this component again in the planning of the territory makes it possible to rewrite the relationship between people and their environment until then neglected. As explained previously, "[…] by the absence or the poverty of biodiversity on these works (of civil engineering), or by the elements of discontinuity of the watercourses that they can generate, they generally present an ecological 'loss'" [12]. This new strategy therefore takes advantage of this desire for nature to enhance the connection of our towns and villages to their natural ecosystems.

These natural ecosystems, their biodiversity and above all the ecosystem services they provide, although too often neglected, are essential to human life. These ecosystem services are defined as "the benefits that human societies derive from functional ecosystems" [15].

Divided into four categories, they include production services, cultural services related to the environment of everyday life and well-being, support services and finally, regulation services such as climate and extreme event regulation. Ecosystems and living organisms therefore contribute in particular to "cushioning natural disasters by limiting the damage caused by floods, hurricanes, tsunamis, landslides or even droughts while ensuring the maintenance of the quality of air, water and soil" [16]. Often invisible, these services are therefore taken for granted by most of us and are still very rarely taken into account in urban policies.

Through this new understanding of risk by absorption, other functions will be able to be questioned again: soil permeability, water cycle, preservation and development of biodiversity, reduction of the effects of heat islands, air quality, safeguarding ecosystem services, …. Nature will ultimately play a leading role in this new management strategy. In addition to promoting flood risk management, it will bring added value to the territory. Conceived and inserted under the prism of the concept of multi-benefits, it will be favorable both for human societies and for biodiversity.

4 CASE OF THE DISTRICTS OF MONS AND VALENCIENNES

The subsequent objective of our study is to be able to develop a strategic absorption plan leaving more room for water as described above, in the districts of Mons (Belgium) and Valenciennes (France). To achieve this and within the framework of this article, we will first attempt to provide a history of the evolution of these territories in their management of the risk of flooding and the space left to water over time. Secondly, a diagnosis of the current territory targeting its vulnerability to the risk of flooding and its resistance infrastructures will be carried out in order to establish possible adaptation leads for an inclusive risk management.

4.1 Presentation of the study area

The district of Mons is located in the province of Hainaut in the Walloon Region in Belgium while the district of Valenciennes is located in the northern department of the Hauts-de-France region in France (Fig. 1). Both forming an integral part of the international hydrographic district of the Scheldt (DHI Escaut in French), the main waterways crossing them are the Scheldt for the district of Valenciennes, the Haine and the Nimy-Blaton-Péronnes canal for the district of Mons.

4.2 History of flood risk management in the territory

The international hydrographic district of the Scheldt has an essentially flat landscape. The rivers making up the district are therefore "of the lowland type with wide valleys and reduced flowrates and water flows" [17]. The changes that have taken place over time can be explained in particular by the fact that it is "one of the most densely populated and industrialized hydrographic districts in Europe" [17].

The Scheldt and its tributaries, mainly included in the Valenciennes district, are therefore low-sloping low-flowrates rivers, flowing into large floodplains with a flat profile and forming meanders. It therefore has a natural propensity to overflow. Over time, a large part of its meanders has been rectified and channeled in order to guarantee protection against flooding on the one hand and to promote navigation on the other hand [17].

In fact, from the 12th century, works to drain wet plains and to protect against flooding were carried out. These took place in "a context of population growth where the need for

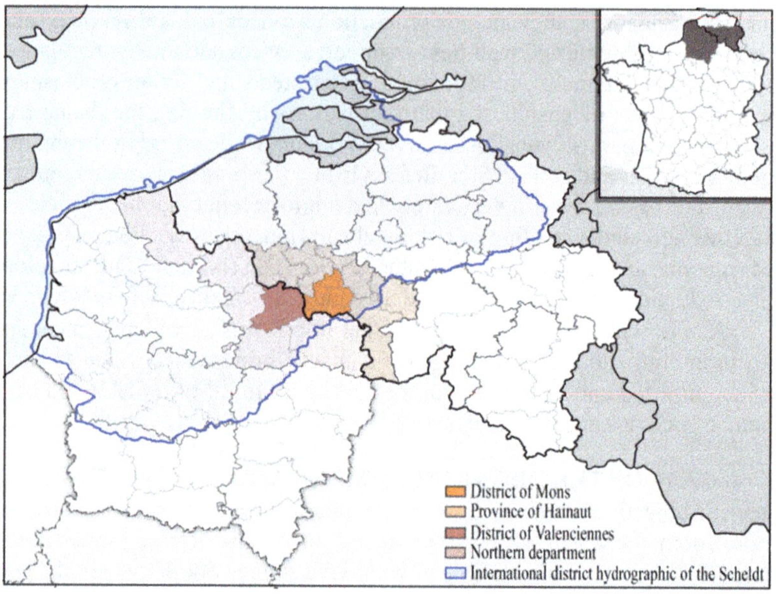

Figure 1: Situation of the districts of Mons and Valenciennes in the international hydrographic district of the Scheldt. *(Source: Personal card produced on the basis of ISC-CIE data.)*

arable land is increasing sharply" [2]. Many canals facilitating the rapid evacuation of water are therefore created, particularly and more frequently in the region of Condé, which is very humid and regularly subject to natural and military floods. The landscape of the Scheldt valley was finally completely crisscrossed in the 18th century by this ever-expanding hydraulic system of canals.

The industrial boom of the 19th century will in this case again lead to major works to modify the course of the Scheldt. Straightening of meanders, widening and deepening of its major bed, embankment, etc. so many actions carried out in order to control the flow of the river and make it navigable. Sounding the death knell for the presence of water in the urban centralities of the Valenciennes district, the gradual covering of rivers and urban canals in the 19th and 20th centuries completes this French containment policy developed in the name of flood protection and public sanitation [2].

The territory of the Mons district is also traversed on all sides by a river which will mark its development, La Haine. Characterized by a steeply sloping flow regime and a narrow valley on its middle and upper parts, navigation is only possible on its lower part, between Condé and Mons, where the landscape is relatively flat and the valley, very wide [18].

Already used in the 13th century for the transport of goods, it was quickly dammed and diverted for the first time in the 18th century in order to increase, just as with the Scheldt, the possibilities of flooding in the event of a siege [19]. Various dyking, damming and canal works will be carried out in the valley of Haine but, although it has always been prone to flooding and the region between Mons and Condé has always been swampy, the main works of engineering will rather be carried out in a commercial logic [20]. In fact, given the growing demand for coal at the dawn of the industrial revolution, the construction of the Mons-Condé canal at the beginning of the 19th century appears to be "an essential condition for allowing the Mons basin to fit into the large-scale waterway projects that emerged at that time" [21]. However, it will be filled a few years later in favor of the Nimy-Blaton-Péronnes canal, suitable for larger boats and also crossing the Montois district from side to side.

These various modifications of the landscape are not the only elements to have considerably complicated the flow of water in these two regions. Intensive mining brought an upheaval in the physical geography and completely changed the hydrographic regime of the two valleys. Sinking basins gradually formed and needed, mainly on Belgian territory, the raising of the banks, the containment of numerous rivers but also the installation of numerous pumping stations in order to raise the sewage, drainage and runoff waters and thus protect the entire Mons valley from flooding [18]. This irreversible phenomenon of the collapse of rocks, initially soaked in water, is called subsidence.

4.3 The "resist" strategy in action on the territory

At present, the major hydrographic network of the Mons and Valenciennes territories is totally artificial or largely modified (Fig. 2). The only natural components are the various minor tributaries of the Scheldt and the Haine. They generally match with a weak urban development in their surroundings. It is therefore the proximity or not of Man which conditions the place of water within the territory and over time. It is also interesting to note that, while many measures have been put in place for the management of the flood risk in the districts of Mons and Valenciennes during the last decades, the DHI Escaut still considers the majority of the area as "Presenting a significant potential risk of flooding" (Fig. 3).

Another alarming observation: the ecological state of the hydrographic network studied (Fig. 3). Indeed, the DHI Escaut classifies rivers according to several categories ranging from "very good" to "bad". It is to be noted that no hydrological component is included in the "very good" category, only one being in the "good" category: the "Marre à Goriaux", a mining subsidence pond has declared a national biological reserve since 1982. While the

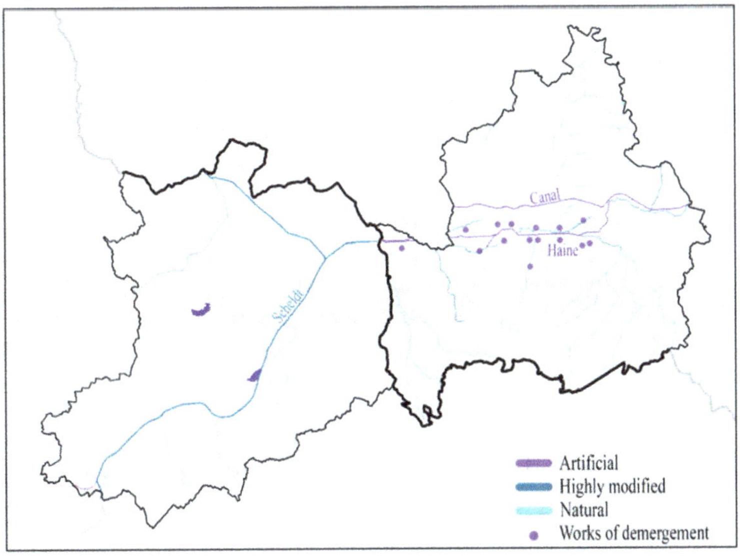

Figure 2: Status of surface water in the district of Mons and Valenciennes. *(Source: Personal card produced on the basis of ISC-CIE data.)*

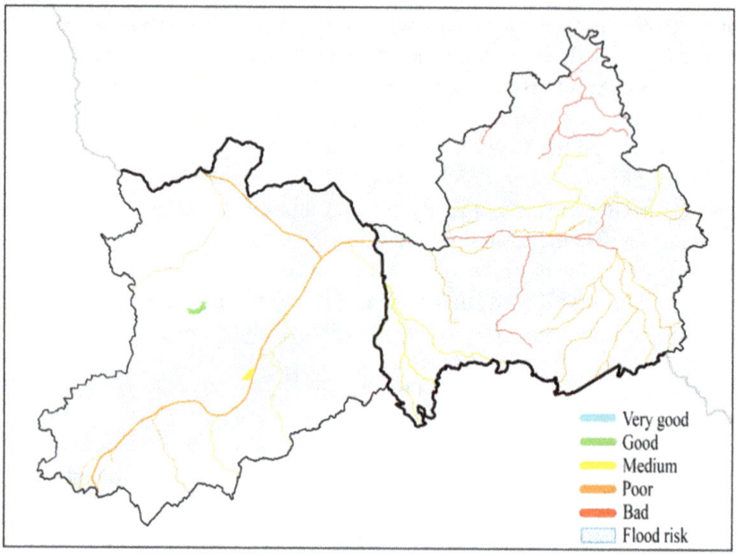

Figure 3: Flood risk and ecological state of surface water in the districts of Mons and Valenciennes. (*Source: Personal card produced on the basis of ISC-CIE data.)*

major watercourses such as the Scheldt and the Haine are included in the "mediocre" and "bad" categories due in particular to their strong waterproofing, one would have thought that the secondary water axes would have may have been of a higher quality than listed.

All the components of the hydrographic network should therefore be studied in order to reduce the potential risk of flooding while improving the ecological state.

5 CONCLUSION

Managing flood risk through adaptation has finally long been rooted in our traditions. From time immemorial, populations have not stopped adapting and reacting to the disasters they have experienced: dikes of earth, wood or stone, canals, ... The objective is always to protect the land as much as possible from invasion or rising waters to allow safe development of our societies.

The objective of this article was initially to be able to analyze the past by making a history of flood risk management over time and its "resist" strategy. The rise of the industrial period and the civil engineering infrastructures that accompany it only amplify this development, always closer to water as well as its damming, plunging into oblivion the inherent risk as well as the possible consequences, often dramatic, of the acts that we carry on the territory. Given the limits and consequences of these resistance structures highlighted throughout this article but also due to the increase in risk under the effect of climate change, we must now integrate the management of flooding risk in our development projects.

To complete this, a diagnosis, at present, has been carried out on the cross-border territory of the districts of Mons and Valenciennes to be able in the future of our study, to propose a more inclusive flood risk management and in full awareness of the risk incurred, by the use of multi-benefit nature both for human societies and for biodiversity. "Observe the present, know the past to anticipate the future" [22]. The diagnosis of the territory carried out within the framework of this article therefore consists of the first essential step in the implementation of a flood risk management strategy under the symbol of absorption on the territory of the districts of Mons and Valenciennes.

The aim is not to completely erase, on the entire territory, especially given the present phenomenon of subsidence, the artificial structures which are so many physical, visual, social and biological barriers. The aim is indeed to design flood risk management that reconciles human societies and their relationship with water and natural ecosystems.

The shift from a policy of "fight against" towards "living with" has already been initiated, to a greater or lesser extent, by certain nations. We just have to follow in their footsteps under the prism of "multi-benefit" nature. The next step in our study will therefore consist in determining what measures can be developed and applied in the cross-border study area to meet this new "living with" flood risk management strategy. Opening up of waterways previously buried underground, de-artificialization of canalized portions or even the creation of floodplains? Only the future will tell us.

REFERENCES

[1] Terrin, J.J., Renouvellement des approches urbaines face aux inondations. *Villes inondables: prévention, adaptation, résilience*, ed. Parenthèses, col. La ville en train de se faire, Marseille, pp. 12–27, 2014.

[2] Deudon, L., Pour une histoire des inondations et de la vulnérabilité dans la vallée de l'Escaut. Histoire et archéologie de la vallée de l'Escaut: Carnet d'Escaut, 2017. https://escaut.hypotheses.org/28.

[3] Bauduceau, N., Risque d'inondation et stratégies d'aménagement en Europe. *Villes inondables: prévention, adaptation, résilience*, ed. Parenthèses, col. La ville en train de se faire: Marseille, pp. 205–216, 2014.

[4] Peil, C., Hydrologie et projets urbains. *Villes inondables: Prévention, adaptation, résilience*, ed. Parenthèses, col. La ville en train de se faire: Marseille, pp. 234–247, 2014.

[5] Comité Français des Barrages et Réservoirs, L'histoire des barrages. Document technique, 2013. shorturl.at/iwHIJ.

[6] Mandy, M., Inondations: une menace planétaire, documentary, Géorama TV film production, 92 minutes, French, 2015.

[7] Barraqué, B. & Gressent, P., La politique de prévention du risque d'inondation en France et en Angleterre: de l'action publique normative à la gestion intégrée. Rapport de recherche final, Université de Marne-la-Vallée, 2004. shorturl.at/egvA8.

[8] Gautier, Y., Inondations. *Encyclopædia Universalis*. shorturl.at/CEHQ6. page consultée le 23/02/2016.

[9] Gralepois, M. & Guevara, S., L'adaptation aux risques d'inondation façonnée par les métiers de la ville. *Développement durable et territoires*, **6**(3), 2015. shorturl.at/fiyzO.

[10] Scarwell, H.-J. & Franchomme, M., Autour des zones humides: espaces productifs d'hier et conflits d'aujourd'hui. *VertigO*, **6**(1), 2005. shorturl.at/ktKQU.

[11] Scarwell, H.-J. & Laganier, R. Chapitre 2: L'intégration matérielle du risque par le biais des mesures structurelles. *Risque d'inondation et aménagement durable des territoires*. Presses universitaires du Septentrion: Villeneuve d'Ascq, 2004. shorturl.at/diILO.

[12] Rey, F., Breton, V., Breil, P. & Mériaux, P., Les solutions fondées sur la nature pour accorder la prévention des inondations avec la gestion intégrée des milieux aquatiques. *Sciences Eaux & Territoires*, **26**(2), 2018. shorturl.at/tMUW2.

[13] Hubert, G., Ville et inondation: une cohabitation délicate. *Villes inondables: prévention, adaptation, résilience*, ed. Parenthèses, col. La ville en train de se faire: Marseille, pp. 218–233, 2014.

[14] Room for the river programme, Dutch Water Sector. shorturl.at/cksO8. Accessed on: 17 Mar. 2021.

[15] Rives, F., Pesche, D., M2ral, P. & Carrière, M.-F., Les services écosystémiques: une notion discutée en écologie. 2016. shorturl.at/syAHZ.

[16] Une biodiversité en plaine santé pour nous protéger des catastrophes naturelles. BeBiodiversity. shorturl.at/hozDH. Accessed on: 15 Mar. 2021.

[17] Le district hydrographique international de l'Escaut, Comission internationale de l'Escaut. shorturl.at/jxGN8. Accessed on: 15 Mar. 2021.

[18] Annales des mines, Le démergement en Wallonie. 1998. shorturl.at/bfnHM.

[19] Hallez, M., Villes et villages de la vallée de la Haine – Mons. 2017. shorturl.at/xDJLM.

[20] Tison, L.-J., La lutte contre les inondations en Belgique. *Ciel et Terre*, **55**, pp. 47–55, 1939. shorturl.at/sRV36.

[21] Honnoré, L., La canal de Mons à Condé: projets, financement et réalisation. *Revue du Nord*, **4**(387), pp. 813–838, 2010. shorturl.at/auQS8.

[22] Delcourte-Debarre, M., Géohistoire du territoire forestier de l'Avesnois: observer le présent, connaître le passé pour anticiper le futur (France). *VertigO*, **16**(3), 2016. shorturl.at/rxILZ.

NATURE-BASED SOLUTION AS FLOOD PROTECTION STRATEGIES: THE CASE OF CARAHÁ RIVER PARK, BRAZIL

RENATA V. DE OLIVEIRA GOBETTI[1], ORNELLA SALIMBENE[2] & DEBORAH PANEPINTO[2]
[1]Technological University Dublin, Ireland
[2]Department of Environmental Engineering, Politecnico of Torino, Italy

ABSTRACT

Following the road paradigm, many cities, at the beginning of the 20th century, experienced a reckless increase in the number of vehicles on the road causing an unsustainable urban and environmental situation. The origin of the problem surely lies in the occupation of the preserved areas. In many South American states, for example, highways generally ignore the presence of environmental obstacles (such as rivers) as they may cause a "delay in modernization". As a result, riparian forests are invaded and turned into highways, water beds are channeled and most rivers are contaminated with wastewater. Due to the occupation of green and riparian areas, urban floods have become more and more and consequently urban spaces have been disqualified. This situation also involved the Carahá River that crosses the city of Lages in southern Brazil, in fact the river embankment has become completely impermeable due to the construction of an urban road. Proposing urban spaces with an emphasis on improving environmental quality and integrating forms of public transport are basic points that could lead to the sustainable development of our cities. In this context, the aim of this research is to propose a sustainable river project for the Carahá River, with the aim of presenting nature-based solutions to prevent floods and improve the environmental, functional and aesthetic quality of the city. To develop this research, first theoretical studies were carried out on the relationship between rivers and cities, secondly three international case studies and related literature were analyzed with the aim of deepening the knowledge of the connection between rivers and population. Finally, an intensive study of the territory was carried out and a social analysis was conducted through a semi-structured interview in which the population of Lages participated. The result of this study contributed to effectively defining the intervention guidelines for the Carahá river park project considering, as an added value, also the main needs of citizens.

Keywords: river revitalization, flood disaster management, nature-based solutions (NBS), Brazil.

1 INTRODUCTION

Nature-based solutions (NBS) are increasingly being adopted as measures to mitigate and adapt to climate change, to reduce flood risks and to improve urban ecosystems [1], [2]. NBS can reduce risks to people and environments as effectively as traditional infrastructure but potentially offer many additional benefits such as the following: improved natural habitat for wildlife, improved water and air quality, improved conditions socio-cultural community [3]. Indeed, over the decades, the occupation of river banks has caused severe flooding and highlighted the enormous lack of proper urban planning of cities in order to avoid these disasters. Some authors have defined the extreme river events as "the revenge of the rivers", highlighting how the race for modernization and strong anthropization, have transformed the life of the people and have profoundly changed the relationship between the city and the river. For example, in the United States, Riley demonstrates that starting in the 1930s, there was no hesitation in channelling the river, digging tunnels or narrowing river beds, and man-made interventions were decidedly invasive. In the same decade, in Brazil, according to Gorsky [4], the river beds were rectified to combat the floods that caused serious epidemics and anthropogenic interventions were justified as a measure of prevention and public health.

WIT Transactions on Ecology and the Environment, Vol 250, © 2021 WIT Press
www.witpress.com, ISSN 1743-3541 (on-line)
doi:10.2495/WRM210181

The consequences and environmental impacts generated by these interventions were manifold. Gorski states that uncontrolled and unsustainable anthropization processes expose aquatic life to altered temperatures, increase the speed of water and shut down the biotic components of the system, eliminating its biodiversity.

Furthermore, corrections of river beds cause more erosion and silting, remove aquatic life, damage water quality and alter the hydrological conditions of the river, its drainage capacity, discharge processes and ultimately its flow. In an attempt to alleviate flood problems in some areas, the Carahà river has undergone important changes to its bed over the years. Despite efforts, flooding in specific areas of the city of Lages has become an increasingly frequent problem especially after extreme weather events. To aggravate the problem, also the lack of basic sanitation, in fact in 2010 only 20% of the wastewater generated in the city was treated, while the rest was discharged into the river (Municipal Water and Hygiene Secretariat (SEMASA)). Observing the precariousness of basic sanitation, the increasing environmental pollution and the disqualification of the river in the urban environment, the main purpose of this work was to develop a landscape redevelopment project based on the NBS (s) to realize a river park and restore the environmental, functional and aesthetic conditions of the banks of the Carahá river.

2 THEORETICAL RESEARCH

2.1 Creating a public space of quality

According to Lucia Maria [5] to carry out corrections of the riverbed and banks in a river, it is essential to know the characteristics of the site and understand the potential it could offer. It is extremely important to conduct an efficient investigation of land use, hydrology, vegetation, natural drainage and flood prone areas. To intervene and create a high-quality public space around the river, Jacobs [6] listed the following seven strategies to follow: experience (the city as a collective environment), identity and control (feeling of belonging to the space), access to opportunities, imagination and distraction (having new experiences), authenticity and meaning (understanding the collective space), public and community life (encouraging people to public participation), urban self-esteem (urban space is self-sufficient), good environment for all (accessible to all). To achieve these goals, Jacobs states that the urban context should have the following prerequisites:

a) streets and neighborhoods must have an active public life and planning should promote sustainable actions to combat mono-functionality (e.g. not promoting exclusively residential areas) and consequently facilitate the diversity of urban functions (e.g. mixed use);
b) integrate activities (life, work, free time) close to each other, the collective space should be understood as the structuring element of the urban form and not play the role of "excess" space (due to the arrangement of the buildings);
c) avoid or eliminate the concept of urban structures organized according to dispersed and distinct buildings with an excessive complexity of arrangements and relationships in space.

In short, for the construction of a quality public space, it is necessary to attribute to the place a diversity of uses, functions and users. For Jacobs, parks are defined as ephemeral places and when they are successful in the urban context, it means that they have no competition with other green areas. The author further states that if the goal of an urban park is to attract the greatest number of types of people at different times of the day, then it is essential to promote a generalization of attendance through the multiplicity of reasons people have for attending the park. For this assumption it is not necessary to have a complex plan of the park but just add some relevant difference or quantity of vegetation that will be accentuated by the different uses in the park. In addition to real river parks, there are alternative drainage systems that reduce the cost of urban infrastructure and also alleviate the problem of flooding. This type of alternative drainage systems are called green infrastructures (GI), an effective example being rain gardens that could be placed near the road curb to receive surface rainwater runoff. Another example is provided by the linear depressions rich in vegetation (earth and other filter elements) which cause the slowing down, the purification of rainwater and increase the runoff time. To adopt a green infrastructure for environmental qualification, the use of native vegetation with high filtering and purifying power is essential. The most effective plants for these purposes are the following: bamboo (bambusa vulgaris), calla (zantedeschia aethiopica), achira (canna edulis) and papyrus (cyperus) [7]. Specifically, the reed bed (phragmites australis) has a rapid growth advantage and has an efficient system of oxygen transport from the leaves to the roots; the water hyacinth (eichhornia crassipes) on the other hand has the ability to absorb heavy metals and organic elements and is able to reduce coliforms and bad smells of polluted water by 77%. Fig. 1 shows the main plants with a purification effect and the main environmental aspects related to them.

Environmental Aspect	
Shading	The shading of the trees has the purpose of softening the thermal rigor of the hot season, reducing the sensation of heat of the users.
Natural Illuminance	In order to obtain a better control of the natural illuminance in the urban area, it is necessary to take into account some criteria in the selection of tree species, such as: size, canopy type, foliage and fundamentally its luminous transmittance.
Air Relative Temperature and Humidity	The trees, especially those of large size, add to the urban area the more thermal capacity (the more mass is included) increasing its inertia and causing a daytime fall in temperature variations. The type of foliage of the tree is important to obtain a control of the humidity of the air in the different enclosures.
Urban Ventilation	The wind significantly influences the ambience of the architectural and urban spaces, and consequently, the thermal sensation of its users. Through an appropriate proposal it is possible to reduce the effects such as very high or low speed, transport of dust, generation of noise and damage of buildings.
Minimizing Noise	Vegetation can help reduce noise contamination in five different ways: by absorbing sound (sound is eliminated), by drifting (changing the direction of sound), by reflection (sound reverts back to its source), by refraction (sound waves change direction around an object by concealment – the unwanted sound is covered by a pleasant sound).

Figure 1: Plants in act in the water treatment and environmental aspect. *(Source: Authors.)*

3 LITERATURE REVIEWES: RELATIONSHIP BETWEEN RIVERS AND CITIES

3.1 Los Angeles (California, USA)

The city of Los Angeles located in California (USA) was developed on the banks of the Los Angeles River. In 1894 the river was rectified and as a result numerous floods began to occur. In 1914, the river was channeled by the United States Army Corps of Engineers. The proposed revitalization of the Los Angeles River only started in 2002 and provided for an intervention cut in the urbanized area of the city corresponding to 59.2 km of river extension [8]. The intervention was divided into nine sections, defined by specific characteristics, deficiencies and potential. Fig. 2 shows the Los Angeles River before revitalization and the Los Angeles River Project for revitalization.

Figure 2: Los Angeles river before revitalization and Los Angeles River Project for revitalization. *(Source: Jao, 2014.)*

In the early stages, arboreal and herbaceous vegetation was added. Subsequently, the focus was on water depollution and the creation of residential areas. These strategies are mainly aimed at saving the ecological functions of the river, attracting the population and reconnecting communities with each other. For this project, four main objectives have been created and specific guidelines have been developed to achieve the expected result, as seen in Table 1.

3.2 Cheonggyecheon (South Korea)

Cheonggyecheon, in Japanese means "clean water", in the past the river was a utility in the daily life of the rural city, it represented the playground for children and laundry for women. In 1958 the river was invaded by the main road which would later become one of the most important urban arteries. In 1999 the municipality started the Cheonggyecheon river recovery project, which involved the demolition of the highway and the construction of a linear park, giving way again to the natural riverbed [9], [10]. The extension of the demolition was 5.4 km and about 75% of the demolition material was reused for the construction of the river park, for the construction of four connection points and new bridges, as well as for the rehabilitation of the stream (Fig. 3) which has been recovered and transformed into a linear urban park of 5.8 km in length, 400 hectares and 80 meters in width.

The main guidelines of the project were to reclaim the Cheonggyecheon River, promote urban renewal by creating a public space for the population, change the growth model of the city through sustainable means of transport in addition to the automobile.

Over the years Seoul has expanded the public transport network and changed the traffic flows that circulate in the city center, so there has been an increase in the number of users

Table 1: Guidelines for the revitalisation. (*Source: Gorski, 2010.*)

Goals	Guidelines
Revitalisation of the river	To valorise the floodplains with the recovery of riparian vegetation; restoring ecosystem functionality (restoring ecological and hydrological functions in the long term); improve water treatment and quality; secure public access.
Creation of green linear system of connection between neighborhoods and the river	Create a continuous green path; connect the neighbourhoods to the river; expanding public and recreational spaces; connect the river to the identity of the neighbourhoods; incorporate elements of public art along the river.
Creation of attraction and opportunities for the community	Make the river a focus of activities: accessible, safe, healthy, green and celebratory place; celebrate the river as cultural heritage; engaging the local population in the process of building and planning the neighbourhood.
Quality of life of the population	Increase the supply of employment, housing, and commerce; create an environmentally appropriate urban design and guidelines and opportunities for land use; concentrate efforts on the requalification of badly areas and poor communities, in social disadvantages, guarantee equal access to leisure areas, work, public transportation, etc.; increase public revenue and increase income distribution.

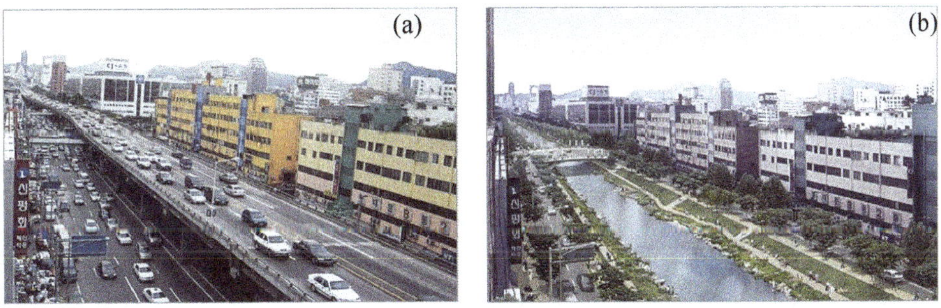

Figure 3: Cheonggyecheon. (a) Before revitalization; and (b) After revitalization. *(Source: Giaretta, 2011.)*

choosing new transport systems. The project promoted the revitalization of the river space so that citizens could spend more time on the banks of the river. In addition, the reduction in traffic has led to cleaner air and reduced heat islands, with an average temperature 3.6°C lower than in other parts of the city. A said consequence of the microclimatic variation has led to the birth of new natural habitats in the city (fish, birds and insects) which have rebalanced some fundamental biological chains. The Cheonggyecheon rehabilitation project has transformed Seoul into a green city with a river of clean and valued waters. One of the main objectives for the development of this project was to transform the city of Seoul into a sustainable city capable of integrating the Cheonggyecheon River into the daily life of the population. However, one of the downsides of revitalization was gentrification and the increase in value from 30% to 50% of all properties along the river.

3.3 Torino city (Italy)

In this third case study, the research team of the Polytechnic of Turin examined the water quality of the Po river in a small stretch between two stations (Castiglione Torinese and Brandizzo) in northern Italy (Fig. 4) as a consequence of the polluting loads introduced and the natural ability of the river to receive these loads without critical degradation of water quality. The main objective was to evaluate the effects of the discharge of a purification plant (SMAT) through a qualitative study of water comparing loads, origins, destinations and effects [11]. To achieve these objectives, the hydrological and qualitative picture of the stretch studied was reconstructed using the mass balance method to define the polluting loads in the selected stations and a one-year sampling campaign was carried out. On the basis of the analyzes, it was possible to conclude that the environmental state of the river Po was only minimally influenced by the discharge of the SMAT purification plant and therefore an intervention on diffused loads was necessary (for example, limitation of the use of fertilizer). These results can be useful for river system planners and administrators in the development of future intervention and remediation policies. Furthermore, the proposed methodological approach can be adopted for other significant assessments of the impacts of the polluting load, both from concentrated and diffuse sources.

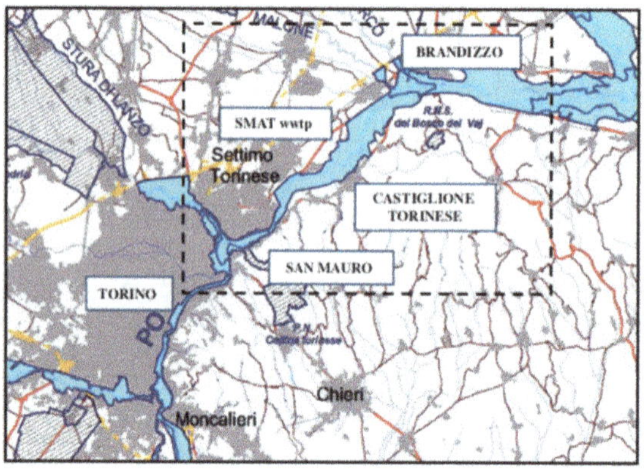

Figure 4: Pò River, main cities and study area (Italy). *(Source: Panepinto et al., 2016.)*

4 STUDY AREA: CARAHA RIVER (BRAZIL)

The city of Lages is located in the mountainous region of the state of Santa Catarina (SC) in southern Brazil and is the largest municipality in the state of SC by territorial extension (Fig. 5(a)). However, the extension of the urban area in the city corresponds to just over 8% of the total area as shown in the image. The average temperature is 14.3°C, the predominant wind in the city is the northeast and its annual rainfall is about 1600 mm with the periods of greatest rainfall in the months of May, June, July and August. Lages is located in the Uruguay watershed, in the Canoas river basin (National Water Agency, 2014), the city is crossed by the Carahá river for 9 km of extension from start to finish. 83% of the basin area is sealed and public roads and footpaths account for 35% of sealed soil in private residential areas contributing to the floods [12]. Fig. 5(b) (1940) shows the beginning of the urban expansion of Lages (highlighted in red) which took place around the cathedral of the city, furthermore the river rectifications are visible.

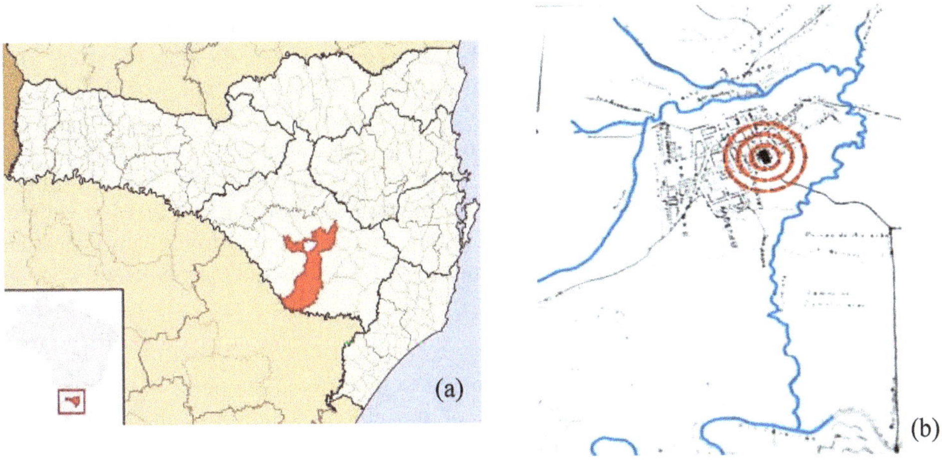

Figure 5: (a) Lages city. *(Source: SIRHESC, 2014)*; and (b) Caraha River in 1940. *(Source: Thiago de Castro Museum.)*

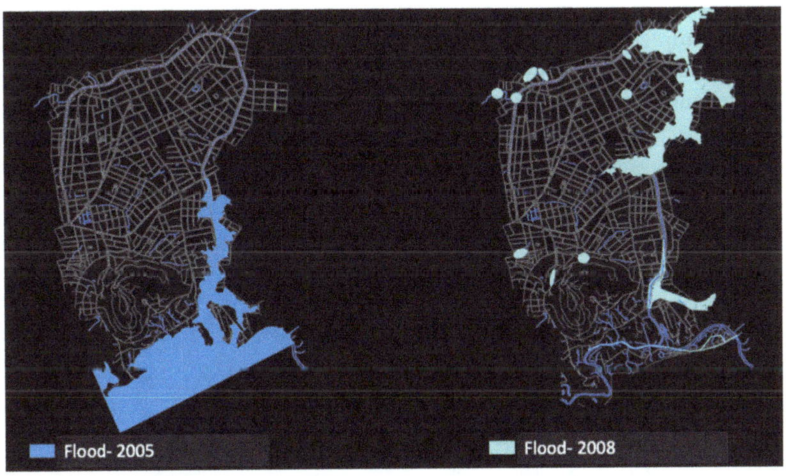

Figure 6: Floodings in 2005 and 2008, consecutively. *(Source: SEMASA, 2020.)*

According to the data available from the Municipal Secretariat for Water and Lages Hygiene (SEMASA), Fig. 6 represents the areas most affected by the rains of 2005 and 2008.

There are two main reasons according to Palma Neto [13] that cause the flooding of the Carahá river in the city of Lages and they are summarized below:

1) The small ciliary band of the river is limited to 8–10 m along the river.
2) Inadequate afforestation in the ciliary forest section where there is a low density of large specimens. Furthermore, another factor that causes flooding is the discharge of wastewater directly into the river without any preventive treatment. In Lages, according to the Ministry of Health (2006), only 37.7% of waste water was collected in 2000, several studies have also indicated that there are no pristine waterways in the urban

perimeter of Lages. In 2013, a legislative motion proposed diverting and suppressing sections of the Carahá River to remedy the floods and improve the flow of water, while neglecting the socio-environmental fragility that this intervention would entail. As demonstrated in this work, transforming a river bed into a "concrete box" involves a brutal intervention in nature and does not address the causes of the flood problem. A project called "Carahá de Cara Nova" has been proposed in the city which aims to preserve the river and raise awareness of its importance. Working concurrently with this existing project, the proposed revitalization of the river park aims to raise awareness and enhance the importance of the river for the city and its consequent conservation.

4.1 Research methodology: Questionnaire

The citizens of Lages were given a questionnaire to understand the perception of the local population in relation to the public spaces of the city. According to the questionnaire presented to a sample of fifty people of various age groups, 82% replied that they felt the lack of public spaces in the city of Lages. In fact, despite the great potential for the creation of a river park, the Carahá River has only a still precarious cycle path along its banks. To better organize the questionnaire, The River Carahá, was divided into six sections (Fig. 7) so as to understand which section was the most frequented and used by citizens. Many of the people interviewed confirmed that they do not use some sections of the river bank due to a lack of urban infrastructure: adequate cycle paths, protective embankments, lighting. However, most people who answered the questionnaire confirmed that they use river banks more than three times a week for physical activity (walking, running or cycling).

Figure 7: Section of river Carahà. *(Source: R.V. de O. Gobetti.)*

The graph in Fig. 8 shows the results of the social survey. The results show which equipment is most awaited by the population for the revitalization of the park. Lawn, footpaths, fruit trees cover the highest percentages, from 30 to 40%, resulting in the greatest needs for the local population. Fishing activities, recreational sports services (skateboarding,

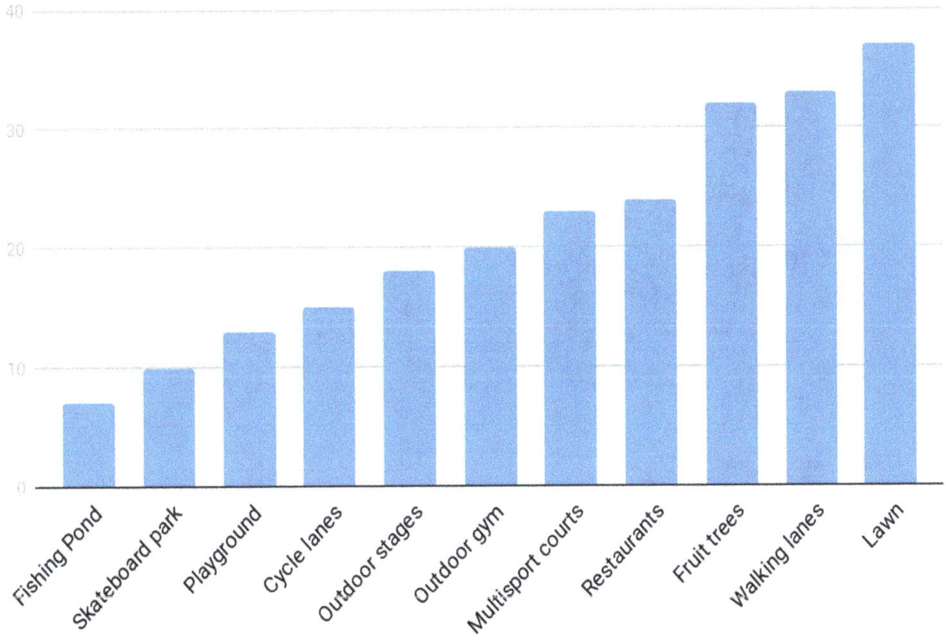

Figure 8: Result of social survey. *(Source: R.V de O. Gobetti.)*

cycle paths, gym) rank between 8% and 20%. Finally, catering and recreational activities cover 23% of the responses, resulting among the greatest needs of the population immediately after lawn, footpaths and fruit trees.

On the basis of this questionnaire and in situ studies, the two areas most affected by floods were chosen as intervention areas and a green infrastructure was proposed as a practical solution. The first area chosen for the redevelopment is located on a more pronounced curve (section 01 in Fig. 9) and is characterized by the congruence of a tributary of the Carahá with inadequate conservation radius. During a technical-environmental analysis that investigates the entire section and the surrounding area, a density of mainly residential buildings was found in this area. Consequently, it is important to intervene on the territory through conservation and development actions of leisure areas for residents. An improvement of the pre-existing recreational space was planned, including in the design the public facilities suggested by the population that participated in the questionnaire. Thanks to its proximity to the city center, the area affected by the redevelopment has a high potential for the provision of services such as bars and restaurants, spaces for hosting events and a market. The second chosen area was completely flooded in 2005. This was caused by the rectification of the river which caused an increase in the speed of the water flow causing a flood. The area of section 02 is mainly residential with isolated infrastructures such as the municipal school and the church. The potential identified for this area is that of free time, being an area further away from the city center, it lends itself more to recreational activities in nature. The project identified as possible equipment: sports fields, skating, playground, gym and green areas for contemplation.

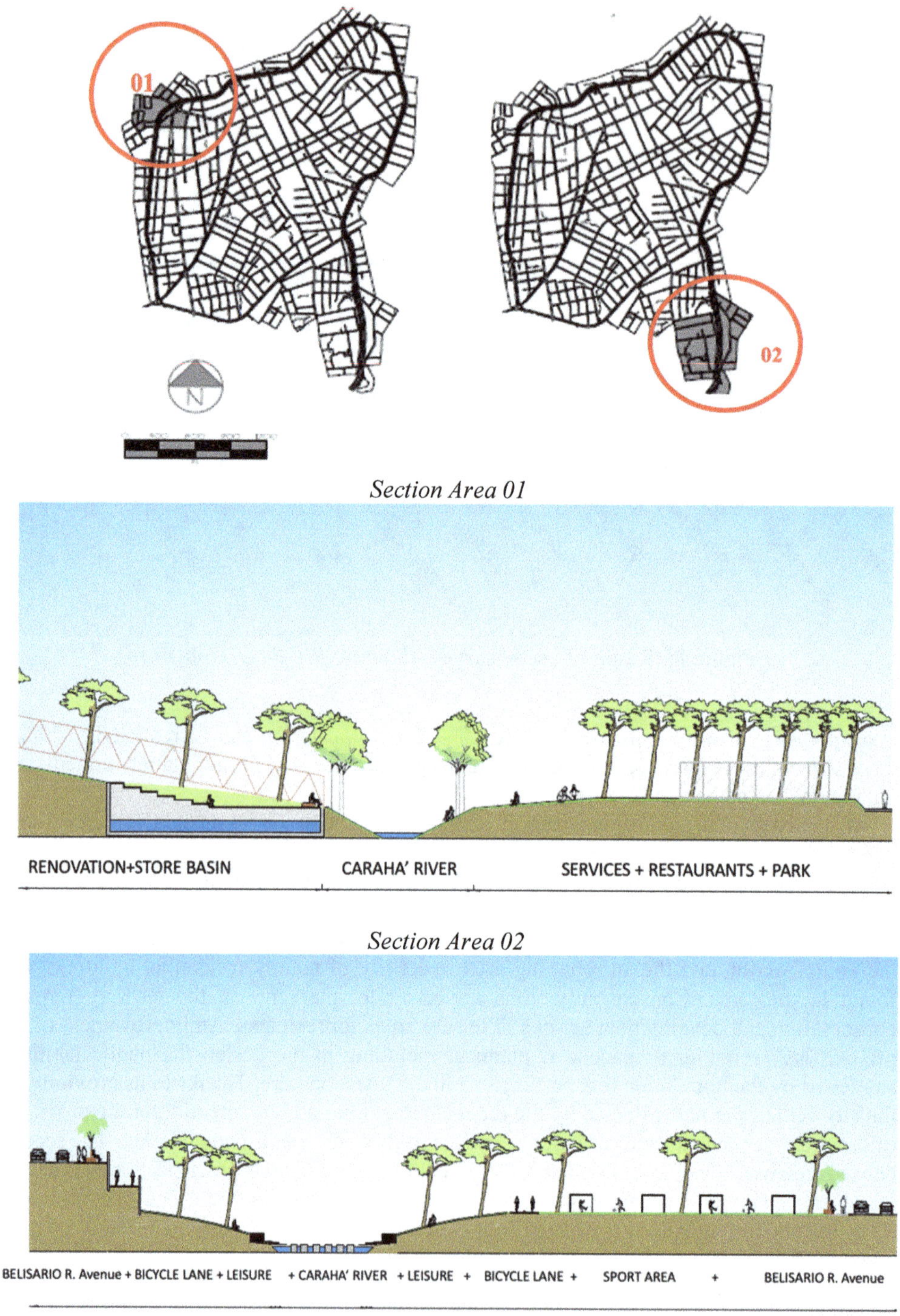

Figure 9: Intervention areas, section 01 and section 02. *(Source: R.V. de O. Gobetti, 2021.)*

5 CONCLUSION

The city of Lages has grown considerably over the years and has received significant proposals which need to be well researched before being implemented. The work presented here proposes a preliminary idea of a river park suitable for the prevention of floods and the requalification of the environmental, functional and landscape aspects of the city. For this reason, reference theories and studies were presented that show the impact caused by the canalization and the closure of the river, thus demonstrating that they are not the best solution to adopt [14]. The theoretical review also aims to understand the importance of the river-city connection and to present the direction that urban planning can follow in Brazil in a sustainable way [15]–[17]. To do this, we have exhibited reference studies that are able to reconcile the urban environment with the natural environment and have contributed to the development of guidelines in various aspects. Through a survey on the rains of 2005 and 2008 provided by the Municipality of Lages, areas of flood and low presence of public spaces on the banks of the river were detected, which was the main constraint for the determination of specific areas of the river park. Furthermore, the questionnaire was carried out with people belonging to different age groups, different social classes and dissatisfied with the public spaces existing in the city. However, throughout the study carried out on the territory, it is perceived that the range of influence of public areas often coincides with each other, so this dissatisfaction can be attributed to the lack of existing infrastructures. So that this does not happen again in the planned proposals, the questionnaire also elaborated a program related to the needs of the citizens. In general, the purpose of the Carahá River Park proposal was to include two basic approaches: that of a circuit (Av. Belisário Ramos) offering well-structured cycle paths and tree-lined streets and secondly green recreational spaces. Green areas should have two specific objectives, namely, to represent an NBS that can contain river floods and also meet the recreational needs of citizens. The creation of a river park structured in this way would allow to offer citizens various opportunities and would develop an integrated connection plan between citizens and the river which is the basis of a river regeneration program.

ACKNOWLEDGEMENT

Thanks to Prof. Me. Arq. Gabriela Morais Pereira, Universidade do Estado de Santa Caterina, UDESC (Brazil) for data support and for supervision of the project work.

REFERENCES

[1] Cohen-Shacham, E., Janzen, C., Maginnis, S. & Walters, G., Nature-based Solutions to address global societal challenges. IUCN publication, 2016.

[2] Benjamin D. et al., Natural Assurance Scheme: A level playing field framework for Green-Grey infrastructure development. *Environmental Research Journal,* 2017.

[3] Dong, X. et al., Enhancing future resilience in urban drainage system: Green versus grey infrastructure. *Water Research,* **124**, 2017.

[4] Gorsky, D. et al., The effects of smolt stocking strategies on migratory path selection of Adult Atlantic Salmon in the Penobscot River, Maine. *North American Journal of Fisheries Management*, 2011.

[5] Lucia Maria, S.A. et al., Environmental restoration of urban rivers in the metropolitan region of Rio de Janeiro, Brazil. EUE. *Environment Urban Journal*, **4**, 2010.

[6] Jacobs, D. E. et al., Linking public health, housing, and indoor environmental policy: successes and challenges at local and federal agencies in the United States. *Environmental Health Perspectives*, **115**(6), 2007.

[7] Zanella, L. *Plantas ornamentais no pós-tratamento de efluentes sanitários: Wetlands construídos utilizando brita e bambu como suporte*. Campinas, São Paulo. 2008.

[8] Jao, C. *Feds Okay $1-Billion Los Angeles River Project.* https://www.kcet.org/shows/earth-focus/feds-okay-1-billion-los-angeles-river-project. 2014. Accessed on: 20 Jul. 2020.

[9] Rowe, P. G. *Os resultados e a história do projeto de restauração do Cheonggyecheon, em Seul.* AU. (Revista online). Ed 234, set 2013. http://au.pini.com.br/arquitetura-urba'nismo/234/restauracao-do-cheonggyecheon-seul-coreia-do-sul-296126-1.aspx. Accessed on: 16 Apr. 2014.

[10] Giaretta, R. (2011). *Uma impressionante renovação urbana em Seul.* http://portalarquitetonico.com.br/uma-impressionante-renovacao-urbana-em-seul. 2014. Accessed on: 16 Apr. 2014.

[11] Panepinto, P., Marchese, F. & Genon, G., Evaluation of Po River water quality in Torino (Italy): Effects of diffuse and local point loads. *Urban Water Journal*, **13**(6), pp. 583–599. 2016. DOI: 10.1080/1573062X.2015.1011669.

[12] SIRHESC. *Portal de informações sobre o Sistema de Informações sobre Recursos Hídricos do Estado de Santa Catarina. Bacia do Rio Canoas*, 2014. http://www.sirhesc.sds.sc.gov.br/sirhsc/conteudo_upload/316/Estado+canoas.JPG.

[13] Palma Neto, C. *Diagnóstico socioambiental da bacia hidrográfica do rio Carahá.* 2013.

[14] Fondation Espace Pour La Vie. *Capturing water, cooling the air: Rain gardens*, 2020. https://fondationespacepourlavie.ca/en/solutions/capturing-water-cooling-the-air rain-gardens/.

[15] Cometti, J. L.S. et al., *Qualidade da água e perspectivas de revitalização de um riacho urbano no Recife, Brasil*, RICA, vol. 11. 2020.

[16] Gorski, M. C. B. *Rios e cidades: ruptura e reconciliação.* São Paulo: Editora Senac. 2010.

[17] Solomon, S. *Climate Change 2007—The Physical Science Basis: Working Group I Contribution to the Fourth Assessment Report of the IPCC*, Cambridge University Press, 2007.

DEVELOPMENT OF FLOOD TRACKING TECHNIQUE FOR FLOOD DISASTER MANAGEMENT IN URBAN AREAS

DONG HO KANG[1], SUK HO LEE[2] & BYUNG SIK KIM[1]
[1]Department of Urban and Environmental Disaster Prevention Engineering, Kangwon National University, Republic of Korea
[2]Risk Management Research Department, UNO Co., Ltd., Republic of Korea

ABSTRACT
The change in precipitation due to climate change has induced local and torrential rainfalls. As a result, protected lowlands have an increased risk of damage caused by river inundation and subsequent flooding. Thus, it is necessary to have the technology to predict the damage due to flooding in a protected lowland. This study developed a distributed flood and inundation model called Simplified Inundation Model (SIMOD) to simulate a flooding path for identifying how flooding caused by river embankment's breakage is inundated and spread in a protected lowland. Two methods were applied to simulate a flood path: the multiple flow direction method (MFDM) and the flat-water assumption (FWA) method. Virtual scenarios were set to calculate an inundation range over time in the Seongseoje, located downstream of the Geumho river, and evaluate the model. In addition, the urban inundation analysis was executed using the SIMOD. The results were compared to results from using the commercial inundation analysis two-dimensional flood routing model (FLO-2D) to verify the model's applicability. The study applied the digital elevation model (DEM) terrain data for conventional flooding analysis and digital surface model (DSM) terrain data to represent urban building groups and road distribution. The 24-hour simulation was done less than 7 minutes using the SIMOD, and the difference in flooded areas was less than 20% compared to FLO-2D. Moreover, the study identified the need for topographical data using DSM for urban areas, as the analysis proves the influence of roads and buildings on flooding.
Keywords: simplified inundation model (SIMOD), two-dimensional flood routing model (Flo-2D), multi direction method (MDM), flat-water assumption (FWA).

1 INTRODUCTION

Global weather and water-related disasters caused by climate change have increased events of flood and drought. The extreme weather phenomenon called "super El Niño" continues to bring record-breaking rainfalls and consequential damages to the United States and Latin America. These severe rainfall events have submerged some urban areas in the mid-western region of the United States by overflowing the Missouri and Mississippi Rivers. Similarly, Paraguay in Latin America has experienced its worst flooding due to this weather phenomenon. The La Niña phenomenon, El Niño's reverse and equally extreme weather phenomenon, increases rainfall in the western Pacific regions, directly impacting the flood and drought South Korea has been experiencing. The country experienced severe drought successively in 2014 and 2015 and is currently suffering from drought. Damages caused by this phenomenon may worsen with floods in a few years due to the increased severity and frequency of extreme precipitation and drought over time [1]. Moreover, the current design of South Korea's drainage-related infrastructure did not consider extreme weather phenomena because its frequency analysis was based on the assumption of stationarity. Thus, the design flood was calculated using the frequency analysis considering the non-stationarity of extreme weather characteristics [2].

The design criteria for levees of main rivers in Korea were structured to protect against 100-year floods. However, considering the extreme weather characteristics of interchanging severe rain and drought, flooding would be less frequent but more severe. Furthermore,

WIT Transactions on Ecology and the Environment, Vol 250, © 2021 WIT Press
www.witpress.com, ISSN 1743-3541 (on-line)
doi:10.2495/WRM210191

because of extreme precipitation, the possibility of floor damage in protected lowlands (urban areas) increases due to possible overflow and levee breaches. In particular, a flood stage increases rapidly in urban streams with a short travel time during torrential rainfalls, putting stream facilities, such as roads, bridges, and buildings in protected lowlands, at a very high risk of damage [3]. Thus, it is necessary to cope with flooding due to extreme weather by establishing an emergency action plan (EAP) at urban streams. A highly reliable model is needed to predict flooding and set up an appropriate measure.

The stormwater management model (SWMM), two-dimensional flood routing model (FLO-2D), and MIKE FLOOD are available as multi-dimensional flood models to simulate flooding. For the SWMM, it takes a long time to develop baseline data, such as storm sewer networks. Likewise, developing baseline data, such as river network configuration and boundary conditions, is time-consuming using MIKE FLOOD. For the FLO-2D model, its simulation time takes much longer as it requires advanced hardware resources and technical skills to develop a high-performance real-time forecast system in a short period using a multi-dimensional flooding model [4]. Thus, it is necessary to create a time-efficient and cost-effective method by simplifying a complex model.

In this study, a distributed flood and inundation model called the Simplified Inundation Model (SIMOD) was developed to simulate a flood routing path in a protected lowland (urban area). The digital elevation model (DEM) or digital surface model (DSM) uses terrain data and simple hydrological data, such as inflow floods. This model was developed to express a flood routing path over time, which has the advantage that it can produce results within a relatively short simulation time, around 10 minutes. The flood routing path is simulated in densely populated urban facilities of Daegu Metropolitan City following a levee (Seongseoje) break and Geumho river basin overflow. A comparative analysis with the commercial FLO-2D model was conducted to verify the developed model's applicability.

2 RESEARCH TREND

The research trend in South Korea and overseas has changed from employing inundation analysis methods using various mechanisms in the past to using strategies applying inundation models easily and rapidly by simplifying models in recent years [5]. Simulated flooding considering land use, buildings, narrow flow path, and changes in underground rainwater drainage to model the runoff in a flood plain in urban areas. Similarly, Pathirana et al. [6] developed a two-dimensional (2D) urban inundation model and linked the model with EPA-SWMM5, which was a one-dimensional (1D) drainage network model. They also simulated urban inundation by simplifying the model's process using the spread and dynamic models. In addition, Nanshan et al. [7] developed an integrated flood and inundation simulation model to predict and evaluate the flood risk using friction parameters to estimate flow data and damage area to provide terrain data, DEM, and model's inflow and outflow boundary conditions. In addition, Yang, et al. [8] conducted a study on developing a real-time flood forecast system using a simplified model with DEM for flood routing in a protected lowland. Likewise, Liu et al. [9] developed and applied an efficient cellular automata model to simulate rainfall-runoff and flood-inundation process with extreme rainfall events. This model can simulate infiltration, runoff, and dynamic flow behavior.

Related studies in Korea were conducted mainly to analyze inundation in protected lowlands for EAP or flooding map creation [10]. Applied Preissmann's method to analyze flooding in a river by considering levee overflow and physical breakage characteristics and propagating the resulting flood wave was to a protected lowland. Their study used 1D and

2D models for inundation analysis in a protected lowland. Subsequently, they developed a numerical analysis model to connect the areas between the stream channel and protected lowland where the two models can interact. In addition, they introduced a related equation for the analysis of submerging effect to accurately calculate the overflow from the river according to the water level rise in the protected lowland.

Lee and Yeon [11] performed an inundation analysis by hourly analyzing the inundation depth and area using XP-SWMM. A different investigation by Lee et al. [12] studied flood and levee collapse characteristics in the Jungnangcheon, a typical urban stream in Korea, to analyze the levee inundation due to torrential rainfall urban stream basins and the damage size due to inundation under various conditions spatiotemporally. Specifically, they analyzed unsteady stream flows and flooding in a protected lowland for each levee breach condition using the Flood Wave routing model (FLDWAV). The subcritical and supercritical flows, or mixed flows where two flow conditions were combined spatiotemporally, could be calculated using this model.

In addition, Park et al. [13] analyzed the inundation and flood based on levees' erosion and scour theory. Their analysis was done during the river levee collapse, changes in levee breakage width over time, estimation of inundation depth and area, and levee collapse while considering the change in river flow to predict the estimated damage caused by inundation. Moreover, they analyzed the water level in urban streams where a levee collapsed, the breach flow calculation, flood map creation, and inundation depth through the developed model. Similarly, Kim et al. [14] calculated a dam break outflow according to various dam collapse scenarios to prepare a method of flood inundation map, which was coupled with the 1D dam-break flood forecasting model (DAMBRK) for dam collapse and geographic information system (GIS). In addition, they compared the maximum flood stage with the right and left levee heights and calculated the peak flow, maximum flood stage, flood-wave travel time, and travel time at the maximum flood stage. Their study was done to show detailed flood inundation map preparation by water depth based on the hydraulic analysis of the river linked with the GIS. Conclusively, their study overcame the limitations of 1D hydraulic analysis models on the river had regarding the floodplain analysis. Moreover, Kim et al. [15] calculated a peak flood discharge over the collapse duration by simulating a virtual dam collapse during the probable maximum precipitation and simulated inundation time, shape, and area by collapse duration.

3 THEORETICAL BACKGROUND

Based on scenarios, the river inundation flow due to past floods can be generated in any arbitrary cell over time. The generated flow has different pathway directions following the downward slope that produces a flow direction slope and the upward slope that creates a reverse direction slope. The multiple flow direction method (MFDM) was used when the flow had a downward slope, while the flat-water assumption (FWA) was employed when the flow had an upward slope. Fig. 1 shows the SIMOD data processing and flow chart. In the MFDM, the total outflow was distributed to the lowest elevation cell among the adjacent eight cells of the inundation starting cell. In contrast, the total outflow was distributed to the neighboring cells when surrounding all eight cells of the inundation starting cell had a higher elevation than that of the starting cell. Thus, the flood at a specific time from the inundation starting cell is transferred to adjacent cells according to two algorithms, MFDM and FWA, explained in the following subsections.

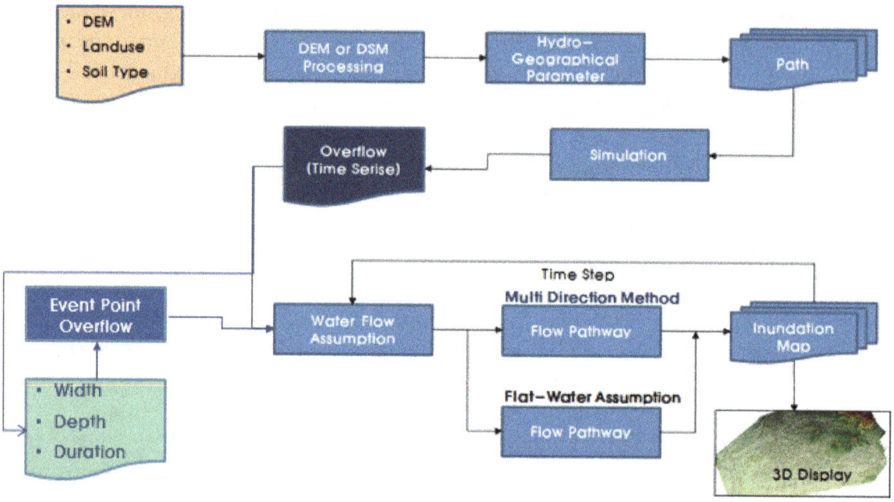

Figure 1: Flowchart of simplified inundation model (SIMOD).

3.1 Multi-flow direction method

The flood scenario determined the river flood flow. The river flood flow's starting cell was then established as the point at which the flood started. When the simulated flood began, the total outflow was distributed and transferred to adjacent cells. Previously, the single-flow direction method (SDM) was mainly used to distribute the flood discharge. In the SDM, the total outflow was distributed to adjacent cells using the D-8 and D-∞ methods, as shown in Fig. 2(a) and (b). The D-8 method distributes the total outflow to the lowest slope cell out of eight adjacent cells of the flood starting cell. In contrast, the D-∞ method, proposed by Tarboton [16], determines the steepest downward slope of all eight adjacent cells as the flow direction angle, thereby creating one or two cells adjacent to the flow direction (Fig. 3(b)). Then, the water flow was distributed to one or two adjacent cells to the flow direction of their adjacent cells. In these two methods, the floodwater flows ultimately in a single direction. However, the flow directions may have many directions depending on the slopes. Accordingly, it is necessary to have a method that distributes the water flow to a cell whose elevation is lower than that of the starting cell.

The MFDM is a method that distributes the total outflow to downslope neighboring cells from the starting cell, where the elevation difference of all adjacent cells is weighted. Assuming that the lowest elevation difference is set to one, the total outflow is distributed differentially to adjacent cells whose elevation difference is lower than the lowest elevation difference according to the ratio of elevation difference. This process continues until the elevations of adjacent cells are higher than that of the starting cell when the distributed floodwater reaches the lowest elevation.

3.2 Flat-water assumption (FWA)

The FWA method, proposed by Yang et al. [4], follows the floodwater flow as it continues downward until it reaches and accumulates in the lowest elevation cell. Because of water

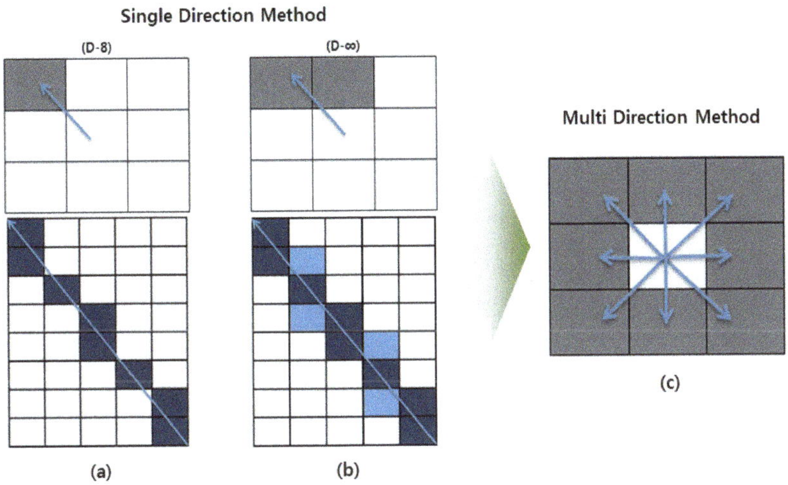

Figure 2: Flow distribution method.

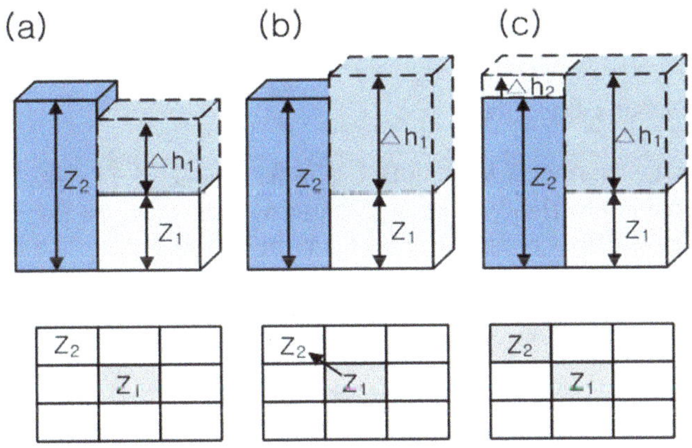

Figure 3: Water spreading using the flat-water assumption [8].

accumulation, elevation rank in the cell is equivalent to that of the adjacent cells; thus, the cell becomes the inundation starting cell where the total flow is accumulated. In Fig. 3(a), the water level in the Z1 cell increases until the water spreading condition [$Z1 + \Delta h1 > Z2$] is met. When the total outflow is added, Z2 should satisfy the flat-water assumption [$Z1 + \Delta h1 = Z2 + \Delta h2$] (Fig. 3(c)), and the flow of the cell is spread to neighboring cells according to this algorithm.

4 TARGET AREA AND APPLICATION

4.1 Select of target area

The use of SIMOD is most suited for areas with a nearby river and a levee of sufficient height, and where the influence of protected lowland facilities, such as buildings and roads, can be analyzed to apply the flood and inundation scenarios. Thus, as shown in Fig. 4, a levee (Seongseoje) located downstream of the Geumho river was selected as the study area.

The Geumho river basin includes a part of one metropolitan city, one province, three cities, two counties, seven districts, four towns, and 20 townships, such as Daegu Metropolitan City, Gyeongsan-si, Yeongcheon-si, Pohang-si, Chilgok-gun of Gyeongsangbuk-do. Its basin area is 2,092.42 km^2 with a channel length of 116.0 km. In the Geumho river basin, there are 52 regional tributary streams, including Jahocheon, Gochoncheon, Sillyeongcheon, and Sincheon, where the total stream length, including the Geumhho river, reaches 590.76 km [17].

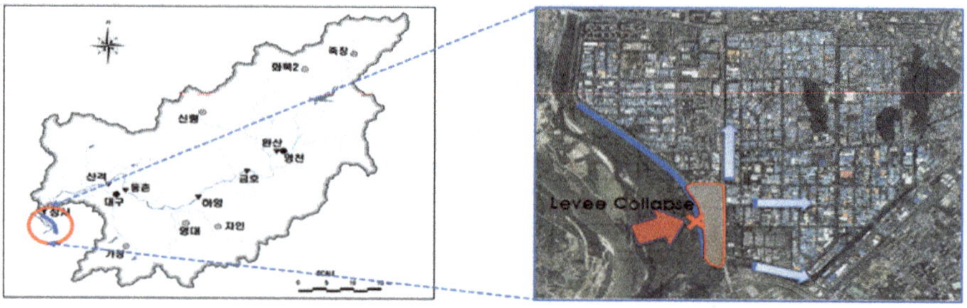

Figure 4: Location of target area and levee collapse.

4.2 Construct of input data

SIMOD requires two input data for simulations to be conducted, namely, inflow flood and terrain data. For the inflow flood, a virtual scenario was set where the flood overflew due to the Seongseoje levee collapse, as shown on the right side of Fig. 5. The inflow water was introduced to the protected lowland (industrial complex) beyond the flood-control reservoir. Fig. 6 shows the flow chart of inflow calculation.

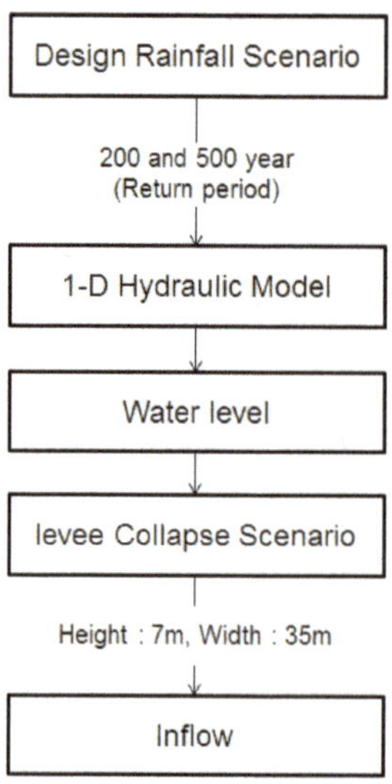

Figure 5: Flowchart for inflow calculation.

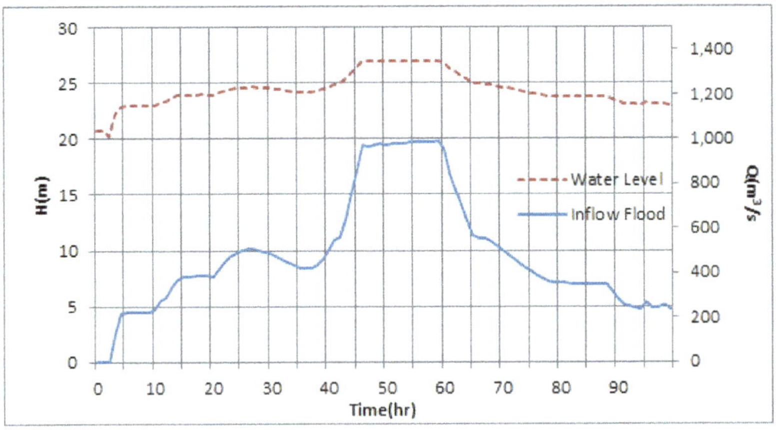

Figure 6: Water level and inflow flood.

The stream's water level was calculated based on heavy rainfall and typhoon scenarios made by the Technical Development Research Division of Integrated Safety Management for Waterside Structures based on Disaster Scenarios in Korea (Fig. 7). The levee collapse scenario assumed that the 7-meter high and 35-meter wide levee was destroyed at the collapse area shown in Fig. 5. If the stream's water level overflows through the collapsed levee, the inflow flood is calculated using eqn (1), assuming a rectangular weir equation. And the total amount of outflow that was calculated is used as input data for the SIMOD model and the Flo-2D model, and the time steps of the two models are the same

$$Q = \frac{3}{2} C \cdot b \cdot \sqrt{2g} \cdot H^{\frac{3}{2}} \quad (1)$$

where Q refers to the flow rate (m^3/s), C refers to the flow coefficient, b refers to the width (m), and h refers to the height (m), and C has a range of 0.6 to 0.66, and the average value of 0.63 is $1.85(m^{\frac{1}{2}})$. If this is applied to eqn (1), it is equivalent to eqn (2)

$$Q = K \cdot b \cdot h^{\frac{3}{2}} \quad (2)$$

where Q refers to the flow rate (m^3/s), K refers to the flow coefficient, b refers to the width (m), and h refers to the height (m). Here, it was assumed that the approaching speed at the stream was ignored, and there was no change in the stream water level. The elevation at the upper end of the levee was 27.63 m (El.). Thus, considering the 7-meter high levee collapse scenario, the elevation at the lower end of the levee was 20.63 m (El.). A flood was introduced to the protected lowland over the entire simulation hours (100 hours). Fig. 6 shows the stream flood stage and inflow flood in the protected lowland according to the virtual scenario, and Table 1 presents the cumulative total outflow by time.

Currently, most studies employ DEM terrain data to analyze inundation. However, urbanization has increased due to the development of civilization, subsequently increasing the factors when analyzing flooding, such as impervious areas of building groups. In this study, existing DEM and DSM terrain data represented building groups and road distribution in urban areas and analyzed inundation in an urban region. The results of DEM and DSM terrain data are shown in Fig. 7.

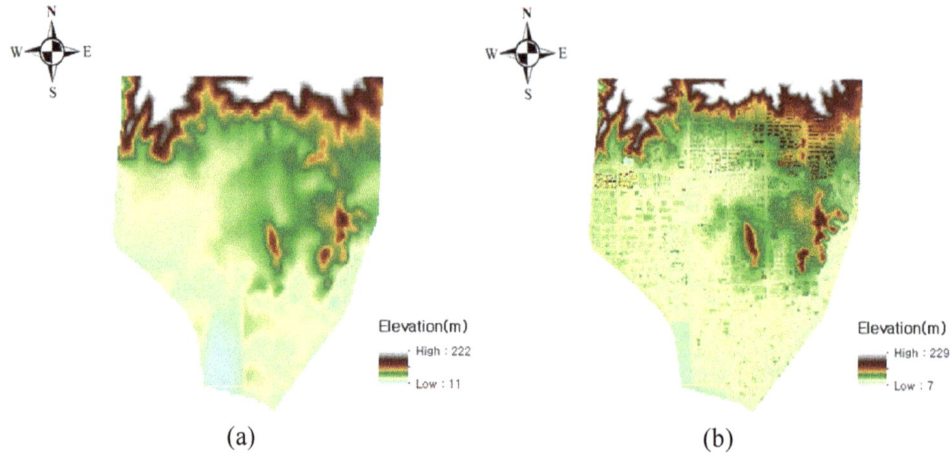

Figure 7: Terrain data (digital surface model, DSM). (a) DEM; and (b) DSM.

Table 1: Cumulative total outflow.

Time (hr)	Total outflow (m^3)	Time (hr)	Total outflow (m^3)	Time (hr)	Total outflow (m^3)	Time (hr)	Total outflow (m^3)
5	2,055,750	30	36,406,264	55	99,849,314	80	154,357,400
10	6,133,322	35	44,462,268	60	117,492,331	85	160,698,147
15	12,077,848	40	52,536,904	65	130,134,271	90	166,652,328
20	19,036,509	45	64,653,536	70	139,754,559	95	171,194,834
25	27,411,181	50	82,179,722	75	147,738,737	100	175,585,297

4.3 Analysis of inundation using DEM topographic data

The analysis was conducted by applying the calculated outflow to the commonly used DEM. The results exhibited around a 10% flooded area difference, except for 38.7% of area difference during the first 60 minutes when the flood started. In addition, the inundation analysis time showed that the simulation execution time for 24 hours using the SIMOD model was 7 minutes, which verified that the analysis was fast.

4.4 Analysis of inundation using DSM topographic data

The flood path following the distributions of building groups and roads is essential when simulating an inundation analysis in urban areas. The inundation analysis was conducted using DSM terrain data to verify building groups and roads in the study area. The analysis results showed that the difference in inundated areas was around 20%. The analysis results using DEM and DSM are shown in Tables 2 and 3. This result confirmed that DSM terrain data had an advantage that could verify urban areas' flood paths. Thus, the need to use DSM terrain data was demonstrated when conducting an inundation analysis in urban areas. Tables 4 and 5 show the difference in inundation area and runtime between two models.

Table 2: Variation of inundation area (DEM).

	Time (hr)	Area (km^2)	Time (hr)	Area (km^2)	Time (hr)	Area (km^2)
SIMOD	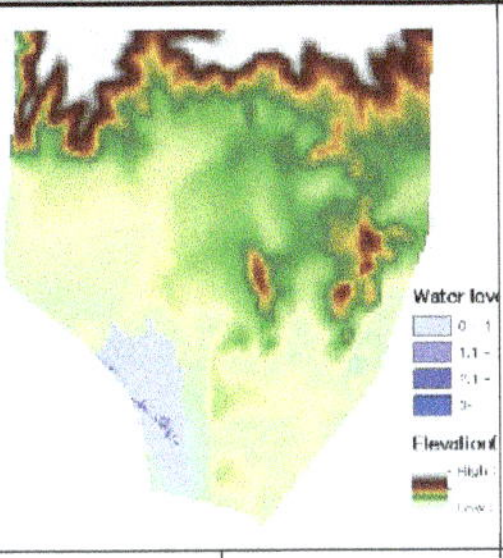		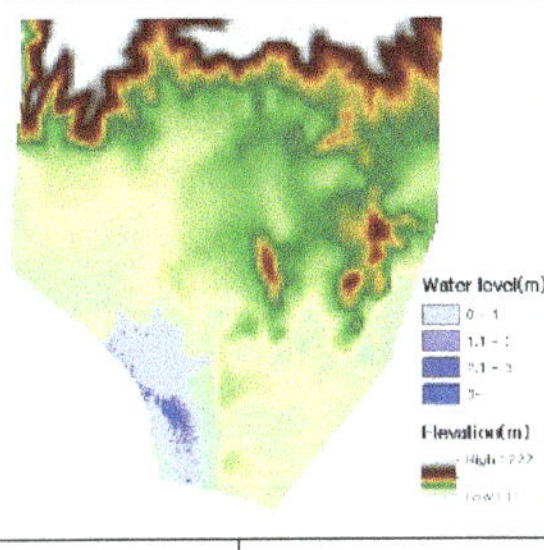		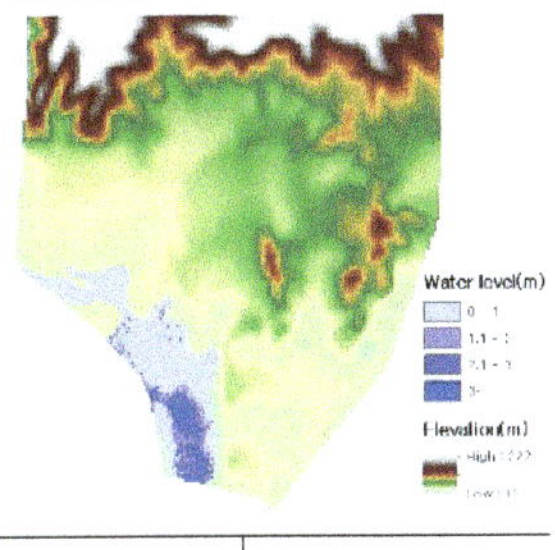	
	1	0.93	2	1.02	4	1.63
FLO-2D	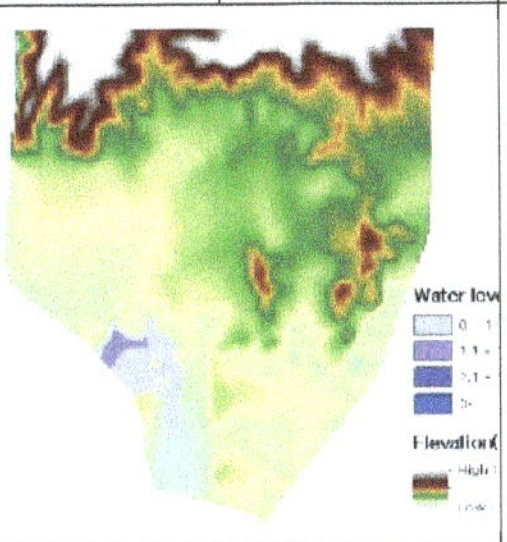		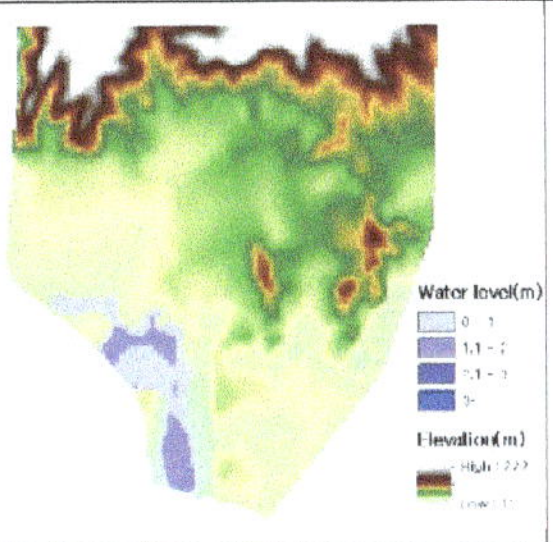		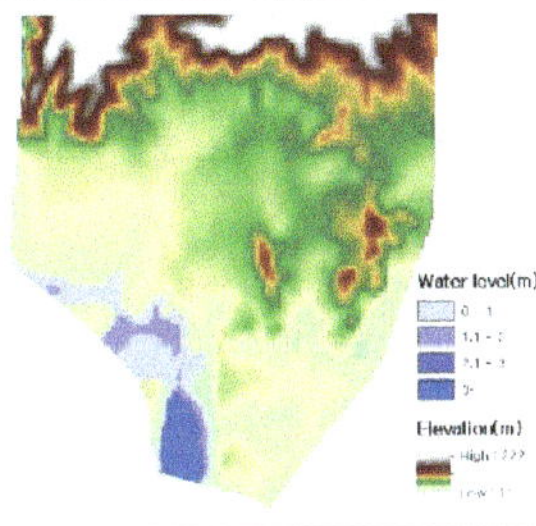	
	1	0.57	2	1.10	4	1.42
SIMOD	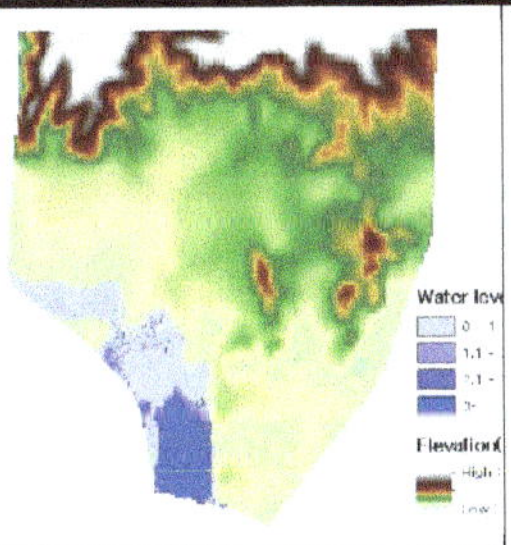		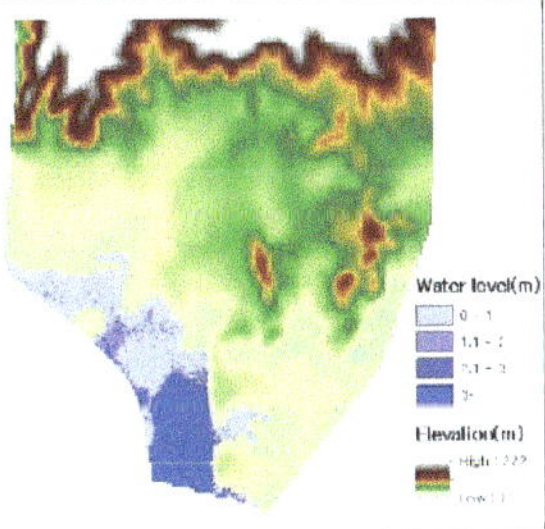		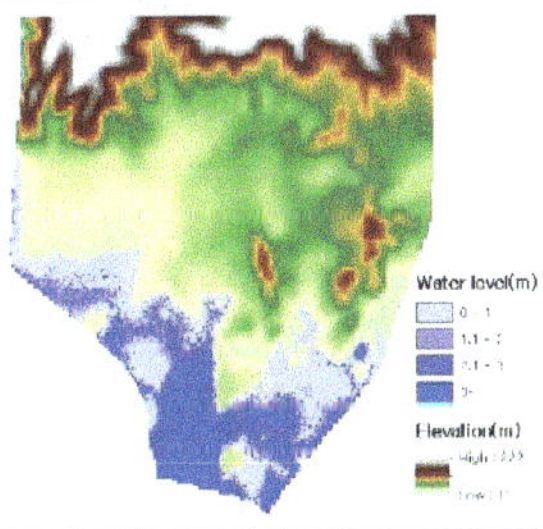	
	8	1.98	12	2.27	24	4.59
FLO-2D	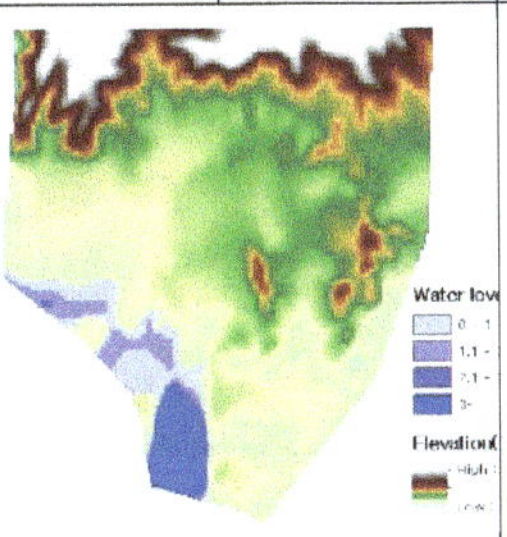		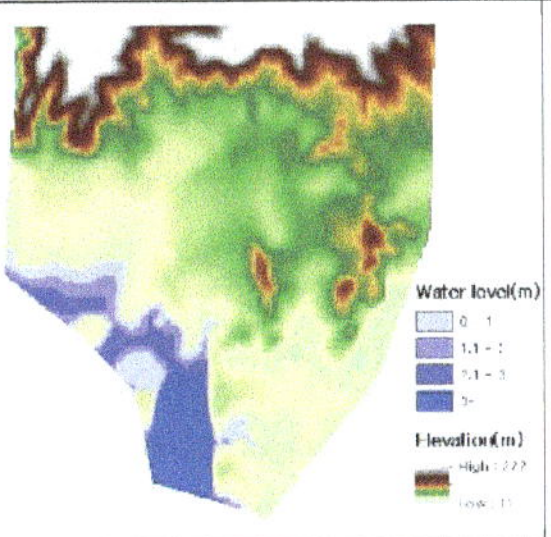		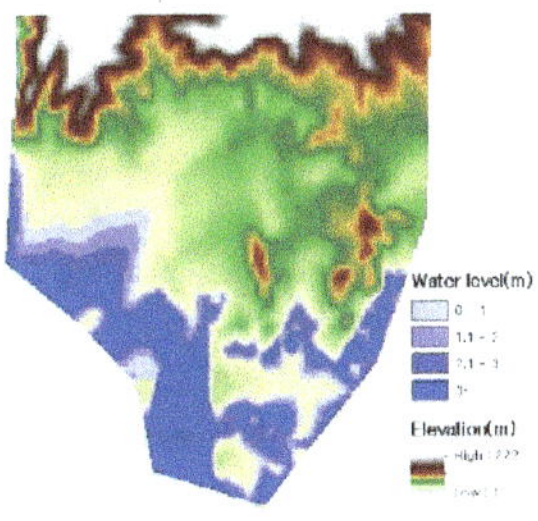	
	8	1.81	12	2.40	24	5.25

Table 3: Variation of inundation area (DSM).

	Time (hr)	Area (km^2)	Time (hr)	Area (km^2)	Time (hr)	Area(km^2)
SIMOD	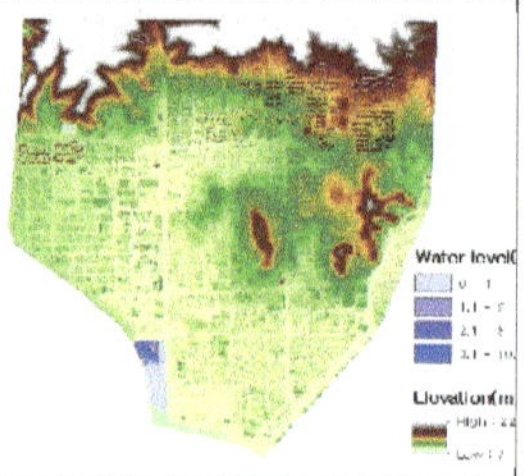		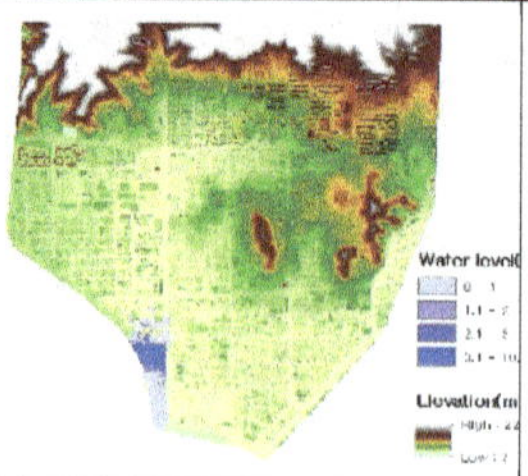		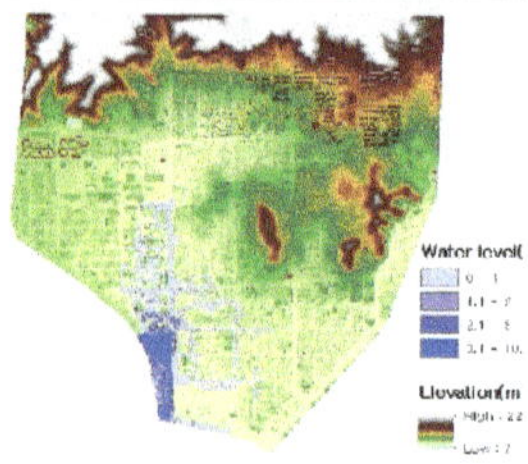	
	1	0.20	2	0.35	4	1.09
FLO-2D	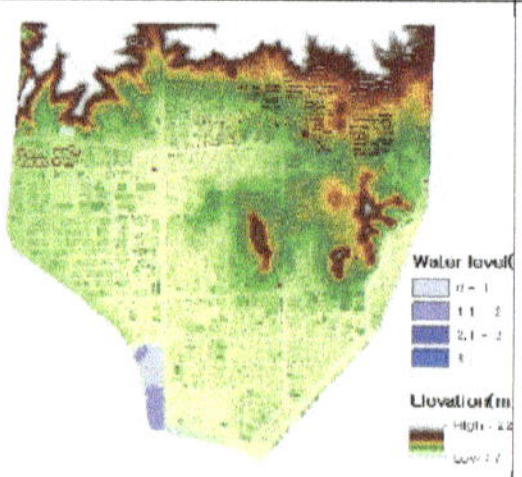		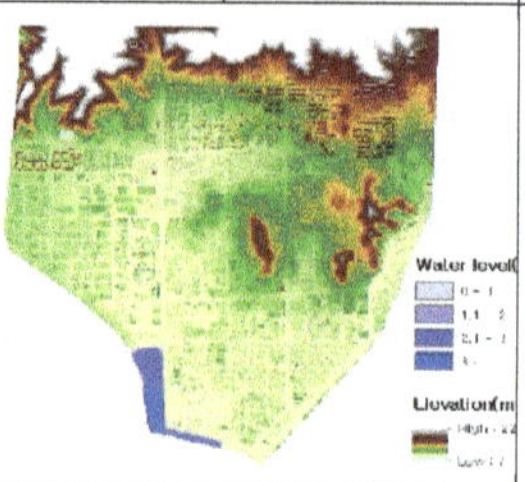		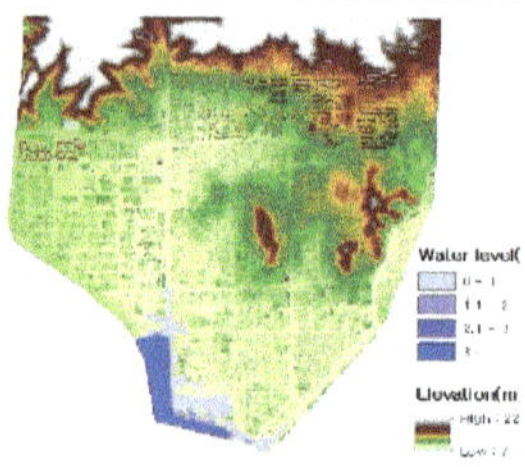	
	1	0.25	2	0.34	4	0.80
SIMOD	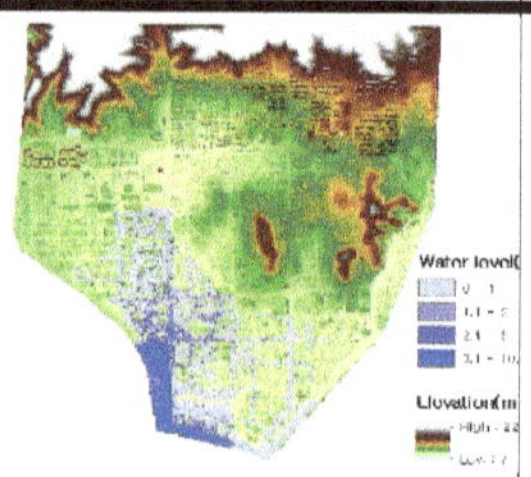		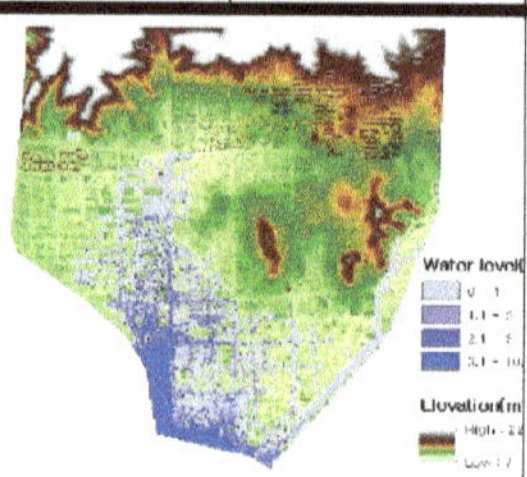		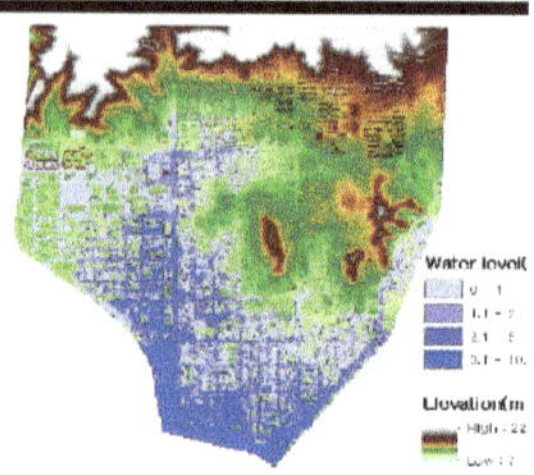	
	8	2.18	12	3.30	24	6.06
FLO-2D	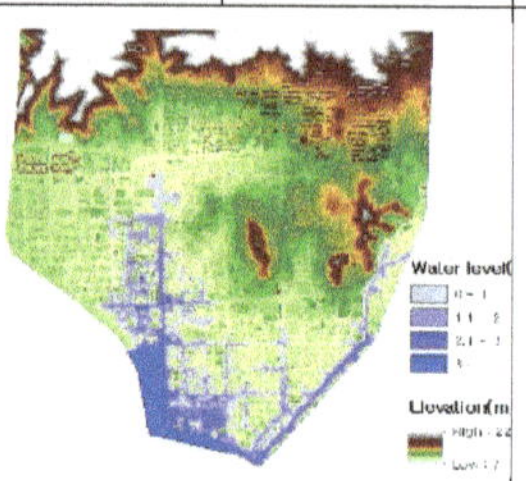		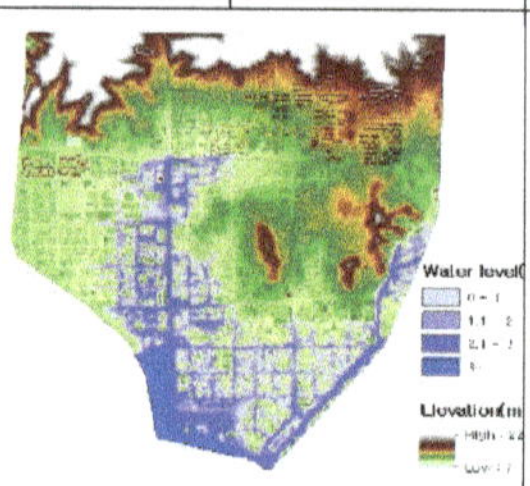		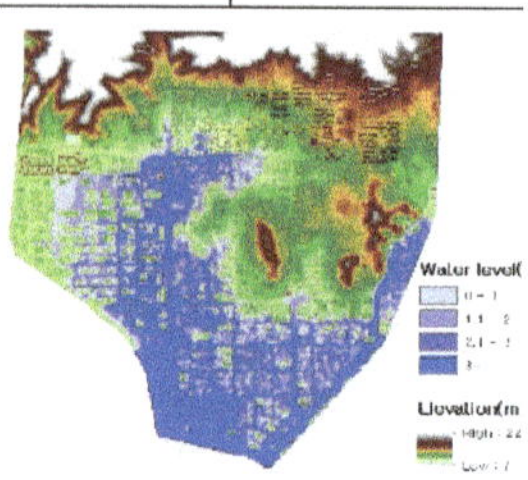	
	8	2.51	12	4.23	24	6.85

Table 4: Result of applying Simplified Inundation MODel (SIMOD) and two-dimensional flood routing model Flo-2D models using digital elevation model (DEM) data.

	SIMOD (km^2)	SIMOD run time	FLO-2D (km^2)	FLO-2D run time	Area difference (%)
1	0.93	1 sec	0.57	18 sec	38.7
2	1.02	1 sec	1.10	42 sec	7.9
4	1.63	10 sec	1.42	1.5 min	13.2
8	1.98	30 sec	1.81	4 min	8.7
12	2.27	2 min	2.40	10 min	5.8
24	4.59	7 min	5.25	25 min	14.3

Table 5: Result of applying Simplified Inundation MODel (SIMOD) and two-dimensional flood routing model Flo-2D models using digital surface model (DSM) data.

	SIMOD (km^2)	FLO-2D (km^2)	Area difference (%)
1	0.20	0.25	26
2	0.35	0.34	4.8
4	1.09	0.80	26.6
8	2.18	2.51	15.4
12	3.30	4.23	28
24	6.06	6.85	12.9

5 CONCLUSION

This study aimed to verify the applicability of SIMOD, a distributed flood routing model that could support decision-making by analyzing urban inundation quickly. SIMOD was a simplified model of many complex hydraulic and hydrologic algorithms required for the urban inundation analysis into two outflow distribution method algorithms: MFDM and FWA. This study aimed to analyze the applicability of the results obtained with the minimum analysis time. Thus, this study compared the analysis results with those using the commercial FLO-2D model. The comparative analysis results were as follows:

1. The flood analysis results using SIMOD and FLO-2D models verified that SIMOD took a shorter time to analyze inundation than FLO-2D. SIMOD took around seven minutes for a 24-hour simulation, whereas FLO-2D took approximately 25 minutes. Even considering the initial input time, SIMOD could take 10 minutes for a 24-hour inundation analysis.
2. The model analysis results using the DEM terrain data exhibited that the error rate of the inundated area between the SIMOD and FLO-2D models averaged around 20%. Because this was flood routing using the total inflow for the fast analysis of SIMOD,

there would be a difference in the degree of error. Nonetheless, sections that exceeded the error rate of 20% meant a difference in the model precision, requiring the future improvement of SIMOD.

3. The model analysis results using DSM terrain data verified that the interference by facilities, such as building groups and roads in the urban region, were expressed. This verification could not be done using DEM terrain data, and the flood routing path over time could be reflected correctly according to the actual topography.

ACKNOWLEDGEMENT

This work was funded by the Korea Meteorological Administration Research and Development Program under Grant KMI [2021]–[000312].

REFERENCES

[1] Kim, B. & Ha, S., The impact assessment of climate change on design flood in Mihochen basin based on the representative concentration pathway climate change scenario. *Journal of Wetlands Research*, **15**(1), pp. 105–114, 2013.

[2] Jeung, S., Lee, S. & Kim, B., Nonstationary intensity-duration-frequency curves under climate change. *Korea Water Resources Association*, 2015.

[3] Lee, J., Jeon, W. & Hur, S., Analysis of inundation characteristics for EAP of highway in urban stream – Dongbu highway in Jungrang stream. *Korean Society of Hazard Mitigation*, **6**(3), pp. 69–76, 2006.

[4] Yang, T.-H., Chen, Y.-C., Chang, Y.-C., Yangand, S.-C. & Ho, J.-Y., Comparison of different grid cell ordering approaches in a simplified inundation model. *Water*, **7**, pp. 438–454, 2015.

[5] Syme, W.J., Pinnell, M.G. & Wicks, J.M., Modelling flood inundation of urban areas in the UK using 2D/1D hydraulic models. *The Institution of Engineers, Australia 8th National Conference on Hydraulics in Water Engineering*, 2004.

[6] Pathirana, A., Tsegaye, S., Gersonius, B. & Vairavamoorthy, K., A simple 2-D inundation model for incorporating flood damage in urban drainage planning. *Hydrology and Earth System Sciences*, **15**, pp. 2747–2761, 2011.

[7] Nanshan, Z., Yasuto, T. & Kaoru, T., Simplified flood inundation model integration with rainfall-runoff processes. *Annuals of Disaster Prevention Research Institute, Kyoto University*, **50** pp. 75–78, 2007.

[8] Yang, T.H., Chen, Y.C., Chang, Y.C., Yang, S.C. & Ho, J.Y., Comparison of different grid cell ordering approaches in a simplified inundation model. *Water*, **7**(2), pp. 438–454, 2015. DOI: 10.3390/w7020438.

[9] Liu, L. et al., Developing an effective 2-D urban flood inundation model for city emergency management based on cellular automata. *Natural Hazards and Earth System Sciences*, **15**, pp. 381–391, 2015.

[10] Jeon, M., Han, D. & Choi, K., Flood inundation analysis from levee failure in Nakdong river. *Korea Water Resources Association*, **38**(4), pp. 259–269, 2005.

[11] Lee, J.H. & Yeon, K.S., Flood inundation analysis using XP-SWMM model in urban area. *Korean Society of Hazard Mitigation*, **8**(5), pp. 155–161, 2008.

[12] Lee, J.T., Hur, S.C., Kim, J.H. & Han, K.Y., Characteristics of the inundation and process of making a flood Map according to the levee break conditions in urban stream – Jungrang experimental basin. *Journal of the Korea Water Resources Association*, **39**(5), pp. 383–394, 2006.

[13] Park, J., Han, K. & Ahn, K., Development of numerical model to analyze levee break. *Korea Water Resources Association*, **38**(4), pp. 259–269, 2009.

[14] Kim, B.H., Choi, S.Y. & Han, K.Y., An analysis method of 1D hydrodynamic model based on GIS for flood inundation mapping. *Korean Society of Hazard Mitigation*, **11**(6), pp. 227–235, 2011.

[15] Kim, D.H., Jang, D.W., Park, H.S. & Choi, G.W., A study on flood impact upon the downstream area of a river by Dam break. *Crisis and Emergency Management: Theory and Praxis*, **10**(2), pp. 117–132, 2014.

[16] Tarboton, D.G., A new method for the determination of flow directions and upslope areas in grid digital elevation models. *Water Resource Research*, **33**(3), pp. 9–19, 1997.

[17] Ministry of Land, Infrastructure and Transport, Kumho River Maintenance Basic Plan, 1997.

Author index

Alvarado J. 109

Barbaro G. 181
Barillà G. C. 181
Becciu G. 131
Becue V. 197
Benalcazar J. 95
Berrezueta E. 95, 109

Carrión-Mero P. 95, 109
Cataldo J. 143
Cavalieri I. 169
Cenci J. 197
Chavez Navarrete E. 61
Coszach E. 197

de Oliveira Gobetti R. V. 209
de Smet I. 197

Elborolosy Y. Y. 143
Escobar-Segovia K. 61
Espinosa L. A. 3

Fiamma V. 181
Foti G. 181

Garcés-León D. 61
Gomez de la Torre M. 47
Gordillo P. 109

Hall J. R. 73
Herrera-Franco G. 95, 109
Hussain M. 157

Janggu J. 157
Jaya-Montalvo M. 95
Jesus J. 35
Jiménez-Oyola S. 61

Kang D. H. 221
Khainaar S. 197
Kiet T. H. V. T. 81
Kim B. S. 221

Laudati P. 197
Lee S. H. 221
Lobo Marchioni M. 131

Malesinska A. 181
Mambretti S. 121
Mancuso P. 181
Mandalari M. 181
Martin G. 73
Matsui K. 23
Merchan B. 95, 109
Monteiro M. 35

Nadya S. 157
Nguyen P. T. 81

Pacheco F. 35
Panepinto D. 209
Paz-Barzola D. 61
Portela M. M. 3
Puntorieri P. 181

Raimondi A. 121, 131
Raimondi F. 131
Rivière C. 197
Rojas J. 47
Romero-Crespo P. 61

Salgado B. 61
Salimbene O. 209
Saltos A. 47
Sanches Fernandes L. 35
Santos R. 35
Sanyal H. 143
Schippa L. 169
Stroppa F. F. 121

Tamai K. 13
Thoa N. T. K. 81
Torres R. 95

Veintimilla L. 109
Villa P. 47

Zakaria A.-R. 23
Zelenakova M. 3

Urban Water Systems & Floods III

Edited by: ***S. MAMBRETTI**, Polytechnic of Milan, Italy and* ***D. PROVERBS**, Birmingham City University, UK*

Flooding is a global phenomenon that claims numerous lives worldwide each year. Apart from the physical damage to buildings, contents and loss of life, which are the most obvious, impacts of floods upon households and other more indirect losses are often overlooked. These indirect and intangible impacts are generally associated with disruption to normal life and longer term health issues. Flooding represents a major barrier to the alleviation of poverty in many parts of the developing world, where vulnerable communities are often exposed to sudden and life-threatening events.

As our cities continue to expand, their urban infrastructures need to be re-evaluated and adapted to new requirements related to the increase in population and the growing areas under urbanization. Topics such as contamination and pollution discharges in urban water bodies, as well as the monitoring of water recycling systems are currently receiving a great deal of attention from researchers and professional engineers working in the water industry. The papers contained in this volume cover these problems and deals with two main urban water topics: water supply networks and urban drainage.

Originating from the 7th International Conference on Flood and Urban Water Management, the included research works include innovative solutions that can help bring about multiple benefits toward achieving integrated flood risk and urban water management strategies and policy.

ISBN: 978-1-78466-379-7 eISBN: 978-1-78466-380-3
Published 2020 / 196pp